A SHORT HISTORY OF

SCIENTIFIC IDEAS

TO 1900

BY

CHARLES SINGER

OXFORD UNIVERSITY PRESS

Oxford University Press, Walton Street, Oxford OX2 6DP

LONDON GLASGOW NEW YORK TORONTO
DELHI BOMBAY CALCUTTA MADRAS KARACHI
KUALA LUMPUR SINGAPORE HONG KONG TOKYO
NAIROBI DAR ES SALAAM CAPE TOWN
MELBOURNE AUCKLAND
and associates in
BEIRUT BERLIN IBADAN MEXICO CITY NICOSIA

ISBN 0 19 881049 0

First published by the Clarendon Press 1959
First issued as an Oxford University Press paperback 1962
Reprinted 1977, 1982

Printed in Hong Kong

PREFACE

THESE pages were written to give an elementary idea of how science came to occupy its distinctive position in the life of our own time. For this it is necessary to have some knowledge also of the Civilizations within which the science of the past has several times waxed and waned. To portray the different settings of science during the ages would be an immense task but I have tried, here and there, to indicate some of those factors in this complex story which seem less widely familiar. Of these a ready general example is the influence of the Arabic-speaking world; a specific case is the importance of the astrolabe.

It is evident that, in the last two generations, there has been a basic change in the status of science. This has affected the whole spiritual climate in which we live. Until the twentieth century the main inner life of civilized societies had been supported, for thousands of years, by activities quite outside the realm of scientific ideas. It is true that for the last few hundred years, and notably since the mid-eighteenth century, science has sometimes come to the aid of industry or agriculture by raising production or otherwise bettering the human lot. But on the spiritual side, save among isolated groups, science did little to exercise the intellect or satisfy the hunger for knowledge, and even less to appeal to the sense of beauty.

In England, until the later nineteenth century, science hardly penetrated the educational system and is still not integrated with it. Science seldom attracted general public interest until the beginning of the twentieth century. The place of science in the older universities remained at best secondary, at worst precarious. Since then the immense and accelerating increase in scientific activity and the resulting mass of real and applicable knowledge has changed every side of life. Having come to control and direct industry, it is now rapidly and manifestly transforming the very face of the earth and the lot of its living inhabitants, whether human, animal, or plant. A true history of science should end by discussing these latter-day metamorphoses. That last chapter must be a very long one and difficult to write. I am incapable of dealing with it and must perforce leave it to others.

I take this occasion to call to mind that science, the great aim of which is to make the world intelligible, or at least describable, has by its very success been riven into many departments. That this is so is due to the sheer accumulation of knowledge and thus ultimately to human limitations. However distressing and artificial these divisions may appear to the philosopher, it must be remembered that they are, after all, merely formal and their frontiers movable at will and according to circumstances. There remain, however, three real gaps which are of a more fundamental kind. One divides the biological from the physico-mathematical sciences. A second divides the psychological sciences from the *Naturwissenschaften*—for which we have no shorter English term. A third separates the sociological sciences from the others. There is evidence that these hiatuses are now occasionally being crossed. When such journeys become more frequent and more secure, it may be possible to bring a history of science 'up to date'. It is these gaps, and not the mere bulk of knowledge, that makes this consummation so difficult. It must, however, be remembered that this achievement would not change the basic difficulty of forming a mental picture of the scheme of nature as a whole. It seems that the substratum of reality, even if expressible in mathematical terms, can never be reduced to the conceptual. The idea contained in that statement is the real bar between the nineteenth and the twentieth centuries.

This book is based on one bearing the title *A Short History of Science*, drafted first in 1929 and published in 1941. It underwent several revisions but the present form is much more than a revision and rather more than a new edition. I have therefore thought best to give it a modified title. It is one-quarter longer than its predecessor, more fully illustrated, brought with fair uniformity to the end of the nineteenth century, stresses ideas more than formerly, pays more attention to simple philosophical implications, and refers more frequently to the historical and economic setting of science. It has been kept firmly on an elementary level and demands no more than the basic information included in a secondary education.

I have to thank several colleagues for generous help. Professor H. Dingle has written the section from p. 418 to p. 460 and has saved the text from some errors elsewhere. Dr. Angus Armitage has written the opening section on Mesopotamia and Egypt (pp. 6–12). Dr. Derek de Solla Price has provided the account of Ptolemy's *Almagest* (pp. 90–95) and of the astrolabe (pp. 149–51). Many helpful suggestions have been made by Dr. E. J. Holmyard, Professor Douglas McKie, and Professor E. G. R. Taylor. As with the earlier work I have had the constant help of Mrs. Singer.

The origins of the figures are given in the list of illustrations. Special thanks are due to the directors of Imperial Chemical Industries for permission to use a number of figures from volumes of *A History of Technology*, the publication of which was completed in 1958. A share in the editing of that work has taught me much which has been used here and for that I have to thank that public-spirited and far-sighted corporation. *A History of Technology* has afforded me a belated apprenticeship which remains among the most intellectually rewarding experiences of my not very short life.

The Library of the Wellcome Historical Medical Museum has been of great assistance. Its librarian, Dr. F. N. L. Poynter, has courteously done some searching for me and provided several useful figures. Those numbered 146 to 154 were kindly lent by Dr. Dorothy Feyer.

Only the traditional anonymity of the staff of the Clarendon Press prevents personal acknowledgment of assistance by several of its members. My affection and respect for the great institution which they serve is co-temporary with my association with it for well over forty years.

The general sources of the book will be evident to any with first-hand knowledge of the history of science. In view of the aim of the book its bulk has necessarily been severely restricted. On this account, and especially because of its elementary character, references seem out of place but none of them is obscure or inaccessible.

A word of general advice to the reader. The book cannot be understood unless the plan be grasped. To do this not only should a beginning be made by examining the *Table of Contents*, but constant reference to it is needed if the narrative is to be followed. In essence the book is a series of about sixty short and elementary essays on aspects of the history and philosophy of science, but neither the titles, nor the sequence, nor the interconnexion of these essays is haphazard. They follow a plan matured over a number of years. Therefore I beg the reader to keep on consulting the *Table of Contents* if he wishes to understand what I have to say.

C. S.

'*Kilmarth*', *Par*, *Cornwall*

1 January 1959

CONTENTS

LIST OF ILLUSTRATIONS

TEXT-FIGURES

PLATE

The original was drawn and engraved by Sebastien Le Clerc (1637–1714), who was himself a trained and experienced man of science, as well as an artist and scientific draughtsman. This is the earliest representation of a meeting of a learned society. It first appeared in the *Mémoires pour servir a l'Histoire Naturelle des Animause*, Paris 1671.

The place is a room attached to the King's Library at Paris in the Rue Vivienne, near where the Bibliothèque Nationale now stands. The occasion is a visit to the Académie in 1671 of Louis XIV who stands in front. On his left is his minister, Jean Baptiste Colbert (1619–83), who always showed a personal interest in science, both pure and applied. On the King's right is Louis de Bourbon, Prince de Condé (1621–86) and the King's brother, Philip, Duc d'Orleans (1640–1701).

Behind these four stand the members, all portraits. Claude Perrault (1613–88, p. 301), the moving spirit of the Académie, is just behind and between the King and Colbert. G. D. Cassini (1625–1712, p. 304) stands just behind the other two royal personages and is speaking over his shoulder. To the extreme right of the picture, Jean Picard (1620–82, p. 301) holds open his *Mésure de la Terre* (p. 319).

In the forefront of the picture is an armillary sphere, behind and to its right a gigantic concave burning mirror, while by its side is a telescope. Further to the right is a terrestrial globe. Behind these a large framed map and another rolled up indicate the Académie's interest in cartography. Behind the framed map is a drawing of a gazelle and its dissected parts. In the right forecorner is a stuffed civet or some such animal. In the front to the left on a table is the air-pump and compound microscope of Christian Huygens (1629–95, p. 302). On the back wall is the skeleton of a lion together with complex stills and other chemical apparatus. Below the open window is a model, perhaps of the water works at Versailles.

Through the open window is seen the garden where the Académie's astronomical work began. There stands the Académie's large quadrant. In the further distance is the Paris Observatory still under construction (begun 1667, completed 1672, p. 301). The artist's licence has brought this building some two miles nearer than its actual site.

An Ancient Picture of the Universe

'Thou hast ordered all things by measure, number, and weight.'

Wisdom of Solomon. An apocryphal work of *c*. 100 B.C.

The Hope of the Determinist

'I wish I could derive all phenomena of nature, by some kind of reasoning, from mechanical principles: for I have many reasons to suspect that they all depend upon certain forces by which the particles of bodies are either mutually attracted and cohere in regular figures or are repelled and recede from each other.'

NEWTON, 1687

A Classical Statement of Classical Physics

'We should regard the present state of the universe as the effect of its antecedent state and the cause of its subsequent state. An intelligence acquainted with all the forces of nature, and with the positions at any given instant of all the parts thereof, would include in one and the same formula the movements of the largest bodies and those of the lightest atoms.'

LAPLACE, 1814

Modern Physics and Mental Images

'The classical tradition had been to consider the world to be an association of observable objects moving according to definite laws of force, so that one could form a mental picture of the whole scheme. . . . It has become increasingly evident, however, that nature works on a different plan. Her fundamental laws do not govern the world as it appears in our picture in any direct way, but instead they control a substratum of which we cannot form a mental picture without introducing irrelevancies.'

P. A. M. DIRAC, 1933

INTRODUCTION

Nature of the Scientific Process

I. WHAT IS SCIENCE?

'WHAT is meant by *science*?' is the question that will naturally be asked on opening this book. Yet this question, if answered at all, can hardly be answered at the outset. In a sense the book is itself an answer.

Science is often conceived as a *body of knowledge*. Reflection, however, will lead to the conclusion that this cannot be its true nature. History has repeatedly shown that a body of scientific knowledge that ceases to develop soon ceases to be science at all. The science of one age has often become the nonsense of the next. Consider, for example, astrology; or, again, the idea that certain numbers are lucky or unlucky. With their history unknown, who would see in these superstitions the remnants of far-reaching scientific doctrines that once attracted clear-thinking minds seeking rational explanations of the working of the world? Yet such, in fact, is their origin. So, too, we smile at the explanation of fossils as the early and clumsier attempts of an All-powerful Creator to produce the more perfect beings that we know ourselves to be. Yet such conceptions were legitimate stages in the development of modern geological theory, just as the scientific views of our own time are but stages in an agelong process that is leading to wider and more comprehensive conceptions of the nature of our world.

It therefore behoves the historian of science to be very charitable, very forbearing, very humble, in his judgements and presentations of those who have gone before him. He needs to remember that he is dealing with the work of erring and imperfect human beings, each of whom had, like himself, at best but a partial view of truth, but many of whom had a sweep of genius far beyond his own.

There is an unquenchable and irresistible thirst of the soul that demands an explanation of the world in which it finds itself. One expression of that eternal yearning is the formulation of religious systems. Akin to such aspiration is that of the historian, who also seeks law and order in the universe. History, like science, like religion, is a constant search for such law, which yet always just eludes the grasp. And if the historian hopes to be judged at all by posterity, he can but echo the epitaph:

> Reader, thou that passest by,
> As thou art so once was I;
> As I am so shalt thou be;
> Wherefore, reader, pray for me.

Time, still, like an ever-rolling stream, bears all its sons away. It is the stream itself and the spirit that dwells therein that we shall seek to study.

Science, then, is no static body of knowledge but rather *an active process* that can be followed through the ages. The sheer validity and success of the process in our own age has given rise to a good deal of misunderstanding of its nature and not a little misapplication of such terms as 'science' and 'scientific'. We hear of the *scientific methods* of some prize-fighter, and a book has been published on the *Science of the Sacraments.* There is nothing in the laws of this or any other country which forbids its citizens from giving to the words of their language such significance as they may choose, but *science* and *scientific* as employed in these connexions have no relation to the great progressive acquisition of knowledge with which we have here to deal. The very form of the adjective 'scientific' might give pause to those who would force the word to cover such topics as the skill of the boxer, or a knowledge of the theory and practice of the sacraments. By derivation *scientific* implies *knowledge making*, and no body of doctrine which is not *growing*, which is not actually *in the making* can long retain the attributes of science.

2. ORIGINS OF THE SCIENTIFIC TRADITION

Science, then, is a process. But when did the process begin? It is as hard to answer this as to answer the question, When does a man begin to grow old? 'Before that I to be begun, I did begin to be undone.' Anthropologists perceive germs of the scientific process in the rudest races of mankind. When a child first begins to observe, he marks the differences of dress and manner in those about him. The savage sees the action of living beings in the sway of the trees or the stir of the waters. Both generalize from imperfect experience. The baby calls every woman 'mummy' and every man 'daddy'. Both make imperfect attempts to deduce general rules or laws. The attempts of both, in their kind and in their degree, are generalizations which are the essence of science.

FIG. 1. Late Palaeolithic drawings of bison with arrows in the heart. Cavern of Maux on the Ariège, S. France, *c.* 25000 B.C.

Man of the Old Stone Age lived on the flesh of the creatures he could slay. His dependence on the chase led him to observe the habits and the forms of the animals that he hunted. The magic in which he believed suggested to him that the mere representation of these animals, in the act of being slain, might result in their falling within his power (Fig. 1). The accuracy and beauty of his paintings rouse the wonder and admiration of those who explore his caves. The exactness of the observations of the palaeolithic artist and the care exerted in the representation of the form, movements, and even the anatomy of animals certainly betray elements akin to the scientific process.

In the tropics, where man first became human, the days do not appreciably lengthen and shorten with change in relation of earth and sun. The most natural and obvious means of calculating time is by changes of the moon. Her recurring appearances are still recalled in our calendars. Our *months* are but *mooneths* altered to fit our newer reckoning of time. Our *weeks* are but quarters of the 28-day cycle of the moon and recall her changes ('week', compare German *Wechsel* = change).

As man spread beyond the seasonless tropical forest he came to inhabit regions where agriculture arose. There was now need for a calendar that should tell him when to sow and when to reap. The movements of the stars were found to bear certain relations to that of the sun and therefore of the seasons. Observations of a very early date that bear on their relationship have come down to us from the civilization that developed in the valley of the Euphrates and Tigris. Thus the demands of agriculture, the first occupation, after hunting, for which man became organized, led to the accumulation of knowledge and to processes of generalization. These, on their level, are certainly scientific.

A settled agricultural civilization demands tools. Technology developed. The age of stone passed into the age of metals. The treatment of ores and the working of metals called for a class with special knowledge. The development of rights in land demanded some sort of surveying. Greek tradition has it that the inundation of the Nile, by obliterating all landmarks, forced on the Egyptians an annual remeasurement of their fields. Thus *geo-metry* (literally *earth-measurement*) was born. The craft of the butcher, as well as the practice of sacrifice and the examination of the entrails of the victims for purposes of divination, led to some knowledge of the structure of the body. In these processes we may see the practical sources of sciences that we now call metallurgy, mathematics, anatomy.

As society became more complex, commerce developed. A system of numerical notation was now evolved. The ancient world presents us numerous such instances of invention fathered by necessity and mothered by experience. All have a like claim to be included in a history of science. Ultimately a work will be written which will include them all.

The older civilizations, which advanced thus far along scientific lines, all developed cultural and religious bonds which united their members into tribal and ultimately into imperial units. Looking back on the past and viewing it from the vantage-point of our own civilization, we are struck with the failure of these ancient cultures to stress human individuality. In the earlier biblical record the punishment or reward of a people for the shortcomings or virtues of a single member passes without remark. Of none of the great primary discoveries which made social life possible has the name of the discoverer come down to us. The inventors and the successive improvers of the means by which fire can be made, of pottery, of the wheel, of the cutting-edge, of the bow, of the metals and their preparation, advanced mankind along the path that led to science. Yet their names, their dates, even their tribal affinities are utterly lost. So with the early thinkers. While we have ample record of the religious and ethical outlook of the peoples of the ancient world, we have little of that peculiarly individual product of the human intellect that in its later development we call *philosophy*, a product of which science is a part. We have no knowledge of those who first set out on the prime task of the philosopher, the individual endeavour to understand and to explain himself and his world. Even when prophet or priest seeks to deliver a message, he is always insistent that it is not his but another's; and not seldom that other is beyond our ken, for he is the Dweller above the Firmament.

Thus it happens that while we may discern science in these more ancient civilizations, no one has yet been able to give a continuous account of the development among them of scientific ideas; still less has it been possible to show how science influenced the modes of thinking of the more ancient peoples. For a clearer view we must consider another and later culture, that of the Greeks. Nevertheless, the ancient peoples of the Mesopotamian and Nile valleys have left remains that suggest, and perhaps prompted, the striking coherence of Greek thought. To certain aspects of the cultures of Babylon and Egypt we therefore turn.

I

RISE OF MENTAL COHERENCE

The Foundations: Mesopotamia, Egypt, Ionia, Magna Graecia, Athens, to c. 400 B.C.

1. FIRST SYSTEMIZATION OF KNOWLEDGE

DURING the past half-century much has been learnt of Babylonian mathematics and astronomy from inscriptions on baked clay tablets found on sites of the ancient Mesopotamian cities. The records are as yet very far from continuous. The great majority belong to two widely separated periods, the 'Old Babylonian' (*c.* 1800–1600 B.C.) and the 'Seleucid' (last three centuries B.C.).

Something of the nature of mathematics must be much older than the earlier of these periods for even in older texts a mathematical technique has attained full development. Numbers were represented on a system which combined a decimal with a sexagesimal notation. It embodied the principle of place-value; shifting a number one place to the left multiplied its value sixty-fold, successive shifts to the right corresponded to repeated divisions by sixty to form sexagesimal fractions. Moreover, the later or Seleucid texts employ a 'zero' to indicate an empty sexagesimal place between two other figures. Remnants of the sexagesimal system survive with us in the 360 degrees of the circle, the divisions of the hour and minute, and in many other current matters.

The mathematical texts usually consist either of tables for multiplication, squaring of numbers, &c., or of worked examples illustrating the solution of geometrical or algebraic problems. The geometry amounts to little more than estimations of the areas of fields, though the special property of the right-angled triangle (p. 16) was known even in Old Babylonian times. The ratio of the circumference of a circle to its diameter, π as we customarily call it, was taken as equal to 3. This is the value adopted in the Old Testament, perhaps under Babylonian influence. The algebra of the Old Babylonians could solve quadratic equations by a procedure equivalent to evaluating the modern formula, which gives the roots in terms of the coefficients, though the known texts nowhere quote or prove this rule. The Babylonians also handled linear equations in several unknowns and even attempted to solve cubic and biquadratic equations. The future may give us the earlier history of all this knowledge.

The astronomy of the Old Babylonian period amounted to little more than recognition of bright stars, arbitrary demarcations of the heavens, and observations, often undated, of striking celestial or atmospheric phenomena. There are also records of omens drawn from these. The significance attached to such celestial omens marks the beginning of astrology (Fig. 2). This study, originally used to predict the fortunes of contending kingdoms ('judicial astrology'), subsequently developed into complex procedures for foretelling the destinies of individuals ('horoscopic astrology').

FIG. 2. Boundary stone showing the god Marduk. Above his head are heavenly bodies including the zodiacal sign of the scorpion. An inscription on it records a donation to his temple in Babylon. *c.* 2000 B.C.

The later or Seleucid texts, on the other hand, embody complicated systems of theoretical astronomy. These were elaborated by the temple priests who were wont to observe the heavens from characteristic stepped watchtowers or temples, of which the biblical 'tower of Babel' (i.e. of Babylon) is a reminder. Their observations were mainly concerned with the motions of the 'seven planets' known from prehistoric times. These were sun and moon and the five bodies which, following the later Roman fashion, we call Mercury, Venus, Mars, Jupiter, and Saturn, but which the Babylonians associated with their own gods.

Each planet appears to move in the heavens, relatively to the background of stars, in its own characteristic figure and period. The movements were all found to lie within a certain belt of the celestial sphere, idealized for computing purposes as a great circle. This is the *Zodiac* (Greek *zōdion*, small figure of an animal). It was divided by the Babylonians into 12 equal portions, or 'Signs', each named after the constellation with which it coincided in those times.

The periodicities exhibited by the planets, and more particularly the revolutions of the sun and moon, were utilized for the measurement of time. The monthly vicissitudes of the moon, being more obvious than the annual travel of the sun, served to regulate the earlier lunar calendar. This way of reckoning time by the moon was (and still is) retained for religious purposes, though the solar year became adopted as a necessary agricultural unit. No natural numerical relation exists between the lunar month and the solar year. Nevertheless, the most compelling motive behind the remarkable development of Babylonian astronomy may well have been the need of a working rule connecting lunar and solar reckoning.

About the fifth century B.C. it was established that 19 solar years are equal to 235 lunar months (125 of 30 days and 110 of 29 days each) to within a fraction of a day. These 19 years, comprising 12 of 12 months each and 7 of 13 months each, were combined in a certain order to form what has come to be called, after its reputed Greek inventor, Metōn (flourished *c.* 430 B.C.), the 'Metonic Cycle'. The sequence in which 29-day and 30-day months followed one another was seen to be affected by variations in the rates of motion of sun and moon, by the latitude of the moon, and by the inclination of the ecliptic to the horizon.

In the tables for predicting the dates of successive new moons, separate columns indicated the corrections to be separately applied for these various factors affecting the length of the month. These corrections were represented as fluctuating discontinuously between upper and lower limits in a characteristic manner. The Babylonian tables, which extend also to the prediction of planetary phenomena, have been classified into two main systems, according to the artifices employed to represent the variation of the sun's rate of motion through the course of the year.

The Seleucid records were the work of a small group of specialists.

They do not show any great refinement of the underlying observations the numerical results of which, in any case, appear to have been rounded off to facilitate computation. The celestial occurrences to which the Babylonians attached most significance—new moons, heliacal risings and settings (first and last appearances of stars before

FIG. 3. Egyptian slit palm leaf and plumb-line, sixth century B.C. The transit of a star is timed as it crosses the plumb-line, being observed through the slit.

sunrise), &c.—are horizon phenomena, and therefore observed under necessarily unfavourable conditions. The most valuable Babylonian contribution to astronomy was probably a list of eclipses dated from 747 B.C. Of this list the Greek astronomer, Ptolemy (p. 89 f.), was to make use nine hundred years later.

On the mythological plane, the Babylonians seem to have conceived the earth as a disk rising into a central mountain and ringed by a moat of ocean. Beyond this ocean was a mountain-wall upholding a solid hemispherical firmament.

By contrast to the mathematical achievements of the Babylonians, those of the Egyptians were primitive. Their system of numeration was a decimal one, numbers being represented by the collocation of conventional symbols for the unit and for successive powers of 10 up to one million. Place-value was unknown, nor was there any zero.

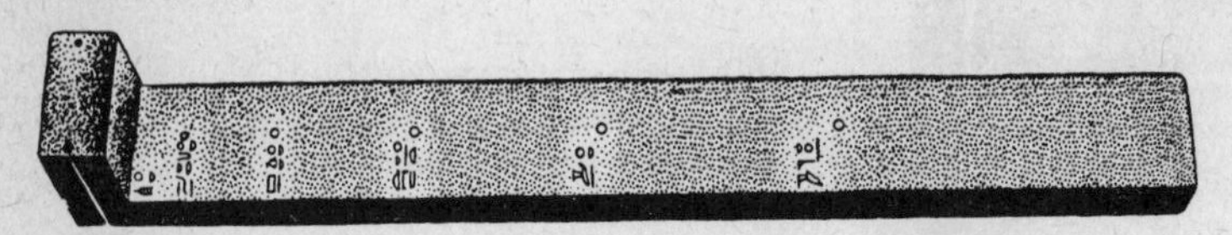

FIG. 4. Modern Egyptian shadow-clock. The pattern goes back to the fifteenth century B.C. at least. The division on which the shadow of the end block falls indicates the approximate hour.

Addition and subtraction were mechanical operations. Multiplication was effected by repeatedly doubling the multiplicand and adding together the appropriate products. Division was the inverse of multiplication. Where a division involved a remainder, fractions were introduced, but an Egyptian fraction (other than $\frac{2}{3}$) always had unity as its numerator. It had therefore to be resolved into a sum of such unit fractions written in juxtaposition. Thus $\frac{3}{4}$ was expressed as $\frac{1}{2}+\frac{1}{4}$.

The initial problems in the best-known Egyptian mathematical papyrus deal with such resolution of fractions. The remaining problems are concerned with solution of equations by trial, mensuration of areas and volumes of simple figures, progressions, and calculations apparently relating to the slope of the side of a pyramid. The area of a circle is assumed to be equal to the square on eight-ninths of the diameter (which makes $\pi = 256/81 = 3{\cdot}16$ approximately). Among Egyptian mathematical documents the most considerable contains an application of the correct rule for calculating the volume of a truncated pyramid with a square base.

In astronomy the Egyptians hardly developed their observational technique nor did their ideas on the nature of the world reach any scientifically significant level. They pictured the heavens as a goddess supporting herself on hands and feet, or the Sun-God as sailing by day upon a celestial river, descending by night to visit the abode of the dead.

The dependence of the Egyptian economy upon the dating of the annual Nile-flood conferred a particular significance upon the length of the year. This was originally determined as the period of recurrence of the earliest appearance before sunrise of the bright star Sothis (our Sirius). Thus was established a conventional year. It consisted of 12 months of 30 days each, with 5 extra days. This scheme later provided classical and medieval astronomers with a serviceable unit of time. The Egyptians also divided day and night each into 12 equal 'hours' which therefore varied in length according to the season. A peculiar feature of later Egyptian astronomy was the

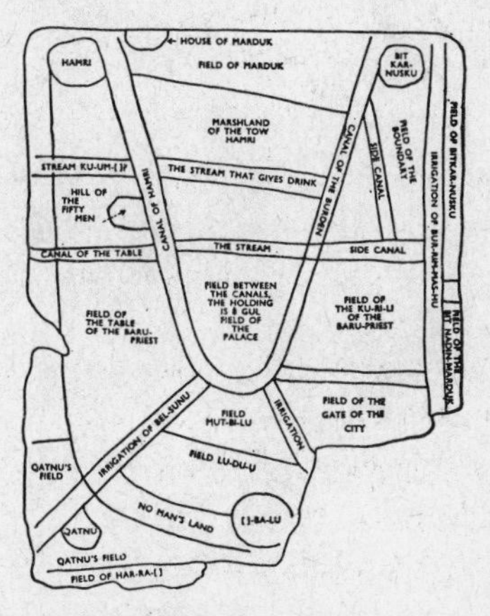

FIG. 5. Babylonian outline map of fields and canals near Nippur in Mesopotamia from a baked clay tablet of *c.* 1300 B.C. Cuneiform writing translated.

concept of the 36 'Decans', a sequence of constellations which rose at regular intervals serving to indicate time by night and presiding astrologically over successive decanal, or 10-day 'weeks'.

The Egyptians had simple apparatus for observing the transit of a star (Fig. 3) and also for timing the divisions of the day by shadows (Fig. 4). Attempts to elicit elements of biological science from Egyptian and Mesopotamian work have displayed high technological development. They have also revealed some quite surprising knowledge of plants and animals and that curiosity about them which heralds the dawn of science. Nevertheless, nothing has yet emerged on these lines that is sufficiently organized or systematized to be discussed here as biological science.

Both in Egypt and Mesopotamia surveying had led to the preparation of rough maps. Specimens have survived from both countries but we cannot estimate their scale (Figs. 5 and 6).

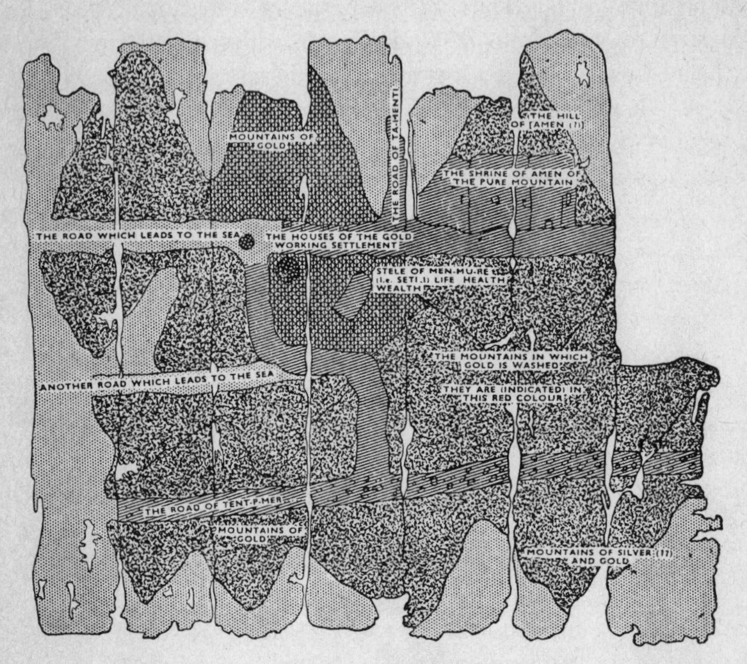

FIG. 6. Egyptian outline map of route through gold-bearing hills in the Wadi Hammamat in the Eastern Desert near Gulf of Suez, from a papyrus of *c.* 1300 B.C. Hieroglyphic writing translated.

2. BEGINNINGS OF IONIAN SCIENCE AND THE EASTERN GREEK SCHOOL

In writing history it is commonly necessary to rely upon written documents. Without such records, the narrative is always imperfect and often incoherent. The earliest scientific documents that we possess that are in any degree complete and continuous are in the Greek language. They were composed after 500 B.C., but we have a few scattered records of Greek science for about a century before that time.

It is certain that Greek science in its origin was dependent on traditions that came from more ancient civilizations, notably from

Egypt and Mesopotamia. On this the Greeks themselves insisted. They have been confirmed by modern discoveries.

The origin of the Greeks is far from clear but we know that tribes, who spoke a language closely similar to theirs, occupied Eastern Mediterranean islands and coasts in the second millennium B.C. Invading streams of them passed to the sea-coasts and islands of Asia Minor. Chief among these Asiatic Greeks were the Ionians, who colonized the shores of the Aegean from Ephesus in the north to Halicarnassus in the south. Yet farther south settled the Dorians (Fig. 7). South Italy and Sicily were colonized secondarily both from Greece and Asia Minor (Figs. 7 and 12). It was among the Ionians that the first great scientific movement arose. Dorian elements, however, crept into it at an early date.

The Ionians were very favourably placed for the reception of foreign ideas. Eastward they were in relations with ancient cultures, Mesopotamian and others. Mesopotamia was invaded in the sixth century B.C. by a people from yet farther east, the Persians, who left a permanent mark on all contemporary civilizations. Their influence is to be discerned in the New Testament where we read of the *Magi* (Authorized Version 'wise men', Matthew ii. 1), a Persian word that has given us our term *magic*. Persia was the most vigorous power of the age and brought new contacts to the Ionians. Further, the Ionians were a maritime and trading people. Through their regular sea traffic suggestions came to them from Egypt, the most ancient and settled of all civilizations. Ionians traded, too, with Phoenicia and reached as far as India whence some of their ideas were derived.

It was, in general, a time of travel, of movement, of the breakdown of old and of the rise of new civilizations. Such was the stage, such the atmosphere of change in which science became first clearly distinguished. We see science emerging into the light of historic day in the person of the Ionian Greek Thales.

Though the son of a Phoenician mother, THALES (*c.* 624–565 B.C.) was a citizen of the Ionian city of Miletus. Tradition tells that he was a man of great sagacity, exhibited no less in politics and commerce than in science. He suggested a federal system for the cities of Ionia and made a fortune as a merchant.

In the course of business Thales visited Mesopotamia and Egypt.

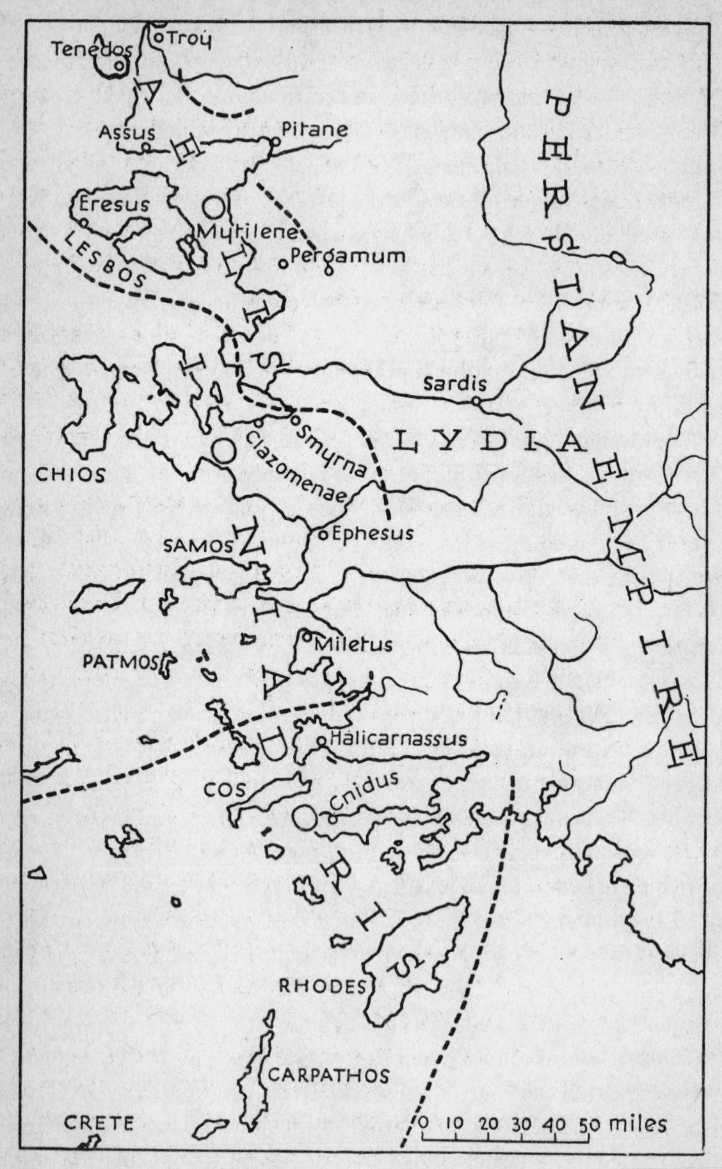

FIG. 7. Western Asia Minor *c.* 550 B.C.

In the former country he learned of the 'Saronic cycle', that is the interval of eighteen years and eleven and one-third days, a multiple of which the observations of ages by Babylonian temple star-gazers had shown to be usual between eclipses of the sun.[1] Knowledge of this is said to have enabled the shrewd traveller to make a lucky forecast of the eclipse visible at Miletus in 585 B.C. His prediction drew much attention. It may well be that the impression thus created directed the attention of the Greeks to the advantages that might accrue from systematic observation of nature. At any rate, they always reputed Thales to be the father of that study.

Further achievements of Thales were chiefly of a geometrical nature. Now it is important here to recall that the Greeks did not invent geometry. They could and did gather some knowledge of the subject from their neighbours in the Nile Valley. The Egyptians, however, had hardly reached beyond an empirical usage of certain special relations of figures, and especially of triangles and rectangles, of pyramids and spheres. Thus, for example, the Egyptians knew that the square on the longest side of a right-angled triangle is equal to the sum of the squares on the other two sides; but they knew it only for such special cases as that in which the sides are in the ratio 3, 4, and 5; thus $5 \times 5 = 3 \times 3 + 4 \times 4$ (Fig. 8). Again, they could estimate the cubic contents of a pyramid, but only of a pyramid of a certain definite type with a certain definite number of sides sloped at a certain definite angle. Thales succeeded in generalizing such special cases. He thus discovered that the angles at the base of an isosceles triangle are equal; that when two straight lines cut one another the opposite angles are equal; that the angle on the circumference of a circle subtended by the diameter is always a right angle; that the sum of the angles of a triangle is equal to two right angles; that the sides of triangles with equal angles are proportional.

Thales, moreover, succeeded in applying such knowledge. He was able, for example, by a simple application of the principle of similar triangles, to determine the distance from the shore to a ship at sea (Fig. 9), and to measure the height of a pyramid by comparing the

[1] *Saros* from a Babylonian word *saru* (Greek *saros*) for the number 3,600, i.e. 60^2 and hence for a period of 3,600 years. The application of the word to the cycle of 223 lunations (18 years 11 and one-third days) is a modern misunderstanding. The word is now firmly fixed in scientific nomenclature.

length of its shadow with that cast by an object of known height. Such problems had been tackled before his time. But Thales not only sought to enunciate them clearly and to solve them demon-

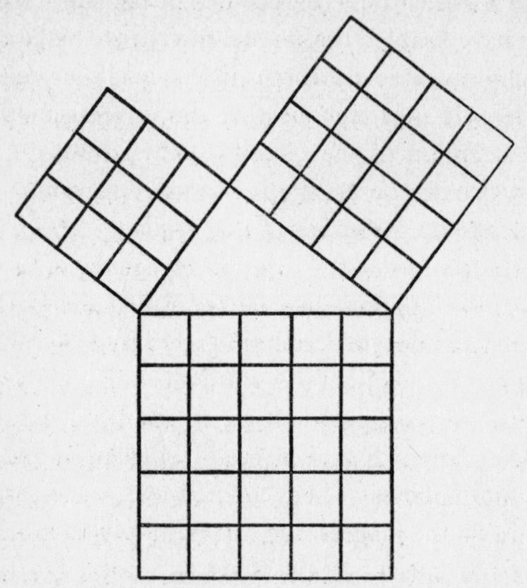

FIG. 8. Special case of squares on sides of right-angled triangle.

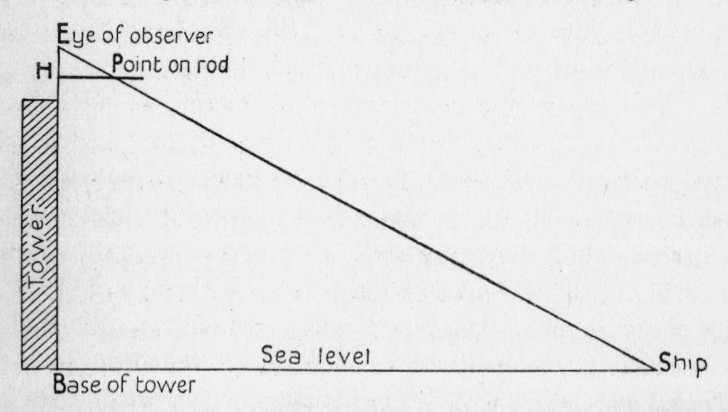

FIG. 9. Thales measures distance from base of tower to ship at sea. Triangle *ESB* similar to triangle *EPH*.

strably but also to widen and generalize them so as to lay bare their essential nature.

As with every Ionian thinker, the ultimate object of the thought of Thales was to find a formula for all things. He thus set himself the task of discerning constancy amidst the diversity and variety of nature. This is but to say that his science was a part of his philosophy. To the general question 'Of what is the world made?' he would answer 'Water', meaning thereby some mobile essence, changing, flowing, without distinctive shape or colour, yet presenting a cycle of existence passing from sky and air to earth, thence to the bodies of plants and animals, and back to air and sky again. But his real place in the history of science is better brought out by the more concrete statement that in his mathematical work we have the first enunciation, as distinct from implicit acceptance, of natural laws.

Following on Thales, a long line of Asiatic Greeks, mostly of Miletus, contributed to the extension of the conception of natural law. Thus ANAXIMANDER (611–547 B.C.), a Miletan pupil of Thales, took much interest in geography. He was the first among the Greeks to represent the details of the surface of the earth by maps. The idea of map-making was known in Egypt and Mesopotamia (Figs. 5 and 6). Anaximander, however, sought to convey a concrete picture of the surface of the earth as a whole. The suggestion doubtless came from Mesopotamia, where simple diagrams of this sort were being made in his time. From Babylon or Egypt also he introduced the sun-dial. It consisted in essence of a *gnomon*, a fixed upright rod, the direction and length of the shadow of which can be measured hour by hour. The records of these make it possible to determine the movements of the sun as well as the dates of the two *solstices* (the shortest and longest days) and of the *equinoxes* (the two annual occasions when day and night are equal). Anaximander was thus led to develop his own astronomical conceptions. He was the first to speculate on the size and distance of the heavenly bodies. The earth was for him a flat disk in the centre of all things. Sun, moon, and stars are enclosed in opaque rings, rotating with the earth as centre. We see them only through vents in these rings.

ANAXIMENES (born *c.* 570 B.C.), another Miletan, extended Anaximander's ideas. The essence of all things he regarded as 'air' rather than the 'water' of Thales. This air was linked with that essence which is essential to life. He called it *pneuma*—literally *breath*—and held that in a sense the universe itself was alive: 'As our soul, being air, sustains us, so *pneuma* and air pervade the whole World.'

At about the same date CLEOSTRATUS of Tenedos, who lived rather outside the Ionian zone, made two important contributions to astronomy. One was an improvement in the calendar, involving a better measure of the solar year. The other was the knowledge of the signs of the zodiac which he introduced from Mesopotamia (p. 8). Zodiacal signs are frequently encountered upon Mesopotamian boundary stones and indicate the time of year at which the stones were erected (Fig. 2).

Among the Greeks of Asia Minor towards the end of the sixth century B.C. there was not only considerable speculative activity, but also the sum of positive knowledge was being systematically increased. The process was encouraged by the roving character of the Asiatic Greeks. Active and daring seamen, they brought back to their homes accounts of many of their adventures by land and sea.

Of these early explorers, the most distinguished was HECATAEUS, also of Miletus (born *c.* 540 B.C.). He visited Egypt, the provinces of the Persian Empire, Thrace, and Lydia. He penetrated the Dardanelles and explored the coasts of the Black Sea. About 500 B.C. he adventured westward to the Gulf of Genoa and as far as Spain, reaching Gibraltar. There he had been preceded by the Phoenicians, who had set up to their god Melkarth a great column on either side of the Strait. Later writers identified Melkarth with Hercules, and the gateway of the Mediterranean came to be called the 'Pillars of Hercules' (Tartessus = Tarshish of the Bible and Tingis = Tangier). Hecataeus collected his experiences into a geographical handbook. He is memorable for that scepticism of the marvellous which is a hall-mark of the man of science. He detested mythology. 'The stories of the Greeks', he says, 'are in my opinion no less absurd than numerous.' We can form some idea of the view of Hecataeus of the general plan of the earth's surface (Fig. 10).

By the fifth century, the character of Ionian thought was modified by contact with Persia which, under its great Emperor Darius I (522–486 B.C.), was advancing westward. The weak and quarrelsome little Asiatic Greek States were coming under its shadow. The Persian service attracted many of their citizens, who brought back to their native homes further knowledge of the world. Among the more typical of these venturers was the physician DEMOCEDES of Cnidus (born *c.* 540 B.C.). The peninsula of Cnidus was the seat of the most ancient medical school of which we have any record (Fig. 7).

After travelling widely in Greek lands, Democedes became the medical attendant of the Persian monarch. Later he was employed

FIG. 10. Conjecture of how the world seemed to Hecataeus *c.* 500 B.C.

as a spy to explore the coasts of Greece. He escaped from this service, however, and settled in the Greek colony of Croton, in the instep of Italy. Here he devoted himself to writing a treatise on medicine, the first Greek work on that subject of which we have tidings. Croton became an important scientific centre (Fig. 12).

Thus, as time wore on, Ionian thinkers came more closely into contact with other civilizations. Their work becomes increasingly sophisticated. Philosophy is no longer the product of the leisure hours of business men, of sailors, or of physicians. Thinking has become a profession.

Amongst the great Ionians who concerned themselves exclusively with philosophy was HERACLEITUS of Ephesus (*c.* 540–475 B.C.). He is specially remembered for his view that 'everything is in a state of flux'. Change is the only reality. 'There's nothing is and nothing was, but everything's becoming.' Fire, the most changeful of elements, is the origin and image of all things. Living creatures are formed of a mixture of the changeful essences of which fire and air are types. Nothing is born and nothing dies. The illusions that we call birth and death are but a rearrangement of these unresting elements.[1]

Very different from the point of view of Heracleitus was that of his younger contemporary, the Miletan LEUCIPPUS (flourished *c.* 475 B.C.), founder of the atomic doctrine of matter. That theory has had a wide influence in both ancient and modern times. It has been associated with the attitude towards the world known sometimes as 'philosophic materialism'.

Leucippus—of whom we know little—is overshadowed by his pupil, DEMOCRITUS (*c.* 470–*c.* 400 B.C.) who was perhaps also of Miletus. This Democritus was a contemporary of Socrates (470–399 B.C.; p. 36), though the outlook of the two men is in the strongest possible contrast. For Democritus, very different from Heracleitus, all things were composed of solid *atoms*, together with the space or *void* between them. We should note that this void has as much claim to be regarded as a primary reality as the atoms themselves. The atoms are eternal, invisibly small, and cannot be divided. (The word *atom* means 'indivisible'.) They are incompressible and homogeneous. They differ from one another only in form, arrangement, and size, that is to say only quantitatively, not qualitatively. The qualities that we distinguish in things are produced by movement or rearrangement of these atoms. Just as atoms are eternal and uncaused, so also is motion, which must, of its nature, originate in preceding motion. As everything is made up of these unchangeable and eternal atoms, it follows

[1] The thought of Heracleitus bears a certain resemblance to that ascribed to the founder of Buddhism who was his contemporary. Whether one derived from the other or both from a common source is a matter which future research may decide.

that coming into being and passing away are but a seeming, a mere rearrangement of the atoms. The beings that you and I think we are, are but temporary aggregations of atoms that will soon separate to enter into the substance of other beings or things. And yet, in ages of time, perhaps, we shall be re-formed, when it may so fall out that our atoms come together again. Thus history repeats herself endlessly.

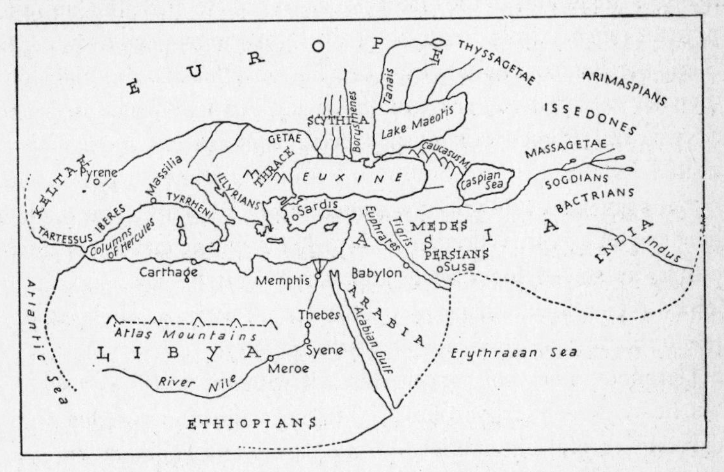

FIG. 11. To illustrate the geographical knowledge of Herodotus.

At first sight the positive teaching of Democritus and the concrete character of his atoms suggest a 'common-sense' philosophy that might be set against the Heracleitan vagueness. It must be remembered, however, that the atoms of Democritus were in no sense the product of experimental investigation. His atoms, like their motion and like the void in which they moved, were alike hypotheses and based on no sort of exact knowledge or experience. His teaching has obvious parallels with nineteenth-century scientific doctrines concerning the 'indestructibility of matter' and the 'conservation of energy', but the parallels are more apparent than real. Democritus also observed animals closely and may even have dissected some. Yet despite the positive trend of his thought, his followers—known as 'Epicureans' after the most distinguished of them, EPICURUS of Samos (342–270 B.C.)—showed little tendency to extend the range of scientific ideas.

Much of the spirit of Ionia is summed up in the life and writings of HERODOTUS of Halicarnassus (*c.* 484–425 B.C.). The native town of this remarkable man was within the limits of the Persian Empire at the time of his birth, and he remained a Persian subject till he was well into his thirties. From an early date his inquiring spirit led him to travel. He explored Greece and Asia Minor thoroughly, visiting many of the islands of the Greek Archipelago. He made the long and difficult journey from Sardis in Lydia, near the modern Smyrna, to Susa, the Persian capital (Fig. 11). He travelled next to Babylon; then he explored the coast of the Black Sea and penetrated into Scythia and Thrace. His journeys were extended westward, and he visited Italy and Sicily. Southward from his home he passed into Syria, sojourned at Tyre, saw something of Palestine, and made a long stay in Egypt. Wherever he heard of anything curious or interesting, he stayed for a while and noted what he saw. Finally he joined a Greek colonizing party that settled in Italy. He spent the rest of his life preparing his delightful *History*.

Herodotus does not concern himself with the world as a whole, but he gives an excellent idea of the geographical knowledge of his day. His careful observations on the nature and habits of different peoples entitle his work to be regarded as the first treatise on the science of man. He is thus the father of anthropology, as he is also the father of history. Many of his allusions to the beliefs and practices of the time help us to check the early records of the history of science.[1]

3. PYTHAGOREANS AND THE WESTERN GREEK SCHOOL

From a very early date Greeks had penetrated westward and had established colonies in Southern Italy and Sicily, *Magna Graecia* as the area came to be called (Fig. 12). The intellectual activity of these western colonies played an important part in the development of Greek science. The most influential of the western scientific movements was that of the 'Pythagoreans'.

The founder of this school or sect, PYTHAGORAS (born *c.* 582 B.C.),

[1] Herodotus is especially responsible for the view that Greek institutions were derived from Egypt.

was a native of Samos in Ionia. He travelled widely. About 530 he settled at Croton, where a Dorian colony had been established. There he founded his brotherhood or sect, which persisted long after him. The veil of mystery which his followers drew over themselves often prevents us from ascribing to their actual originators the scientific advances which they made.

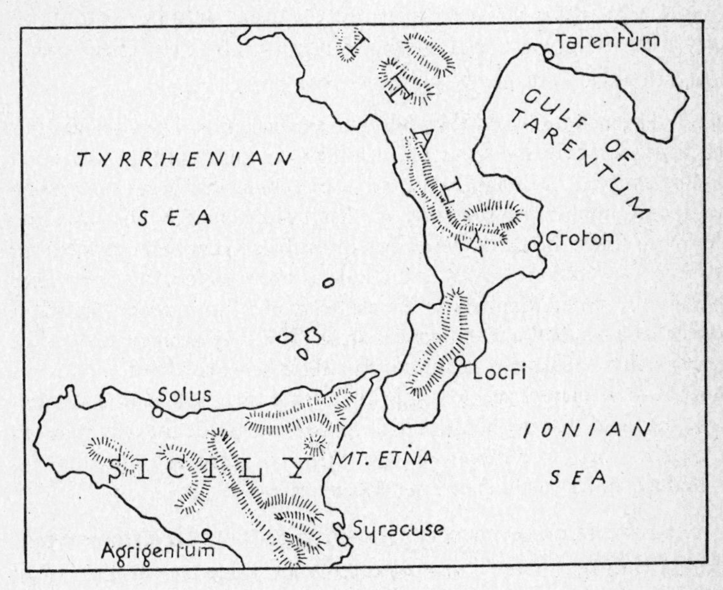

FIG. 12. Western Greek Colonies.

From the hazy philosophical outlook of the Pythagoreans there emerge certain ideas which have exerted a profound influence. Foremost is their peculiar teaching on the subject of numbers. These were held to have a real and separate existence outside our minds. The use by the Greeks, as by the Hebrews, of letters to express numbers gave an especial currency to this conception, which was capable of, and received, all sorts of mystical and magical application. An example will readily come to the mind in connexion with 666 'the number of the beast' in the book of Revelation (xiii. 18).[1] There was

[1] It is held that 666 represents the numerical value of *Nero Caesar* spelt in Hebrew letters.

a similar Pythagorean tendency to ascribe an objective independence to the divisions of time. Again a biblical illustration is to hand:

> Job cursed the day.
> Let that day perish wherein I was born,
> Let it not be joined unto the days of the year.
> (Job iii. 1–6.)

The very word *mathematics*[1]—which means simply 'learning'—was given its special relationship to numbers by the Pythagoreans. Aristotle tells us in his *Metaphysics* that

> the Pythagoreans devoted themselves to mathematics. They thought that its principles were the bases of all things. In numbers they saw many resemblances to the things that exist and are coming into being—one modification of number being *Justice*, another *Reason*, another *Opportunity*—almost all things being numerically expressible. Again they regarded the attributes and ratios of the musical scale as expressible in number. They therefore regarded numbers as the elements of all things, and the whole heaven as a musical and numerical scale. The very arrangement of the heavens they collected and fitted into their scheme. Thus, as 10 was thought to be perfect and to comprise in itself the whole nature of numbers, they said that the bodies which move through the heavens were ten in number; but since the visible heavenly bodies are but nine, they invented a counter-earth. (See Philolaus, p. 27.)

The conception seems very fanciful to us now. Nevertheless fancies of this type have been repeatedly of value in the history of science. The human mind, it must be supposed, is somehow attuned to the processes of nature. We live in a world that is susceptible of mathematical expression. Thus the theoretical investigations of mathematicians correspond in some degree to the findings of the physicists and astronomers. Such is the nature of things, though why this should be so is a mystery. Perhaps it is not even the business of science to discuss this mystery. But consciousness of a correspondence between the workings of our minds and the workings of nature is illustrated by this doctrine of the Pythagoreans. Their conception of the 'harmony of the spheres'—on which Aristotle touches in the

[1] Greek *mathēsis*, 'learning', *mathētēs*, 'disciple', so used in New Testament, *mathēmatikos*, 'fond of learning', so used by Plato and Aristotle. The word *mathematics* did not enter the English language till the late sixteenth century. The curious plural form is an elliptical expression for 'mathematic sciences' and has no foundation in Greek.

above passage—was related to an interest in music. It proceeded from the observation that the pitch of musical notes depends on a simple numerical ratio in the length of the cords struck. This numerical ratio, it was held, corresponded to the distances of the heavenly bodies from the centre of the world.

The beautiful conception of a world bound together in a harmony has captivated the imagination of poets in every age. There was a time

When the morning stars sang together
And all the sons of God shouted for joy.
(Job xxxviii. 7.)

It is the dullness of the ear of flesh, so the Middle Ages would have had us believe, that prevents us from hearing still these glorious tones. The Christianity, which set off body against spirit, at times would claim to catch the heavenly music;

soft stillness and the night
Become the touches of sweet harmony.
. . . .
There's not the smallest orb which thou behold'st
But in his motion like an angel sings,
Still quiring to the young-eyed cherubins;
Such harmony is in immortal souls;
But, whilst this muddy vesture of decay
Doth grossly close it in, we cannot hear it.
(*Merchant of Venice*, Act V, Sc. i, ll. 56–65.)

The Pythagorean habit of giving character and qualities to numbers becomes more intelligible to us if we remember that for the Greeks mathematics was, in effect, geometry. Thus, to take a prominent example, the Pythagoreans distinguished the series

$$1, 1+2, 1+2+3, 1+2+3+4, 1+2+3+4+5, \ldots$$

as *triangular* numbers, and they exhibited geometrically the interesting fact that the sum of any two consecutive *triangular numbers* is a *square number* (Fig. 13).

The so-called 'Pythagorean theorem', that is that the square on the hypotenuse of a right-angled triangle is equal to the sum of the squares on the other two sides (Fig. 8), was referred by the ancients to Pythagoras himself. The Pythagoreans erected a system of plane geometry in which were formulated the principal theorems which concern parallels, triangles, quadrilateral and regular polygonal figures and angles. They discerned many important properties of prime numbers and progressions. In particular they worked out a theory of proportion which involved both commensurables and incommensurables. This was of great importance as providing the link

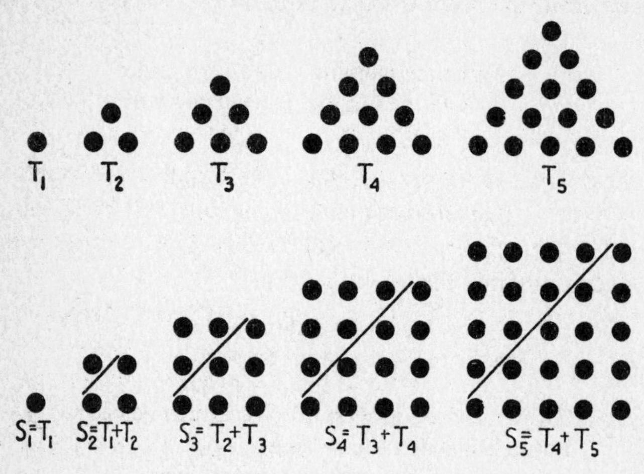

FIG. 13. Triangular and square numbers.

between arithmetic and geometry. They recognized at least three types of proportion. Thus:

arithmetical proportion $a-b = b-c$
geometrical proportion $a:b = b:c$
harmonic proportion $a-b:b-c = a:c$

The most striking mathematical achievement of the Pythagorean thinkers is perhaps their attainment of a conception of *irrational quantities*, such, that is, as are not expressible by ordinary numbers. With the imperfect mathematical notation of the time, however, great algebraical advance was impossible, and irrational numbers could not be algebraically represented (compare p. 122). Greek mathematics was thus forced to preserve its geometrical bias. The Greeks, in fact, constantly resorted to geometric methods when we should prefer algebraic. A very simple example will suffice. The equation

$(x+y)^2 = x^2+2xy+y^2$ was geometrically proved by reference to such a figure as that adjoining (Fig. 14).

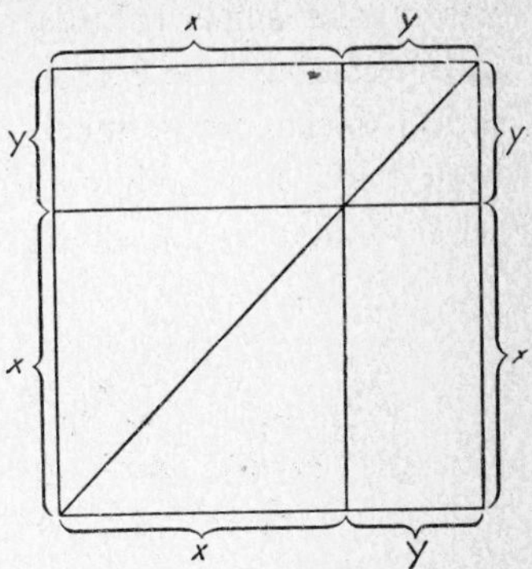

FIG. 14. Pythagorean presentation of equation $(x+y)^2 = x^2+2xy+y^2$.

Led by their mystical view that the sphere is the perfect figure, just as 10 is the perfect number, the Pythagoreans introduced the conception that the earth as well as the planets are spheres. This important advance is among the many in the history of science in which the formation of general ideas on theoretical grounds has preceded and not followed practical observation.

An interesting astronomical hypothesis was put forward by the Pythagorean PHILOLAUS of Tarentum (*c.* 480–400 B.C.). He abandoned the theory that the earth is the mid-point of the universe, and supposed that it is similar to the other planets in its movements, and that all revolve round a central fire. This fire, he held, is invisible to us, since the part of the earth which we inhabit is ever turned away from it. To balance his system he invented a *counter-earth*, bringing his spheres of the movable heavenly bodies up to the sacred number 10, that is to say, sun, moon, earth, five other planets, counter-earth, and sphere of the stars. Philolaus was the first to publish a book on Pythagorean doctrine. It was used by Plato in the composition of the *Timaeus* (p. 40). The conception by Philolaus of a moving earth and central fire influenced Copernicus (pp. 212 f).

Another Pythagorean development was destined to influence thinkers in after ages in a very curious way. Manipulating equilateral triangles and squares in three dimensions, the Pythagoreans discerned four 'regular solids', that is figures with all their sides and angles equal. These four were the regular 4-sided pyramid (*tetrahedron*), the 6-sided *cube*, the 8-sided *octahedron*, and the 20-sided *icosahedron*. They were taken to represent the four elements of the physical world, earth, air, fire, and water. Later was discovered the geometrical mode of constructing regular *pentagons* or 5-sided plane figures. One of the Pythagoreans found that these could be built into a fifth regular solid, the 12-sided *dodecahedron*. In the absence of a fifth element this was taken to represent the universe. The five possible regular solids became later known as the 'Platonic bodies'. They played a large part in subsequent philosophical and mathematical development, much of it very fanciful. Kepler's thought about the Platonic bodies in the sixteenth century suggested the first modern unitary theory of the universe (p. 236) (Figs. 15 and 104).

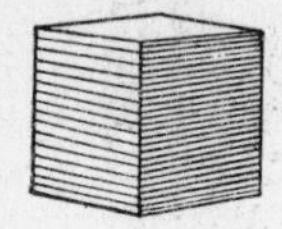

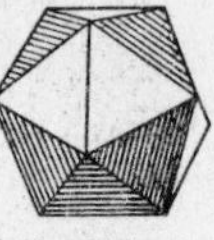
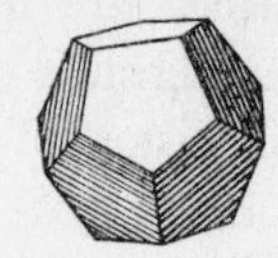

FIG. 15. The five Platonic bodies.

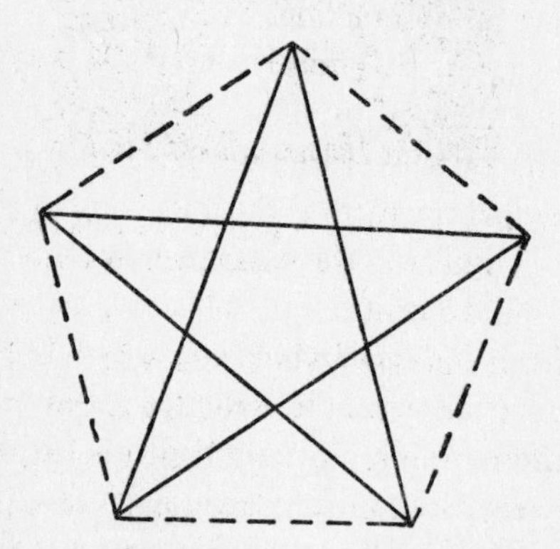

FIG. 16. Magic pentagram, a continuous line or 'endless knot' formed by producing the sides of a regular pentagon or by joining its alternate angles. Tying a ribbon in a loose knot, flattening and turning in the ends yields a regular pentagon.

From the regular pentagon it was easy to pass to the 5-pointed star or *pentacle*, forming an endless line by prolonging the sides of a pentagon. The Pythagoreans used the pentacle as a secret sign of recognition. It thus started on its career of mystery, passing into magic and humbug. For Pythagoreans and Platonists it expressed completeness, health, well-being. Among lesser souls it degenerated into the commonest and most banal of charms. No evil could pass it! Faust has a pentacle on the threshold of his study which prevents Mephistopheles from leaving it. The history of the pentacle provides a type of degradation that science has repeatedly suffered (Fig. 16).

It was not only in cosmical and mathematical speculation that the western colonies exhibited their intellectual activity. During the fifth century B.C. there developed among the Greeks in Italy and Sicily a remarkable naturalistic art. Painters closely observed and represented the parts and structures of animals (Fig. 17). This naturalistic

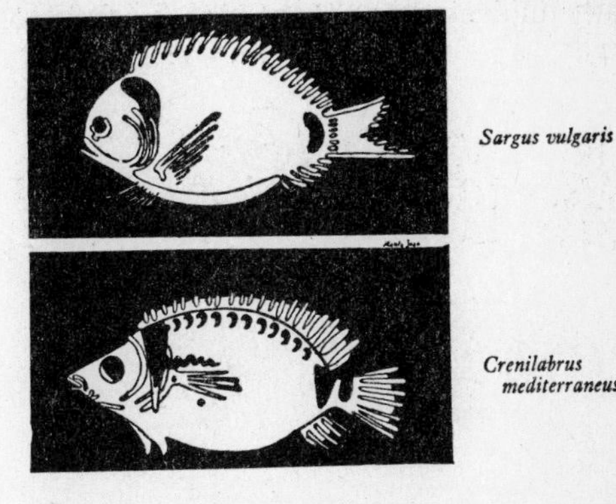

FIG. 17. Paintings of fish on plates from Magna Graecia of fourth century B.C. They are well drawn and of identifiable species.

tendency is reflected by the Italo-Greek scientific thinkers. Among them, ALCMAEON of Croton (*c.* 500 B.C.), a pupil of Pythagoras, extended the scientific field to living things. He began the practice of scientific dissection. He discovered the nerves that proceed from the brain to the eyes. He described those passages connecting mouth and ear, through which, if the nose be pinched and the cheeks blown out, air is driven into the ear-drums. These tubes were next investigated by the anatomist Eustachi, after whom they are now called *Eustachian tubes*. Eustachi lived in Italy more than twenty-two centuries after Alcmaeon! Alcmaeon believed that these tubes carried the *pneuma* (see Anaximenes, p. 17).

An important Western thinker, upon whom Pythagoras had influence, was EMPEDOCLES of Agrigentum in Sicily (*c.* 500–*c.* 430 B.C.). He held that the blood is the seat of the mysterious *innate heat*, an idea taken from folk belief that 'the blood is the life' (Deuteronomy xii. 23). This innate heat he closely identified with the soul. He held the heart to be the centre of the system of blood-vessels through which the innate heat, or essential factor of life, is distributed to the bodily parts. Thus for the followers of Empedocles the heart was the special seat of life. This idea passed to Aristotle (p. 50).

The teaching of Empedocles led to curiosity as to the distribution of the blood-vessels. Our first coherent account of these is in a fragment of the work of DIOGENES of Apollonia in Crete (*c.* 430 B.C.), who was greatly influenced by the thought of Empedocles and his school (Fig. 18).

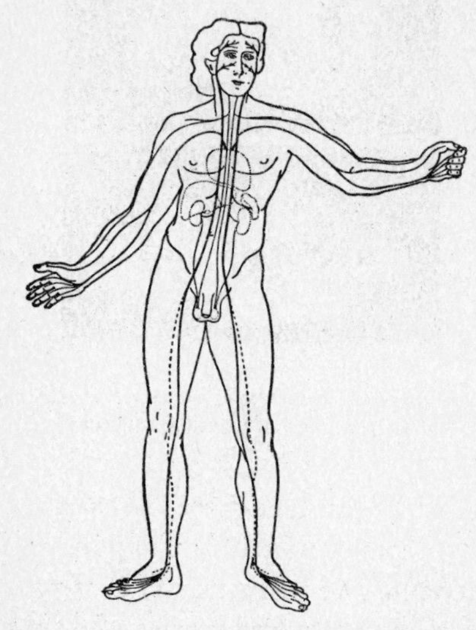

FIG. 18. Plan of blood vessels as described by Diogenes of Apollonia *c.* 440 B.C. He assumed a system of vessels penetrating the whole body and proceeding from great vertical trunks. He distinguished arteries from veins in form, function, and distribution.

Empedocles supposed that Love and Strife alternately held sway over all things. Everywhere there was opposition and affinity. In matter itself the so-called *four elements* could be distinguished as exhibiting these relationships. All matter was held by him to be composed of the four essential elements—*earth*, *air*, *fire*, and *water*. These were in opposition or alliance to one another. Thus water was opposed to fire, but allied to earth. Each of the elements was, moreover, in its turn compounded of a pair of the four 'primary qualities', heat and cold, moisture and dryness (Fig. 19). These qualities exhibit affinity and opposition as do the elements.

It must not be imagined that such philosophers as Empedocles thought that the 'elements' were the substances that we know by the names of earth, water, air, and fire on our earthly sphere. Here we find the 'elements' only in combination. Thus the substance we know as water contains, according to the theory, a preponderance of elemental water, but contains also small amounts of the other three

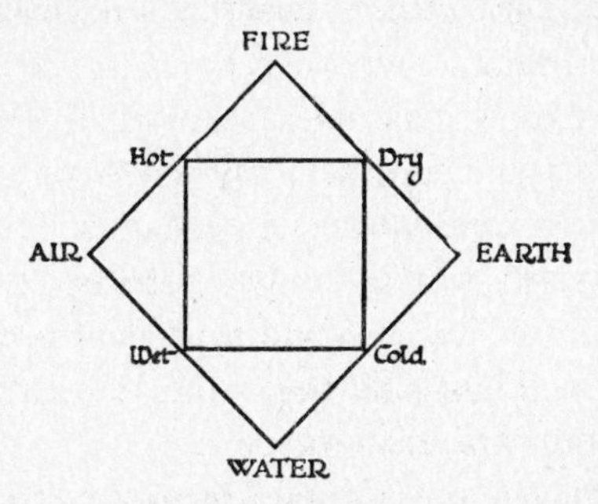

FIG. 19. The four Elements and four Qualities of Empedocles.

elements. The 'element' water forms only the essence of water, an essence that we human beings can never apprehend. And so for the other 'elements'.

This doctrine has left its mark on our language. We still speak of a storm as 'the raging of the elements'; we wear coats 'to protect ourselves from the elements'; and we think of 'elemental forces'. We still read the passage in Galatians in which St. Paul adjures us not to 'turn again to the weak and beggarly elements' (Galatians iv. 9); nor have we difficulty in understanding references to a 'fiery nature' or to an 'aerial spirit'. These things come to us from Empedocles, and they come through Aristotle (p. 54) and the Athenian School.

4. FATHERS OF ATHENIAN SCIENCE

By the middle of the fifth century B.C. both the Eastern and the Western schools of Greek thought were overshadowed by Athens, now the intellectual centre of the Greek world. An important factor in this concentration was the Ionian ANAXAGORAS (488–428 B.C.) of Clazomenae. He came to Athens (464 B.C.) burning with scientific zeal, and attracted the attention and friendship of the statesman Pericles (490–429 B.C.) and of the poet Euripides (480–406 B.C.), both of whom he inspired with his own love of science. Socrates (p. 36) differed profoundly from him. Much of the course of thought in later ages may be traced to this divergence, for Plato was the philosophic heir of Socrates while Aristotle took much from Anaxagoras.

Anaxagoras developed an obscure and difficult philosophic system which involved rational theories concerning many celestial phenomena. He gave scientific accounts of eclipses, meteors, and rainbows. The sun was a vast mass of incandescent metal, the light of the moon was reflected from it, and other heavenly bodies were stones rendered white hot by rotation. Such interpretation outraged religious opinion, and he, like Socrates, was prosecuted for impiety. Defended by Pericles and acquitted, he yet found it prudent to withdraw to his native Asia Minor. Thus early began the persecution of scientific doctrine opposed to current views.

The intellectual conditions in the Athenian metropolis were very different from those in the colonies of Ionia and Magna Graecia. In Athens the greater complexity of life was making itself felt. The systematic accumulation of knowledge was beginning to render a little old-fashioned those who 'took all knowledge to be their province'. The eloquence of the popular educators known as 'sophists' entertained and attracted the volatile Greeks beyond anything else. But many of the sophists were little but professional talkers, and few or none had any direct acquaintance with scientific matters, which were left to another class. Thus something in the nature of scientific specialization became necessary. The movement affected especially two departments, medicine and mathematics. By a curious chance, the two typical exponents of these disciplines bore the same name and came from neighbouring and somewhat similarly named islands. They were the physician, Hippocrates of Cos, and the mathematician, Hippocrates of Chios.

HIPPOCRATES THE PHYSICIAN was born about 460 B.C. on the island of Cos just inside the Dorian Zone. He came of a family of physicians. Both on his own island and on the opposite peninsula of Cnidus (Fig. 7) had been established medical schools whose destiny it was to transform the tradition into a scientific procedure. The change afterwards became very closely associated with the name of Hippocrates.

Hippocrates led a wandering life, following his profession in Thrace, in the neighbourhood of the sea of Marmora, on the island of Thasos, at Athens, and elsewhere. He had many pupils, among whom were his sons and sons-in-law. He is said to have died in his hundredth year, an appropriate age for a great physician. This is almost all we know of his personal history. Yet it is impossible to exaggerate the influence on medicine of the picture that was early formed of him. Learned, observant, humane, with a profound reverence for the claims of his patients, but possessed of an overmastering desire that his experience should benefit others; orderly and calm; anxious to record his knowledge for the use of his brother physicians and for the relief of suffering; grave, thoughtful, and reticent; pure of mind and master of his passions; such is the image of the father of medicine as it appeared to his successors.

While the philosophers developed the conception of a rational world, it was the physicians, typified by Hippocrates, who first put the rational conception to the test of experience. It was they who first consciously adopted the scientific procedure which, in its relation to medicine, is sometimes called the 'Hippocratic Method'.

The method of the Hippocratic writers is that now known as 'inductive'. Without the vast scientific heritage that is ours today; with but a small number of recorded observations and those from scattered and little organized experiences; surrounded by all manner of bizarre religious cults which recognized no adequate relation of cause and effect; above all, constantly urged by the exuberant genius for speculation of their own people whose intellectual temptations they shared, the 'Hippocratic' physicians remained, nevertheless, patient observers of fact, sceptical of the marvellous and the unverifiable, hesitating to theorize beyond the facts, yet eager to generalize from actual experience. There are few types of mental activity known to us that cannot be paralleled among the Greek writings. Careful and repeated return to verification from experience, expressed in a record of actual observations, has been rare at all times in history. It is wonderful that so many Greek works have come down to us expressing this attitude. A large proportion of these are by 'Hippocratic' authors.

It is true that the Greeks had scientific forebears (pp. 6–12). It is probable that they borrowed, more frequently than we know, from other civilizations. But the 'Religion of Science' of these early physicians, the belief in the constant and universal sequence of cause and effect in the material world, was theirs before all other men. The first prophet of that religion was Thales. The first writings on that religion bear the name of Hippocrates. The first great exponent of that religion whose works are still substantially intact is Aristotle (pp. 45 ff.).

The Hippocratic writings, important for the history of medicine, are even more significant for the conception that they contain of the nature of science itself. This conception is beautifully expounded in a treatise on the *falling sickness*, or epilepsy. In those days the affliction was regarded as a divine visitation, a 'sacred disease'. A Hippocratic writer composed a book on it, in which he sets forth the proper attitude of the scientific man towards such claims. It is a monument of the rational spirit, and is perhaps the first book in which there is clear opposition between the claims of science and of religious tradition.

In our own time natural events are not always treated, even by educated men, in the spirit of the Hippocratic writers. Both leases and insurance certificates have still sometimes a clause as to the type of accident to which the lawyers refer as an 'act of God'. The type of these acts of God has altered in the course of ages. They used to include, for instance, infectious disease. Our word 'plague' is from a Latin word meaning a *blow* or *stroke* which comes to us from the days when the 'plague-stricken' were held to be stricken by God himself. The legal term 'act of God' still includes the action of tempest and of lightning. Yet the attitude of the Hippocratic work called *The Sacred Disease*, written more than 400 years before the birth of Christ, is very different:

As for this disease called divine, surely it has its nature and causes, as have other diseases. It arises—like them—from things which enter and quit the body, such as cold, the sun and the winds, things ever changing and never at rest. Such things are divine or not—as you will, for the distinction matters not—and there is no need to make such division anywhere in nature, for all are alike divine or all are alike natural. All have their antecedent causes which can be found by those who seek them. [Slightly paraphrased.]

We have spoken of the belief in the constant sequence of cause and effect as a 'religion' (p. 34), since it was—and perhaps still is—essentially a matter of faith. In Hippocratic times there was as yet no large body of exact observations by which the operations of nature could be exactly forecasted, save only the astronomical record. Thus the regularity of the astronomical sequences was, by an act of faith, set forth as the type to which all nature should accord. The heavenly bodies herald those regularly recurring changes of season which determine the lives of men. It is but a step to regard them as causes of those changes and to treat them as gods. The step was often taken. The planets still bear the names of deities and even Aristotle is touched by this idea (p. 54).

HIPPOCRATES OF CHIOS, the mathematician (*c.* 430 B.C.), was the first to compile a work on the *Elements of Geometry*. This title has made a household word of his successor, Euclid (pp. 63 f.). Hippocrates of Chios is the first known mathematical 'specialist'. He began life as a business man. Chance brought him to mathematics. He came on a law-suit to Athens. That city was rapidly becoming the centre of learning, and the provincial Hippocrates had now an opportunity to consort with philosophers. His real abilities rapidly asserted themselves, and he began to devote himself with ardour to mathematical pursuits.

The work of Hippocrates of Chios may be illustrated by one of his most acute investigations. It gives an idea of the standard which mathematics had attained in Greece about 400 B.C. Hippocrates discovered that the *lune* bounded by an arc of 90°, and by a semicircle upon its chord, is equal in area to the triangle formed by the corresponding chord with the centre as its apex (Fig. 20). The lune--a figure bounded by curves—being thus equated with a figure bounded by straight lines, its area can be ascertained. He discovered two other lunes of which the areas could be similarly expressed. Finally, he discovered a particular lune which, when added to a circle, enables the

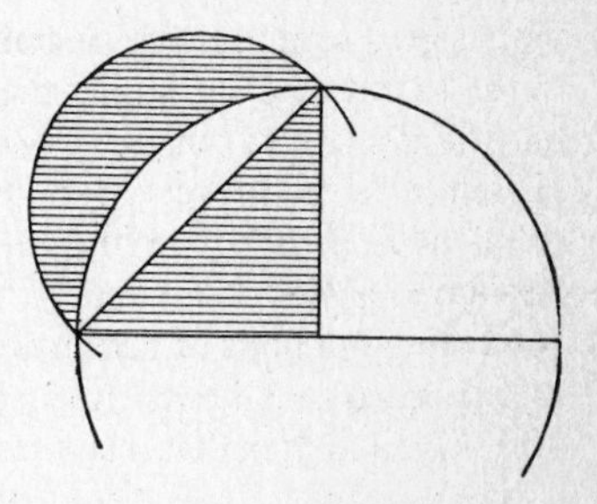

FIG. 20. A lune of Hippocrates of Chios.

whole to be represented geometrically as a square. This lune by itself cannot, however, be squared, and so the method cannot be used for squaring the circle. These remarkable researches became misrepresented and tradition told that Hippocrates *had* succeeded in the impossible geometrical task of squaring the circle. His proofs, in fact, imply great familiarity with advanced geometric methods. They are based on the theorem, which he himself proved, that circles are to one another as the squares of their diameters.

Thus by the end of the fifth century not only had philosophical thought taken a scientific turn, but science itself had emerged as a preoccupation of men set aside from their fellows. Two departments, medicine and mathematics, had become well differentiated. Astronomy had been the special interest of such philosophers as Pythagoras (pp. 22 ff.), Philolaus (p. 27), Empedocles (p. 29), and Anaxagoras (p. 31). This earlier phase of Greek thought terminated in the fifth century with a thinker of a very individual type.

The name of SOCRATES (470–399 B.C.) is associated with a great intellectual revolution, perhaps the greatest that the world has seen. His overwhelming preoccupation was with conduct. For him 'Knowledge is Virtue'. The attitude of Socrates towards the sciences of his day has been set forth by his pupil Xenophon (430–350 B.C.), who tells that

> with regard to astronomy Socrates considered a knowledge of it desirable to the extent of determining the day of the year or of the month and the hour of the night; but as for learning the courses of the stars, occupying oneself with the planets or inquiring about their distance from the earth or about their orbits or the causes of their movements, to all these he strongly objected as a waste of time. He dwelt on the contradictions and conflicting opinions of the physical philosophers . . . and, in fine, he held that speculators on the Universe and on the laws of the heavenly bodies were hardly better than madmen.

There is an aspect of the thought of Socrates that had a profound, though quite indirect, effect on science. He laid great stress on the soul and its persistence after death. This view carried with it a contempt for the body as a mere temporary habitat thrown off by the soul. It was this disregard of the body that, a hundred years later, made human dissection possible and hence the rise of anatomy in Greek circles at Alexandria (pp. 66 ff.).

The triumph of the Socratic revolution depressed for a while both science and physical philosophy. But out of the conflict between the Socratics and the physical philosophers arose the main streams of later Greek thought. These two streams derive their titles and their tendencies from two gigantic figures of the fourth century, the age of Plato and Aristotle.

II

THE GREAT ADVENTURE

Unitary Systems of Thought: Athens, 400–300 B.C.

I. PLATO AND THE ACADEMY

THE thought of PLATO (427–347 B.C.), like that of his master Socrates, was dominated by the ethical motive. Convinced, like Socrates, that Truth and Good exist and that they are inseparable, he embarked on an inquiry which had as its object to expose, account for, and resolve into one comprehensive theory the discrepancies of ordinary thinking. During this process he developed a doctrine destined to be of great moment for the subsequent relation of scientific thought to religion and philosophy. It is the so-called *Doctrine of Ideas.* The nature of this doctrine and the manner in which Plato reached it have been briefly set forth by his pupil, Aristotle.

In his youth [says Aristotle] Plato became familiar with the doctrine of certain philosophers that all things perceived by the senses are ever in a state of flux and there is no knowledge concerning them [see Heracleitus, p. 20]. To these views he held even in his later years. Socrates, however, busied himself about ethical matters, neglecting the world of nature, but seeking the universal in conduct. He it was who fixed thought for the first time on definitions. Plato accepted his teaching but held that the problem of what was to be defined applied not to anything perceived by the senses but to something of another sort. His reason was that there could be no real definition of things perceived by the senses because they were alway changing. Those things which could alone be defined he called *Ideas*, and things perceived by the senses, he said, were different from these *Ideas* and were all called after them. (Aristotle's *Metaphysics*.)

Thus concepts, things of the mind, became for Plato something more concrete, while our impressions of the material universe, percepts, became something more vague. It is as though the word 'horse' were to suggest to the mind not Ned or Dobbin or even a cart-horse or a carriage-horse but a generalized being that is approximately expressed by the biologist's definition of the *species* horse. Further this 'Idea' of the species was more truly an entity than any individual horse. The Platonic 'Idea' contained in it the conception of form, for only in the Idea was the form separated from matter. The conception is put epigrammatically by Plato in the phrase 'the Soul is the place of forms',[1] that is, of those forms which can be defined.

Plato expresses a great admiration for mathematical principles, and he regards mathematics as exhibiting that type of certitude and exactness to which other studies should conform. Mathematics indeed relies for its material upon something of the nature of Plato's Ideas. It might be expected, therefore, that mathematics would appeal to him. Many of Plato's thoughts assume a mathematical guise. He exhibits at times a view which seems to approach that of Pythagoras, who had attached a moral and spiritual value to numbers (pp. 23 ff.). Plato thus tended to respect a science in the degree to which it had progressed in the mathematical stage of its development. The heavenly bodies evinced, in the opinion of those Pythagorean days, the exemplars of perfect geometric forms (p. 25). For astronomy—especially on its theoretic as distinct from its observational side—Plato had therefore a high regard. Indeed, for many of his Greek followers mathematics became identified with astronomy. We think of astronomy as a field for the *application*, the Platonists rather for the *exemplification* of mathematics.

The attitude of Plato was less favourable to those sciences, other than astronomy, to which we nowadays habitually apply our mathematics. On the non-mathematical sciences he smiled even less. He repudiated the theories of such thinkers as Democritus, who not only denied the existence of mind as a separate entity but also assumed the universe to be the result of accident (p. 20). Such a universe was hardly susceptible of exact presentation. In ultimate analysis the position of Democritus was a denial of the validity of philosophy. On the other hand, Plato speaks with respect of Hippocrates the physician, the very type of the scientific man in antiquity—Hippocrates of whom a follower said 'he was the first who

[1] The phrase is not found in the extant works of Plato but is quoted by Aristotle in the *De anima* (p. 47).

separated science from philosophy'.[1] Plato's respect for Hippocrates, however, did not tempt him to follow in his footsteps. Nor is this surprising, for, firstly, Plato assigned a relatively unimportant place to phenomena and, secondly, his mind was too full of a greater vision to enable him to lend himself to the tedium of the pursuit of the inductive method.

Nevertheless, the greatest of thinkers could not refrain from producing some general theory of the universe of phenomena. The work in which this appeared, the dark and difficult *Timaeus*, is under strong Pythagorean influence (pp. 22 ff.). Its spokesman is a member of that sect. Its very darkness and difficulty provide an unintentional appeal for that patient, impartial objective process of observation and record that is the very foundation of science. The *Timaeus* demonstrates how knowledge can be degraded, even by Plato, in the relentless endeavour to explain the universe rather than to describe it. The work displays the Platonic mood at its weakest.

The trend of Platonism in general and of ancient Platonism in particular has normally been away from observational activity, even when friendly to mathematics. There are, however, many and evident exceptions and, moreover, Platonism has often been helpful to science in the presence of an entrenched and static Aristotelianism (pp. 133 f., 138, 145, 155 f., 164 f., 191, 209, 217, 219 f. &c.)

It has been said that 'everyone is by nature a disciple either of Plato or of Aristotle'. There is much truth in this. Aristotle himself set forth the difference between the two attitudes, reduced to its simplest expression. In his great work, the *Physics*, Aristotle discusses the use of mathematical formulae. The objects studied in the physical sciences, he says, do present, of course, planes, lines, and points. Such planes, lines, and points are the subjects also of mathematical study. How, then, are we to distinguish the procedure of mathematics from that of the true physical sciences which often invoke mathematics?

To this, Aristotle answers that the mathematician does indeed study planes, lines, and points, but he studies them as mental abstractions and not as the 'limits of a physical body'. The objects of mathematics, though in fact inseparable from a physical, movable,

[1] Aulus Cornelius Celsus, *De re medica*, Introduction.

and therefore changeable body, are studied in abstraction from that change to which all material things are subject. This process of abstraction necessarily involves error. The mistake made by Plato's theory of Ideas, says Aristotle, is that of attempting to exclude from his consideration of matter those conceptions in which are involved the very nature of matter, though not that of mathematical objects. Thus odd and even, straight and curved, number, line, figure—all these can be studied wholly out of connexion with the change or movement inseparably connected with material things. They are subjects for the mathematician. Such things as flesh, bone, man, nay, even inorganic nature, minerals and earths, sounds and colours, heat and cold, cannot be so studied. They are subjects for the man of science. Change is indeed an essential part of nature, fundamental to real existence, as Thales, 'the father of science', had seen (p. 13) and Heracleitus with his 'being as becoming' had emphasized (p. 20). Yet change has to be ignored in pure mathematical investigation.[1] This principle of change or movement prevents nature from ever really repeating herself, while in mathematical conceptions one unit is exactly like another.

We may see the contrasted effects of the Platonic and the Aristotelian attitudes in the scientific works of the two great philosophers. So far as science is concerned, it is by their fruits that we must know them. Plato has shrouded his views in the *Timaeus*. From the deceptive shadows seen in the twilight of that work he has elevated into picture form, from an 'Idea', a mechanism that never was on sea or land. On the other hand, in the great biological works of Aristotle we have a magnificent series of first-hand observations and positive studies to which, in each succeeding generation, naturalists still return with delight, with refreshment, and with respect (pp. 45–51).

The importance of Plato, so far as the subsequent development of science is concerned, is thus to be sought chiefly in the department of mathematics. Plato was, in fact, an accomplished mathematician and had had Pythagorean teachers. The 'Platonic bodies', the five regular solids which have equal sides and equal angles, were known to the Pythagoreans (p. 28). Plato describes them in the *Timaeus*, exhibiting full understanding of them. There are many other passages

[1] Ignored, that is, in mathematical investigation as then understood.

in his writings which show mathematical penetration, nor is it easy to overrate his influence upon later mathematical developments. We may consider it under four headings:

(*a*) It is through Plato that mathematics obtained, and retains, a place in education. In the abstractions of mathematics he saw an instrument for the training of logical thought. The study of mathematics was thus for him the portal to philosophy. 'Let none who has not learnt geometry enter here' was inscribed over the entrance to his school, the Academy.

(*b*) The hand of Plato may be traced in the actual course of mathematical development. To his logical teaching the body of mathematical knowledge owes the systematic structure and logical finish that have since distinguished it. This factor exhibited itself in his pupils and his spiritual descendants. Such a work as Euclid's *Elements* is in essence a product of Plato's thought and of Plato's school (p. 63). It is certainly no overstatement that, through Euclid, as presented in elementary geometry, every schoolboy is nowadays a student of Plato.

(*c*) The inspiration of Plato can be traced very clearly also in the history of astronomy. He early came to regard the irregularities of planetary motion as inconsistent with his view of the essential perfection of the universe. These movements had, in his opinion, to be explained as somehow compounded of simple circular movements, a conception that he derived from his Pythagorean teachers (p. 24). Plato accordingly set his pupils to seek out rules by which the movements of the heavenly bodies could be reduced to a system of circles and spheres. This was the main task of astronomers from his time to Kepler (pp. 240 f.)—a stretch of two thousand years. During all those centuries the hand of Plato ruled astronomy. Here Aristotle (pp. 52 f.), like Copernicus (pp. 212 f.), is but a pupil of Plato as Plato is of Pythagoras.

(*d*) Plato may be said to have made a positive contribution to science of first-class importance. It cannot be said that this is wholly his creation, since the germs of it are to be found among the Pythagoreans, but its formal introduction is Plato's work. It is the method of assuming that a problem is solved and working back from it until a statement is reached, the truth or falsehood of which is already known. Thus may be discerned whether the problem is, in fact, soluble or not, and indications may be forthcoming as to the general direction of the solution and whether there are any limitations to it. The method is set forth in the *Meno*. Euclid (p. 63) used this method as did Archimedes (pp. 69 f.) and it is current in modern geometry.

There is a curious Platonic conception that is perhaps a mere by-product of Plato's thought but was yet fraught with consequences for after ages. The Pythagorean Timaeus, in Plato's dialogue of that name, pictures the universe as a living thing with a soul penetrating its body. The passage is well summarized by Aristotle:

> Timaeus tries to give a physical account of how the soul moves its body. The soul is in movement and the body moves because it is interwoven with it. The Creator compounded the soul-substance out of the elements and divided it according to the harmonic numbers (p. 26) that it might have an innate perception of harmony and that its motion might be with movements well attuned. He bent its straight line into a circle. This he divided into two circles united at two common points. One of these he divided into seven circles [that is the orbs of the seven planets] in such wise that the motions of the heavens are the motions of the soul (*De anima*).

This view of the universe gave a framework for the neo-Platonic conception that the structure of the universe foreshadowed that of man. Thus arose the doctrine of the intimate relation of *macrocosm* ('great world') and *microcosm* ('little world', that is, Man). This doctrine permeated medieval Christian thought (pp. 133 f.).

Plato's school, under the name of the Academy, persisted for many centuries, but was chiefly occupied with metaphysical discussion. One of his first disciples to distinguish himself in science was EUDOXUS (409–356 B.C.) of Cnidus, the founder of observational cosmology. Eudoxus had also studied with the Pythagoreans. Under the stimulus of Plato he made advances in mathematical theory, but occupied himself chiefly with examining the heavens. Among his achievements is his remarkably accurate estimate of the solar year as 365 days and 6 hours. His most influential contribution was his view that the heavenly bodies move on a series of concentric spheres, of which the centre is earth, itself a sphere. Eudoxus had observed the irregularities in the movements of the planets. To explain these he supposed each planet to occupy its own sphere. The poles of each planetary sphere were supposed to be attached to a larger sphere rotating round other poles. The secondary spheres could be succeeded by tertiary or quaternary spheres according to mathematical and observational needs. For sun and moon Eudoxus found three spheres each sufficient. In the explanation of the movements of the other planets, four spheres each were demanded. For the fixed stars one sphere sufficed. Thus twenty-seven spheres in all were demanded. These spheres—save that of the fixed stars—were treated by Eudoxus not as material but in the manner of mathematical constructions.

CALLIPUS of Cyzicus, a pupil of Eudoxus and friend of Aristotle,

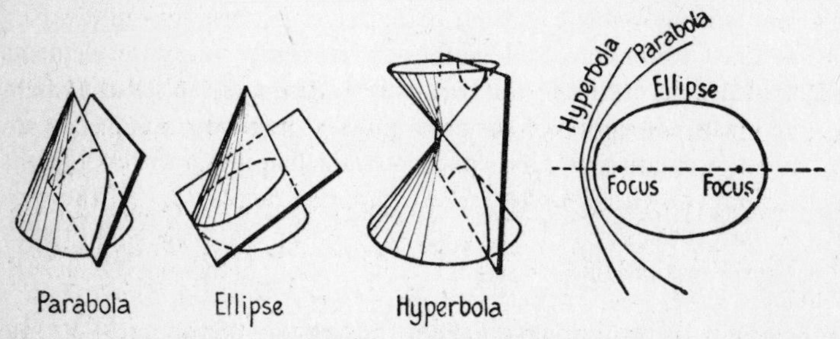

FIG. 21. Types of curve obtained by section of cones by planes. Compare Figs. 30, 99, 103, 105, 127, 128 and 175.

observed movements of the heavenly bodies and irregularities unknown to his master. To explain these he added yet further spheres, making thirty-four in all. The Eudoxan theory thus modified was adopted by Aristotle (pp. 52 f.).

HERACLEIDES of Pontus (*c.* 388–315 B.C.), a pupil of Plato, contributed to astronomy a suggestion that the earth rotates on its own axis once in twenty-four hours, and that Mercury and Venus circle round the sun like satellites. His teaching led on to that of Aristarchus (p. 65).

Important for subsequent mathematical developments was MENAECHMUS, another pupil of Eudoxus. Menaechmus initiated the study of conic sections. He cut three kinds of cone, the 'right angled', the 'acute angled', and the 'obtuse angled', by planes at right angles to a side of each cone. Thus he obtained the three types of conic section which we now call by the names allotted to them by his Alexandrian successor Apollonius (p. 75) (Fig. 21).

Many others of Plato's followers made contributions to pure mathematics, and, in the sense which we have discussed (p. 42), all subsequent mathematicians are Plato's spiritual heirs. There is also evidence of a certain amount of botanical activity in the Academy, and some physiological theories which became popular in later centuries may be traced to Plato. Platonism passed into Christianity early, mainly through St. Augustine, so that the Christian Middle Ages, until the twelfth century, were mainly Platonic. The later school of philosophy known as 'neo-Platonism' also profoundly influenced St. Augustine and Christianity (p. 134).

2. ARISTOTLE

ARISTOTLE (384–322 B.C.) was born at Stagira, a Greek colony a few miles from the northern limit of the present monastic settlement of Mount Athos. His father was physician to the monarch of Macedon. At seventeen Aristotle became a pupil of Plato at Athens. On his master's death in 347 he crossed the Aegean Sea to reside in Lesbos, an island off the coast of Asia Minor. In 342 he became tutor to the young prince Alexander of Macedon. He remained in Macedon till 336 when Alexander started his career of conquest that was to alter the face of the world. Aristotle then returned as a public teacher to Athens. There he owned a garden known as the *Lyceum*, whence the word has derived its special significance. In it he established his famous school afterwards called the *Peripatetic* (Greek, 'walking around'), for he had there his favourite *Peripatos* or cloister where he lectured.

Aristotle's writings cover the whole area of knowledge. The earliest are biological. These were written, or at least drafted, during his residence in Asia Minor (347–342). Most of his other works were produced during his second period at Athens (335–323), in the twelve years that preceded his death. We must always remember that the whole of Aristotle's science, and indeed the whole cast of his mind, was deeply influenced by his biological experience.

Regarded from the modern scientific standpoint, Aristotle appears at his best as a naturalist. His first-hand observations are on living things, and his researches on them establish his claim to be regarded as a man of science in the modern sense. In his great work *On the Parts of Animals*, he sets forth what he regards as the relation between 'physics'—which is for him a general description of the universe—and the study of living things.

Of things constituted by nature [he says] some are ungenerated, imperishable, eternal; others subject to generation and decay. The former are excellent beyond compare and divine, but less accessible to knowledge. The evidence that might throw light on them, and on the problems which we long to solve respecting them, is furnished but scantily by our senses. On the other hand, we know much of the perishable plants and animals among which we dwell. We may collect information concerning all their various kinds, if we but take the pains.

Yet each department has its own peculiar charm. The excellence of celestial things causes our scanty conceptions of them to yield more pleasure than all our knowledge of the world in which we live; just as a mere glimpse of those we love is more to us than the grandest vista. On the other side we may set the certitude and completeness of our knowledge of earthly things. Their nearness and their affinity to us may well balance the loftier interest of the things of heaven, that are the objects of high philosophy.

But of a truth every realm of nature is marvellous. It is told that strangers, visiting Heracleitus (p. 20) and finding him by the kitchen fire, hesitated to enter. 'Come in, come in', he cried, 'the gods are here too.' So should we venture on the study of every kind of creature without horror, for each and all will reveal something that is natural and therefore beautiful. Absence of haphazard and conduciveness of all things to an end are ever to be found in nature's works, and her manner of generating and combining in ever-changing variety is of the highest form of the Beautiful. [Somewhat paraphrased.]

Though it cannot be claimed that Aristotle was an evolutionist in the sense that he regarded the different kinds of living things as actually related by descent, yet there can be no doubt that he fully realized that the different kinds can be arranged in a series in which the gradations are easy. His scheme was a 'Ladder of Nature' (Fig. 22) as it came to be called by later naturalists. Thus he writes in his *History of Animals*:

Nature proceeds by little and little from things lifeless to animal life, so that it is impossible to determine the exact line of demarcation, nor on which side thereof an intermediate form should lie. Thus, next after lifeless things in the upward scale, comes the plant. Of plants one will differ from another as to its amount of apparent vitality. In a word, the whole plant kind, whilst devoid of life as compared with the animal, is yet endowed with life as compared with other corporeal entities. Indeed, there is observed in plants a continuous scale of ascent toward the animal.

The peculiar principle that Aristotle invoked to explain living phenomena we may call 'soul', translating thereby his word *psyche*.

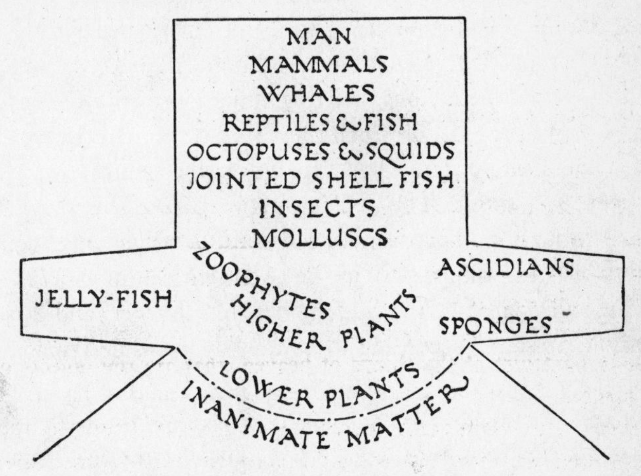

FIG. 22. Aristotle's 'Ladder of Nature'.

His teaching on that topic is to be found in his great work *On the Soul* usually cited by its Latinized title *De anima*. He thinks of things as either 'with soul' or 'without soul' (*empsychic* or *apsychic*). His belief as to the relationship of this soul to the matter in which it is embodied is difficult and complicated, but he tells us that 'Matter is identical with potentiality, form with actuality, the soul being that which gives the form or actuality in living things'. Thus for Aristotle 'soul' is not a separate existence. In this he differs from his master Plato and no less from early Christianity which, through St. Augustine (pp. 134 f.), borrowed much from Plato. Aristotle believes, too, that the soul works ever to an end, and that

As every instrument and every bodily member subserves some partial end, some special action, so the whole body must be destined to minister to some fuller, some completer, some greater sphere of action. Thus an instrument such as the saw is made for sawing, since sawing is a function, and not sawing for the saw. So, too, the body must somehow be made for the soul and each part thereof for some separate function to which it is adapted. [*Parts of Animals*, somewhat paraphrased.]

Aristotle is thus a *vitalist* (Latin *vita*, 'life') and a *teleologist* (Greek *telos*, 'end', 'object'), that is to say, he believes that the presence of a certain peculiar principle is on the one hand essential for the exhibition of any of the phenomena of life, while on the other hand it serves to integrate all such phenomena towards the emergence of the perfect living individual. The Democritans, to whom Aristotle was opposed, believed that all the actions of living things were the result of the interaction of the atoms of which they were composed (p. 20). Thus life, for the Democritans, was capable of mechanical expression. They were *mechanists*. The division between *vitalist* and *mechanist* extends throughout the history of science and has ceased to separate students of living things only in the twentieth century.

Living things are for Aristotle the type of existence, and existence as a whole presents, according to him, evidence of design.

Everything that nature makes is a means to an end. For just as human creations are the products of art, so living objects are manifestly the products of an analogous cause or principle. ... That the heaven is maintained by such a cause, there is, therefore, even more reason to believe than that mortal animals so originated. For order and definiteness are even more manifest in the celestial bodies than in our own frame. ... Thus Nature is marvellous in each and all her ways. [*Parts of Animals*, greatly abbreviated.]

Aristotle attempted to analyse the nature of generation, of heredity, of sex. His are the first presentations of many such topics which are today discussed by naturalists. There is an amazing variety and depth in his biological speculations. These have a permanent value and are constantly cited by biologists of our own time.

Aristotle's psychological studies are only partly within our purview. The psychological questions with which we are concerned come mostly into his discussion of the nature of life. 'Of natural bodies,' he says, 'some possess life and some do not; where by life we mean the power of self-nourishment and of independent growth and decay'. It should be noted that in the Aristotelian sense the egg or germ is not at first a living thing, for in its earliest stages and before fertilization it does not possess 'soul' even in its most elementary form.

In a famous passage from his work *On the Soul* Aristotle says:

> The term life is used in various senses. If life be present in but a single one of these senses, we speak of a thing as alive. Thus, there is intellect, sensation, motion from place to place and rest, the activity concerned with nutrition, and the processes of decay and growth. Plants have life, for they have within themselves a faculty whereby they grow and decay. They grow and live so long as they are capable of absorbing nutriment. In virtue of this principle [the vegetative soul] all living things live, whether animals or plants, but it is sensation which primarily constitutes the animal and justifies us in speaking of an *animal soul*. For, provided they have sensation, creatures, even if incapable of movement, are called animals. As the nutritive faculty may exist, as in plants, without touch or any form of sensation, so also touch may exist apart from other senses.

Apart from these two lower forms of soul (*a*) the *vegetative*, or nutritive and reproductive, and (*b*) the *animal*, or motile and sensitive soul, stands (*c*) the *rational* or conscious and intellectual soul that is peculiar to man.

The possession of one or more of the three types of soul, vegetative, animal, and rational, provides in itself a basis for an elementary form of arrangement of living things in an ascending scale. In fact, the basis of Aristotle's 'Ladder of Nature' (p. 47) is really psychological, depending on the character of soul or mind. It is characteristic of Aristotle's method that the various departments of investigation should thus interlock.

In the closest possible association with Aristotle's biological views stand his innumerable and admirable observations. Among the more striking are the following:

(*a*) A series of records of the life and especially the breeding habits of a large variety of animals. About 540 species are discussed.

(*b*) Embryological investigations of the developing chick, which has ever since been the classic object for such investigations.

(*c*) Accounts of the habits and development of the octopuses and squids which have, in some cases, been surpassed only in modern times.

(*d*) Anatomical descriptions of the four-chambered stomach of the ruminants, of the complex relationships of the ducts and vessels in the mammalian generative system and of the mammalian character of the porpoises and dolphins, all unsurpassed until the sixteenth century.

(*e*) Accounts of exceptional modes of development of fish. Among them is one of a species of dogfish of which the young is linked to the womb by a navel cord and placenta, much in the manner of a mammal. Nothing has contributed more to Aristotle's scientific reputation in modern times than the rediscovery of this phenomenon.

(*f*) As a result of his embryological investigations Aristotle attached very great importance to the heart and vascular system. He came to regard the heart as 'the first to live and the last to die',[1] a conception which passed to the Middle Ages and was current until the eighteenth century.

(*g*) A lasting addition to the technique of scientific instruction was made by Aristotle in introducing diagrams to illustrate complex anatomical relations. Some of his diagrams can be restored from his descriptions (Fig. 23).

Most of Aristotle's biological work reads like that of a modern naturalist, for his methods are closely similar to those of our own time. But when we turn to examine Aristotle's view of the universe we encounter not only a different method of work but a mode of thought so diverse from ours that we can neither understand nor sympathize with him without some special study. The intellectual revolution of the insurgent century (Ch. VII) resulted in complete destruction of the Aristotelian physical philosophy. Modern science is the product of that revolution, and it is difficult for us to go behind it in our thinking.

We are all of us brought up from early years with the idea of the 'uniformity of nature', that is that the same causes always and everywhere produce the same results. Thus, for instance, we think of astronomers exploring the heavens and discovering new facts about worlds other than our own. We assume, and we are justified in assuming, that in the starry spaces there rule the general physical laws which we have learned on our earth. On this principle astronomers

[1] This phrase is often given as a quotation from Aristotle. It occurs, however, nowhere in his writings, though the idea is to be found there.

deduce, for instance, the exact chemical constitution of many of the stars. Did we question ourselves on this matter, we might, perhaps, ask, if the physical laws that we know on earth did not prevail in the stars, how could astronomers make discoveries at all? But this law

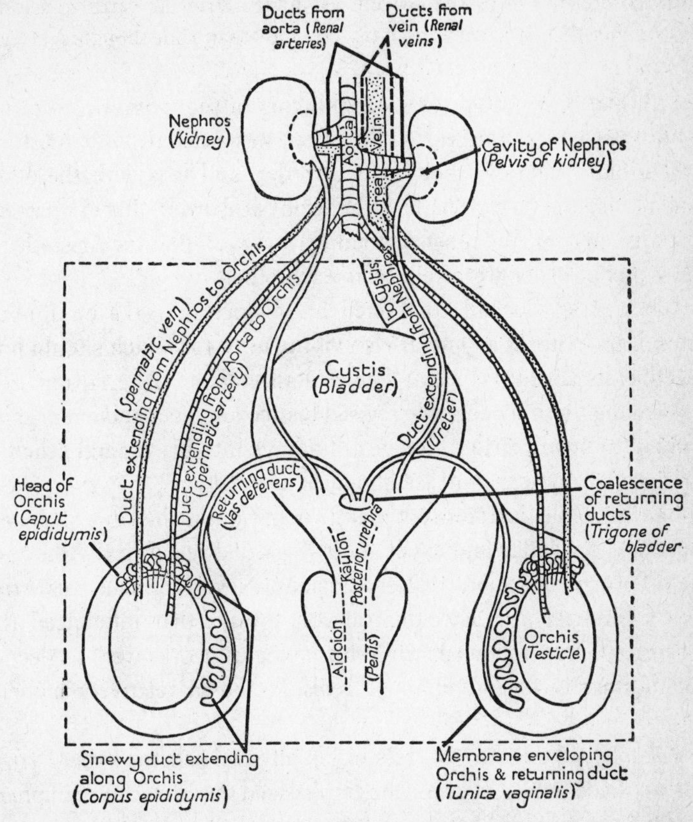

FIG. 23. Diagram of generative and excretory systems of a mammal as described by Aristotle in his *Historia animalium*. Words in brackets are modern and the others transliterations or translations of Aristotle's terms. The dotted line represents the limit of Aristotle's diagram.

of uniformity that we take for granted was by no means obvious to Aristotle. To him heaven was not only different from earth, but its ways were incommensurate with the ways of earth.

Aristotle knew nothing of the book of Isaiah. But his philosophical distinction between the rules of heaven and of earth made a special appeal to the Church fathers and to his medieval followers who had read that book. It was brought nearer to them by a superb and oft-quoted passage,

> My thoughts are not your thoughts, neither are your ways my ways, saith the Lord. For as the heavens are higher than the earth so are my ways higher than your ways, and my thoughts than your thoughts. (Isaiah lv. 8, 9.)

Isaiah, like Socrates (p. 36), was thinking of the moral order in his contrast of heaven and earth. So, often, was Aristotle. But Aristotle was thinking also of other kinds of order, and it is with the other kinds of order, and especially with the physical order, that our present work has to deal. We must remember, however, that for Aristotle all the kinds of order were related to each other.

When Aristotle had completed his biological works he applied himself to set forth a general view of the universe which should link together its various aspects. The structure of the material universe was among these aspects. He revised his account over and over again, seeking to fit his earlier biological findings into his general scheme. Aristotle's physical and astronomical conceptions, however, were unlike his biological conceptions in being untouched by profound personal knowledge and experience. Regarded scientifically they are far inferior to his biological conclusions. Nevertheless it was Aristotle's physical and astronomical conceptions that influenced the centuries which followed, while his biological works were neglected and ultimately forgotten, to be rediscovered in relatively modern times.

Aristotle, like Plato, exhibits in his physical scheme some Pythagorean tendencies. Especially he emphasized the circle and the sphere as the most 'perfect' figures and therefore those on which the world is modelled. Thus he was led to regard the heavens as a series of concentric spheres arranged round our earth as a central body (Fig. 24). These spheres he described, however, as crystalline, mechanizing them from the mathematical scheme of Eudoxus (p. 43). Around our earth was the sphere of the atmosphere and around that spheres of pure elemental nature, being, from within outward and in order of density, earth (or rather earthy exhalation), water, air, and fire (Fig. 68). These spheres of pure elements are as inaccessible to us as the heavens themselves. Next, beyond the sphere of elemental fire, lies the region of a yet more mysterious substance, the *ether* (Greek 'shining') which enters into the composition of the heavenly bodies. Yet farther out are in succession the seven spheres, each of which carries a planet, while beyond is the eighth sphere which bears the fixed stars. Finally, beyond all others, is the sphere whose divine harmony causes the circular revolution of the whole celestial system.

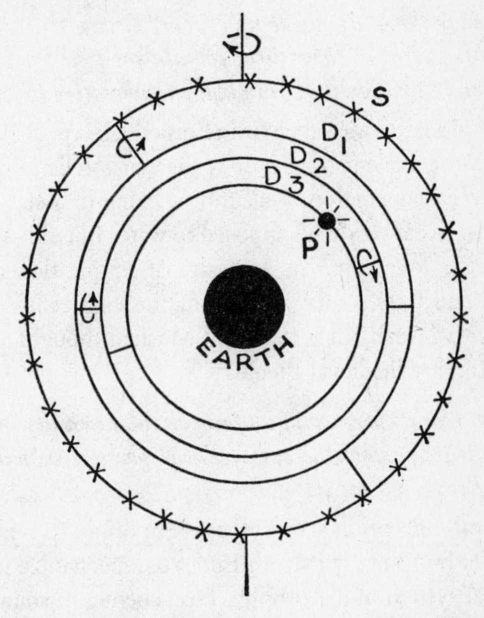

FIG. 24. Eudoxian or concentric planetary scheme, later modified by Callippus (p. 44) and adopted by Aristotle. The figure shows the scheme for one planet only, e.g. Saturn. The earth is stationary at the centre. The starry sphere S rotates about the centre of the earth. In this sphere and carried about with it are the axes of a second sphere D_1 rotating at a different rate: in this arc are the axes of a third sphere D_2 which carries a fourth D_3. To this last is attached the planet whose motion is therefore a combination of the rotations of all four spheres.

Such was the basis of the system that was to control for two thousand years the view that men took of Nature. We may thus summarize the system, its history, and its fate:

(*a*) *Matter is continuous*

In taking this view Aristotle opposed Democritus and sided with

Socrates and Plato. The followers of Democritus and of his disciple Epicurus, who took an atomic view of matter (pp. 20–21), were associated with doctrines which were peculiarly abhorrent to the early and medieval Church. The atomic theory was the only alternative to Aristotle's conception of matter. Thus criticism of Aristotle on this point drew theological odium on itself. The atomic theory, we shall see, therefore passed into the background for many centuries.

(*b*) *All mundane things are made up of four 'elements', earth, air, fire, and water, which, in their turn, contain the four 'qualities', heat, cold, dryness, and moisture, in binary combination* (Fig. 19).

This view of matter was taken from Empedocles (p. 29) and is probably of yet more ancient origin. It is the Aristotelian expression of the Pythagorean conception of all things being in a state of love or hate—fire, for instance, being opposed to water but allied to air. The doctrine of the four elements was almost unquestioned until the seventeenth and lasted until the end of the eighteenth century. It fitted well with Jewish, Christian, and Moslem thought and became a part of orthodox medieval theology.

(*c*) *Stars and planets move with uniform circular velocity in crystalline spheres, centred round the earth. Each sphere is subject to the influence of those outside it.*

This general conception is of Pythagorean origin (pp. 22 ff.). Aristotle did little but borrow it from Eudoxus, mechanize it, and fit it into a general system of philosophy. His scheme, or some modification of it, held its ground till the time of Kepler in the seventeenth century (pp. 240 f.).

(*d*) *Circular movement is perfect since the circle is the perfect figure. Circular movement represents the changeless, eternal order of the heavens. It is contrasted with rectilinear movement which prevails on this our changing and imperfect earth.*

Where imperfection ceaseth, heaven begins

Here again are Pythagorean influences. The basis of the conception is that while heavenly bodies appear to circle round us, bodies on earth tend to fall or rise. Newton at the end of the seventeenth century succeeded in expressing the movements of the heavenly bodies

in known and experimentally demonstrated terms. Until his time the differences between the behaviour of earthly and heavenly bodies remained a puzzle or paradox or both.

(*e*) *The universe is limited in space in the sense that it is contained within an outer sphere. It is unlimited in time in the sense that it is subject neither to creation nor destruction as a whole.*

The finiteness of the universe both in space and time became necessary to all the theological systems of the Middle Ages and notably to that of the Western Church. It was effectively unquestioned till the time of Bruno (died 1600). Thus Aristotle himself could not be completely accepted. The philosophical return to the conception of a universe infinite both in space and time is a landmark in the history of science (pp. 218 f. and 290).

It has been urged against Aristotle that he obstructed the progress of astronomy by divorcing terrestrial from celestial mechanics, for he adopted the principle that celestial motions were regulated by their own peculiar laws. He thus discouraged astronomical observation, placed the heavens beyond the possibility of experimental research, and at the same time impeded advance in the knowledge of mechanics by his assumption of a distinction between 'natural' and 'unnatural' motion. For two thousand years the general outline of the world as set forth by Aristotle remained the orthodox view. It was dangerous even to question it. How far was Aristotle responsible for this intellectual tyranny? To this question there are many answers, of which we shall adduce but four.

(*a*) It was not Aristotle who introduced the distinction between celestial and terrestrial physics. Such distinction had been taken for granted by his predecessors. The Pythagoreans, for example, had made much of it. In fact by his exposition of a positive and tangible scheme he gave a new interest to the study of nature.

(*b*) It is unfair to bring his own greatness as a charge against Aristotle. All our conceptions of the material world—'scientific theories' as we call them—should be but temporary devices to be abandoned when occasion demands. This is a proposition which Aristotle himself puts forth. In expounding the motions of the planets he advises his readers to compare his views with those that they themselves reach. That his scheme lasted for two thousand years without effective criticism is no fault of his. It is rather evidence that the men who followed him were dwarfs compared with 'the master of those who know'.

(*c*) Some of Aristotle's reasons against what we now regard as the form of our world are, in fact, valid. Thus he argues against the motion of the earth. Such movement, if it existed, should, he considered, produce *apparent motion* among the fixed stars. This is a just objection. It was only met in the nineteenth century by the demonstration of such motion. The reason that this was not previously detected is that the vast distance of the heavenly bodies from us makes this apparent motion so small that excessively delicate instruments are needed.

(*d*) We need to remember that the rigidity of the Aristotelian scheme lay not in itself but in the interpretation given to it, especially in the Middle Ages. By linking the theories of Aristotle with their own religious views, men of those times introduced a prejudgement into the debate concerning the validity of the Aristotelian scheme that had nothing to do with its philosophical or scientific value.

3. PERIPATETICS, STOICS, AND EPICUREANS

It is improbable that his connexion with Alexander was of any service to Aristotle himself.[1] There can be no doubt, however, that the great conqueror was a friend of learning and that important investigations were initiated by him. Thus he made an attempt to survey his empire by employing a special force whose duty it was to maintain the condition of the main roads. The services of these men were available for scientific purposes, such as the collection of data bearing on the natural history of the districts where they were at work. Investigations were also made by certain of Alexander's commanders, notably by his admirals, NEARCHUS and ANDROSTHENES. Portions of their botanical and geographical works are preserved.

Aristotle's own work was continued by his school, the Peripatetics, of whom the best known was the long-lived THEOPHRASTUS (372–

[1] A number of statements to the contrary can be found in writings of later classical antiquity. None, however, bears critical scrutiny.

287 B.C.) of Eresus in the island of Lesbos. Though a pupil of Aristotle he lived to be contemporary with the first generation of Alexandrian science (pp. 62 ff.). He made important botanical researches and continued Aristotle's work in Aristotle's spirit. It is interesting to observe that he exhibits the same 'evolutionary' bias that characterizes the biological work of his master (p. 46). In one of his botanical treatises Theophrastus observes that 'where there is growth there is life. Wherefore we should observe these things not for what they are but for what they are becoming. And, moreover, though some be peculiar, yet the general plan can everywhere be traced and is never lost.'

Ancient science suffered from lack of a scientific terminology. This defect Theophrastus attempted to remedy in his own chosen department of botany. For his technical terms he did not rely, as do we, on an ancient and classical language, but sought rather to give special meanings to words in current use. Among such words were *carpos*, 'fruit', and *pericarpion*, 'seed vessel'. From Theophrastus are derived the modern botanical definitions of *carpel* and of *pericarp*. Many Theophrastan plant-names also survive in modern botany.

The botanical works of Theophrastus are the best arranged biological treatises that have survived from antiquity. They contain many acute and accurate observations. Among these are his clear and exact distinction between monocotyledons and dicotyledons. Interesting, too, is his attempted distinction of sex in plants, an attempt which is only successful in the case of the palms. Of these plants, as Herodotus tells us, the ancient Babylonians had the same idea.

Another younger contemporary of Aristotle was AUTOLYCUS of Pitane (*c*. 360–*c*. 300). He worked at his native town and at Sardis, and expounded the geometry of the sphere for astronomical and geographical purposes. A pupil of Aristotle who worked on somewhat the same lines was DICAEARCHUS (*c*. 355–*c*. 285). He employed himself on physical geography and wrote a description of the world accompanied by a map. He, too, worked on information derived from Alexander's officers and was the first to draw a parallel of latitude across a map. This was a convenient reference line. It extended from the Pillars of Hercules (p. 18) due east to the Taurus and 'Imaus' (Himalaya) ranges (cf. Fig. 32) and on to the Eastern Ocean.

It is appropriate to mention here the explorer PYTHEAS of Marseilles (*c*. 360–*c*. 290 B.C.) though he was not of the Peripatetic school. The itinerary of his remarkable voyage can be traced with some exactness. He left Marseilles about March 320 B.C. and made for Spain, followed the coast through the Pillars of Hercules to Cadiz and then along the Atlantic seaboard as far as Cape Ortegal. From there he struck across the ocean to Ushant and on to Cornwall. He next sailed round Great Britain and, returning to Kent, crossed to the continental side of the English Channel and followed the North Sea coast to the mouth of the Elbe. From there he turned north following the Scandinavian coast as far as Trondheim at about latitude 63°. After having put forth thence into the open sea, he turned back along the way he had come and reached Marseilles towards the end of October of the same year.

Pytheas was a good astronomer, and made a number of observations of latitude, among others of his native place Marseilles, which he fixed with remarkable accuracy. He was the first of the Greeks who arrived at any correct notion of the tides, indicating their connexion with the moon and its phases.

One of the best-known of the earlier Peripatetics was the Thracian, STRATO of Lampsacus (*c*. 300 B.C.). He reduced the formation of the world to the operation of natural forces. He recognized nothing beyond natural necessity and, while retaining opposition to atomism, he sought to explain all the functions of the soul as modes of motion.

After the first generation the Peripatetic school devoted itself to preserving or to commenting upon the work of its founder. It exhibited no scientific originality, and from about 300 B.C. onward Athens ceased to be a great scientific centre. Two of the later Peripatetics are, however, of some importance for the history of science. One, ANDRONICUS of Rhodes, was about contemporary with Christ. He prepared a critical text of the works of Aristotle which was probably closely similar to that which we now possess. The other was the Cilician ALEXANDER of Aphrodisias (*c*. A.D. 200). He was an industrious commentator whose writings, much used by the neo-Platonists (p. 133), were the foundation of the Arabian commentaries (pp. 140–156 et seq.) and through them of many of the Latin Aristotelian commentaries.

Contemporary in origin with the Peripatetics was the philosophical school called *Stoic*, from a *stoa* or corridor of the market-place at Athens, where its members used first to meet. The Stoics stressed the operation of natural forces in the manner of Strato the Peripatetic (p. 58). They differed from the Peripatetics, however, in emphasizing the interaction of all different parts of the material world. Thus, while there are reasons for everything in nature, it is also true that everything in nature is among the reasons for the rest of nature. All existence is capable of acting or being acted upon so that 'force' the active and 'matter' the passive principle pervade each other. With this doctrine of 'universal permeation' there is no real difference between matter and its cause. The conception of Deity becomes indistinct and blended with that of 'reason' or 'law' which is but an aspect of a pantheistic system.

Inimical for the future of science was the Stoic cosmology. From 'primitive being' or *pneuma* there separated the four elements in succession, fire first, earth last. The remaining *pneuma* is the 'ether' (p. 53). From these five factors arose a universe on the Aristotelian model. In the world which has thus been formed we, who are parts of it, must obey the inevitable laws. But this world will again decay and dissolve into elements and finally into primitive being or *pneuma*. Our individual souls are part of the universal *pneuma*, temporarily separated therefrom. In the embryo the soul is still in the 'vegetative' stage. It becomes successively 'animal' and 'rational' (p. 49) but joins, in the end, the universal *pneuma*.

So far as human relations and human conduct go, the key to Stoicism is *fate*. The Stoic schooled himself to disregard the inescapable, the nature of which came to be tested by astrology (p. 128). He devoted himself to the development of his own soul through duty, awaiting inevitable absorption into the world-soul.

The Stoic school maintained itself in Athens, Rhodes, and Alexandria. It attained no great importance till Roman Imperial times, but then became the prevalent faith of the upper class (p. 103). Among its exponents were the meteorologist Aratus of Soli (271–213, p. 126) and the Bithynian scholar POSIDONIUS of Apamea (135–50 B.C.). The latter, as an exponent of Stoicism, was anxious to demonstrate the interrelations of different parts of the universe. He was thus attracted to the discussion of the influence of the moon on the tides. He also made estimates of the size of the sun in excess of those of any other ancient writer. Posidonius was a friend of Cicero and thus links Greek with Roman Stoicism.

A rival sect to Peripatetics and Stoics was that of the Epicureans refounded in 307 B.C. by EPICURUS of Samos (342–270). The thought of Epicurus was based on the atomism of Democritus (p. 20) and to a less extent on Anaxagoras (p. 31). The school exhibited little interest in phenomena, and Epicurus himself deprecated scientific pursuits.

Epicurean philosophy spread rapidly and widely in Asia and Egypt. About 150 B.C. it established itself at Rome where its ablest exponent was Lucretius (*c.* 95–55 B.C., p. 104). He has been extolled as a scientific pioneer but in truth there is nothing of that in him. Despite his rank as poet, he is neither a careful nor orderly observer, nor does nature interest him except in its dramatic aspects of volcano, earthquake, thunder, plague, and the like.

The warring of these sects—Peripatetic, Stoic, Epicurean—seem trivial incidents as against the great constructive thought of Plato and Aristotle. With Aristotle we have parted with the first and most active stage of ancient scientific thought. Summarizing his place in the history of science we may say that

(*a*) he represents the final stage of the 'Great Adventure', the attempt to represent the world, including the moral world, as a whole and as a unitary system;
(*b*) he provided a philosophic synthesis which, in more or less modified form, satisfied intellectual aspirations from his own time until the seventeenth century.

In that philosophical system there remained two great breaks in continuity. One hiatus was between celestial and terrestrial physics. This first began to be filled by the workers of the 'Insurgent Century' from Bruno (p. 218) to Newton (p. 288). The other gap was between the world of the living and of the not-living. The Epicurean philosophy attempted to fill the breach in ancient times by the introduction of a 'mechanist' system (p. 104). The Christian Church in medieval times, repudiating with vigour the Epicurean solution, accepted the breach as part of the divine order of the world. The physiologists in modern times, beginning in the seventeenth century with van Helmont (p. 269), Descartes (p. 258), Borelli (p. 278), and Sylvius (p. 280), have been seeking to resolve it ever since.

In leaving the heroic age of Greek science we would again emphasize the 'universal' character of the philosophical attempt that we call the 'Great Adventure'. The scientific activity of the age partook of the nature of what we should now term 'philosophy'. The object of each investigator was to fit his observations and the laws that he deduced into some general scheme of the universe. From their day to ours philosophy has continued her attempt thus to storm the bastions of heaven. But with the new age that we have to discuss, there was a failure of nerve in that great frontal attack. Science, becoming gradually alienated from philosophy, begins to proceed by her own peculiar method of limited objectives. The first series of these attempts resulted in the 'Great Failure', the story of which we shall trace through two thousand years (Chs. III, IV, V). Nerve fails first, as with the Alexandrian school and the separation of science from philosophy in the second adventure (Ch. III), next Inspiration falters under the Roman Empire (Ch. IV), lastly Knowledge itself fades in the Middle Ages (Ch. V). At length there is a rebirth. The science of the Renaissance began again to proceed by the method of limited objectives (Ch. VI).

III

THE SECOND ADVENTURE

Divorce of Science and Philosophy: Alexandria (300 B.C.–A.D. 200)

I. EARLY ALEXANDRIAN PERIOD (300–200 B.C.)

WHEN Alexander died (323 B.C.), his Empire broke into fragments (Fig. 25). Egypt was seized by one of his generals, named Ptolemy, and the Ptolemaic dynasty endured for three hundred years. Its members were mostly able and intelligent men and women. The first

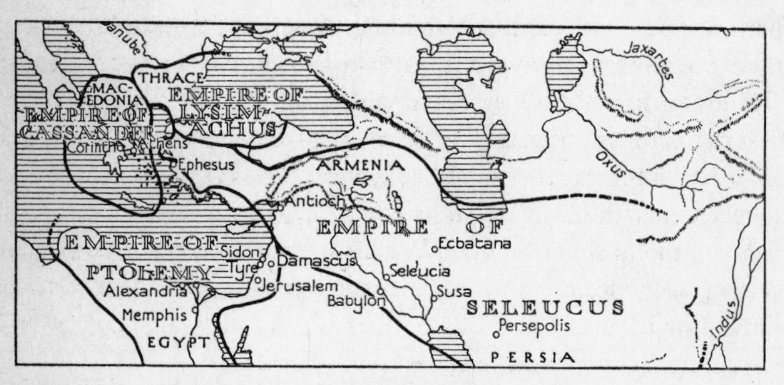

FIG. 25. Break-up of Alexander's Empire.

of the line established the tradition of learning. The second founded a library and museum at Alexandria. That city became the centre of the scientific world. Learned men flocked to it and were supported by funds provided by the Ptolemaic rulers. The school continued very active for a couple of centuries. By 100 B.C., however, it was beginning to languish, and by A.D. 200 it was in rapid decay, though there was spasmodic scientific activity until about 400.

The Alexandrian library in its earlier stages had many distinguished curators. Most were literary men, but some, e.g. Eratosthenes (p. 76) and Apollonius of Perga (p. 75), were also men of science. From 300 B.C. to A.D. 200 most eminent men of science were teachers at Alexandria. A few, notably Archimedes and Galen, were less intimately linked with the Egyptian metropolis. Yet even they were pupils of the school and corresponded with Alexandrian teachers. Greek science from about 300 B.C. onward is thus not inadequately described as 'Alexandrian science'.

Alexandria was not, however, entirely without rivals as a seat of learning. The most prominent were the island of Rhodes and the city of Pergamum in western Asia Minor. Of the rivalry between Alexandria and Pergamum there is an interesting reminder in our language. The Alexandrian books were written on rolls prepared from *papyrus* reeds, whence our word *paper*. To prevent Pergamum from acquiring copies of their literary treasures, the jealous Ptolemies are said to have put an embargo on its export. The Pergamene kings, cut off from a valued import, sought to improve the preparation of skins, the Asiatic medium for writing. Thus was developed the *membranum pergamentum* which has reached our language as *parchment*.

It is characteristic of Alexandrian science that it developed along the lines of 'specialities'. These came to lose their relation to general philosophic thought with which they had hitherto been linked. It is convenient to consider Alexandrian science in three chronological divisions; an *early period* containing the first and second generations of the school to nearly 200 B.C., a *middle period* to about the birth of Christ, and a *late period* to the complete decline of the school. Archimedes (pp. 69 ff.) demands individual discussion.

The early Alexandrian period is noteworthy for the fact that mathematics at once assumed a prominent and independent position. Among the first to be called to the Alexandrian Academy was the illustrious mathematician EUCLID (*c.* 330–*c.* 260). He was trained at Athens, probably by a pupil of Plato. His famous *Elements of Geometry* has determined all subsequent geometrical teaching. Perhaps no book save the Bible has been so much studied. For the next twenty-two centuries parts of the *Elements*, and especially the first six of its thirteen books, were the customary introduction to geometry. Even though the work has now been superseded in the schools, the newer forms of geometrical teaching are based on their Alexandrian predecessor.

To what extent was Euclid's work original? Elementary works on geometry had already been written by other authors, notably by Hippocrates of Chios (p. 35). Before Euclid, it had been generally agreed to base geometry on the straight line and circle. The properties of the right-angled triangle and the doctrine of proportion for both commensurables and incommensurables (p. 26) had been investigated. Some properties of conic sections were known (p. 44). Philosophers were familiar with the five 'Platonic bodies' (pp. 28, 238). The solution of such problems in solid geometry as the relation between the volume of a cone or pyramid and that of the cylinder or prism circumscribed around it had been attained. To all this mathematical activity Euclid certainly added advances in arrangement, in logical sequence, in form of presentation, and in completeness. His treatise displaced all that had gone before it, and rapidly assumed the position which it has since held.

Although Euclid's great work is called the *Elements of Geometry*, its subject-matter extends far beyond what is now regarded as geometry. Thus three of its thirteen books are devoted to the theory of numbers. In particular they contain the proof that no limit can be set to the number of prime numbers. This is a matter of importance in view of the great attention focused on the prime numbers by previous mathematicians such as the Pythagoreans and Plato and by subsequent mathematicians, notably by Eratosthenes (pp. 76 ff.), Euler (p. 310), Lagrange (p. 310), and Gauss (p. 325).

Euclid's tenth book expounds the dominating concept of *irrational quantities*, thus opening up a thought-world of which the facts cannot be given tangible expression. The Pythagoreans (pp. 22 ff.) had already broken into that world, and of it both Plato and Aristotle had had a Pisgah sight, but Euclid was the first to attempt any systematic exploration of it. It should be noted, however, that Euclid and his Greek successors distinguished sharply between *irrational quantities* and *irrational numbers*. In the theory of proportion as developed in Euclid's fifth book, the basis of the theory of irrational numbers is laid but is not developed. For its exposition the world had to wait until Descartes (pp. 226 f.) showed the deep unity of the long separated fields of number and form.

Euclid wrote also on optics. Many of his other works are lost, or survive in Arabic translation or in corrupted versions.

Of those lost we should particularly like to have his work, *On Fallacies*, which dealt with the causes of error in geometrical research. Other of his works dealt with astronomy, optics (p. 86), and music.

ARISTARCHUS of Samos (*c.* 310–230 B.C.) taught at Alexandria soon after Euclid. He was himself the pupil of a disciple of Strato (p. 58). The peculiar views of Aristarchus on the position of the earth among the heavenly bodies have earned him the title of the 'Copernicus of

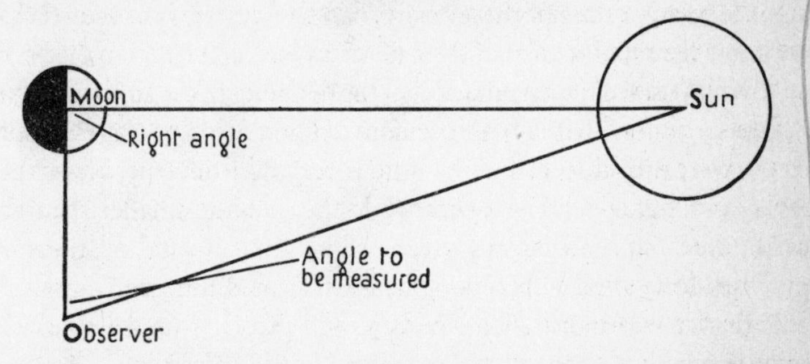

FIG. 26. Aristarchus measures relative distances of sun and moon from earth.

Antiquity'. He extended the view of Philolaus that the earth revolves around a central fire[1] (p. 27) by maintaining that the sun itself is at rest, and that not only Mercury and Venus but also all the other planets, of which the earth is one, revolve in circles about the sun. It is interesting to observe that this view of Aristarchus brought on him the same charge of impiety as had descended on the head of Anaxagoras (pp. 31 f.) two centuries earlier.

We owe to Aristarchus the first scientific attempt to measure the distances of the sun and moon from the earth, and their sizes relative to each other (Fig. 26). He knew that the light of the moon is reflected from the sun. When the moon is exactly at the half, the line of vision from the observer on the earth to the centre of the moon's disk M must be at right angles to the line of light passing from the centre of the sun's disk S to the centre of the moon's disk M. Now the observer can measure the angle that the sun and moon form at his own eye O. With a knowledge of the two angles at M and O the relative lengths of the sides OS and OM can be determined. This gives the relative distances of sun and moon from the observer.

[1] The exact nature of the theory of Philolaus is disputed.

The difficulty lay in determining exactly the angle at *O*. A very small error here makes a very great difference in the result. Aristarchus estimated this angle as 87 degrees when the reality is 89 degrees 52 minutes. In the resulting calculation he estimated the sun as 18 times more distant than the moon, instead of about 390 times more distant.

If we have the relative distances of sun and moon from the observer, the relative sizes of these bodies can be estimated, provided that we know the relative sizes of their disks, as they appear to an observer on earth. Aristarchus calculated that the diameter of the sun was about nineteen times that of the moon. Here further observational errors were introduced, and the ratio is very far from true. Nevertheless, Aristarchus perceived that while the moon is smaller than the earth, the sun is enormously greater. This fundamental relationship may well have affected his thought, for it seemed inherently improbable that an enormously large body would revolve round a relatively minute one.

Contemporary with Aristarchus at Alexandria were other astronomers who recorded the positions of stars by measurements of their distances from fixed positions in the sky. Thus they defined the positions of the more important stars in the signs of the zodiac, near to which pass all the planets in their orbits. Their observations were used later by Hipparchus (pp. 82 ff.).

The philosophy which was the parent of science among the Greeks interested itself in three main aspects of the material world: (*a*) number and form and their relation to each other and to material objects, (*b*) the form and workings of the universe, and (*c*) the nature of man. In Alexandria, where science had freed itself from philosophy (p. 63), it was thus to be expected that the systematization of mathematics and astronomy would be accompanied by a similar development in the basic studies by which alone medicine can continue its progressive scientific tradition. Thus at Alexandria we first hear of open dissection of the human body. This procedure was not only against Greek tradition but also contrary to deep-seated feelings as old, perhaps, as the human race itself. How could this step be taken at this time and place?

No complete answer can here be given but it may be recalled that at Alexandria about 300 B.C. three peculiar conditions converged. (*a*) A host of immigrant Greeks, uprooted physically from homeland and spiritually from their ancient folk-ways. (*b*) Certain of these had a natural curiosity as to the structure and working of the body roused by the biological work of Aristotle (died 322 B.C.) and his still living pupil Theophrastus (died 287 B.C.). (*c*) These immigrant Greeks were in full possession of the works of Plato (died 347 B.C.) wherein contempt of the body as a mere shell of the soul is expressed in terms of most moving beauty in words he ascribes to his master, Socrates (died 399 B.C.) in the hours before his death (*Phaedo*, written about 367 B.C.).

Many of these Alexandrians were thus freed from the servile bonds of superstition but the earliest whose name reaches us is that of HEROPHILUS of Chalcedon (*c*. 300 B.C.), a contemporary of Euclid. He dissected the human body publicly. In describing the anatomy of man he compared it with that of animals. He recognized the brain as the centre of the nervous system, and he regarded it as the seat of the intelligence. The name of Herophilus is still attached to certain parts of the brain. One is called by modern anatomists the 'winepress of Herophilus'. It is the meeting-place of four great veins at the back of the brain. Their arrangement reminded him of the handles of a press. Herophilus was the first to distinguish clearly between veins and arteries. He observed that arteries pulsate. Their movement, however, he did not ascribe to the heart's action, but wrongly considered that it was natural to the arteries themselves.

A little younger than the anatomist Herophilus was the physiologist ERASISTRATUS of Chios (*c*. 280 B.C.), who also taught at Alexandria. He was an atomist and a follower of Democritus (p. 21), but his physiology was based on the idea that every organ is a complex of a threefold system of vessels—veins, arteries, and nerves—extending by ever more minute branching beyond the reach of vision. In those days, and for long afterwards, the nerves were regarded as hollow. Their imaginary cavities were thought to convey the hypothetical 'nervous fluid', much as the veins carry blood. Erasistratus paid particular attention to the brain. He distinguished between the main brain, or *cerebrum*, and the lesser brain, or *cerebellum*. He observed the convolutions in the brain of both man and animals, and associated their greater complexity in man with his higher intelligence. He made experiments on animals which led him to distinguish between the posterior nerve-roots of the spinal cord, which convey sensations and the anterior which convey motor impulses. This was forgotten till rediscovered in the nineteenth century (p. 494).

Erasistratus also observed the lacteals, those lymphatic vessels that convey the white, milk-like fluid—the so-called 'chyle'—derived from the food in the intestine, to the liver. The lacteals were seldom mentioned again until the Italian Gasparo Aselli (1581–1626) recorded them in the seventeenth century.

A word on the views of Erasistratus on the general working of the animal body. He supposed that air (i.e. the *pneuma* of the world) is taken in by the lungs and passes to the heart. Here, as he held, it enters the blood and is changed into a peculiar kind of *pneuma* or spirit—the 'vital spirit'—which is sent to the various parts of the body by the arteries. It is carried to the brain, among other parts, and is there further altered into a second kind of *pneuma*, the 'animal spirit'. This animal spirit reaches different parts of the body through the nerves. This physiological system of Erasistratus was further developed by Galen (p. 99).

After the first generation anatomical enthusiasm at Alexandria waned. We may refer to three special points concerning it and concerning Alexandrian science in general:

(*a*) The names of Herophilus and Erasistratus are linked with the terrible charge of having dissected living men. Historians who have investigated the charge are satisfied that it is false.

(*b*) Erasistratus considered the *pneuma* that circulates in the body to be ultimately drawn from the air, or *pneuma* of the great world. This gave a physiological basis to the philosophical conception of the spirit of man as part of the world-spirit. This conception is encountered in later writings, as in works of the Stoic school of the second century A.D. such as those of the Emperor Marcus Aurelius or in the so-called 'Hermetic' writings of the third century A.D. Physiology and philosophy thus reacted on each other.

(*c*) In the third century B.C. Alexandria was an important Jewish centre. Parts of the Old Testament had been rendered from Hebrew into Greek by about 250 B.C. Greek contacts went far toward rationalizing the Hebrew view of nature. Thus, while earlier biblical literature contains many references to divine intervention in the course of nature, the *Wisdom Literature* of Alexandrian date equates natural law with divine ordinance. In some passages the various types of Greek philosophy are set over against the old Hebrew view. Among the Greeks various 'first principles' had been adopted. Thales had proposed 'water' (p. 17), Heracleitus 'fire' (p. 20), Pythagoras the 'circling stars' (p. 27), Anaximenes 'air' (p. 17), yet other philosophers vague essences that may be rendered 'wind' or '*pneuma*' (p. 59). Finally the new astrological science coming in from Babylon suggested the complex mathematical order of the heavenly bodies, which *signalled* the seasons, as *controlling* the seasons and through them men's lives. This appealed specially to the Stoics. A Jewish work written in Alexandria about 50 B.C. inveighs against all these views:

> Surely vain were all men in their natures, and without perception of God who could not, from the good things that are seen, know Him that is; neither by giving heed to the works did they recognise the Workmaster, but either fire [Heracleitus] or pneuma (the Stoic) or the swift air [Anaximenes] or circling stars [Pythagoras] or raging water [Thales] or the lights of heaven [the Astrologers] they deemed the gods that govern the world. (*Wisdom of Solomon* xiii. 1–2.)

The influence of Greek science can similarly be traced into the domain of Hebrew physiological conceptions. Thus, for instance, the seat of the understanding throughout the *Wisdom Literature*—which is Alexandrian—is usually placed in the heart. This is Aristotelian and contrary to Herophilus and Erasistratus who placed the seat of intelligence in the brain. It is also opposed to the older Hebrew view (e.g. Psalm xvi. 7) which placed it in the liver. In several places, too, *Wisdom Literature* as well as the New Testament writings (e.g. 2 Peter iii. 10, Galatians iv. 8–9) set forth the Greek doctrine of the four elements. And lastly from the melting-pot of pre-Christian Alexandria emerged the homunculus of European alchemy (pp. 283–4 and Fig. 119).

2. ARCHIMEDES. RISE OF MECHANICS

ARCHIMEDES (287–212 B.C.) of Syracuse in Sicily was the greatest mathematician of antiquity. His life was entirely devoted to scientific pursuits, and his work is so fundamental that it affects every department of science. He was himself the son of an astronomer. He visited Alexandria, where he met successors of Euclid. His whole work is instinct with a human element. Moreover, despite his absorption in science, he was not above applying his knowledge to practical matters. Thus his name is remembered in connexion with the Archimedean screw for raising water (Fig. 27). It is said that he invented it during

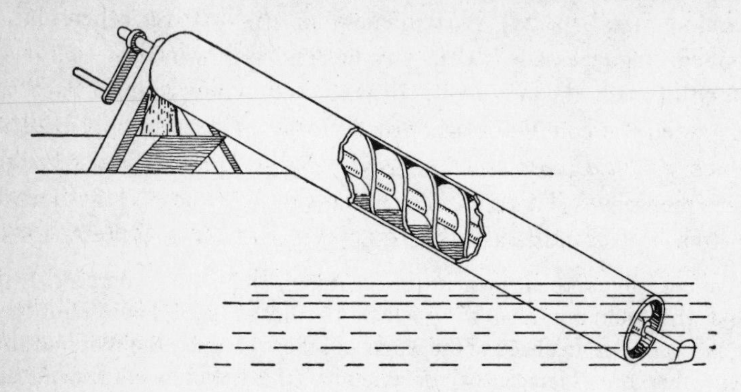

FIG. 27. Screw of Archimedes.

a visit to Egypt, and it is still in wide use there. The use of the screw as a means of applying mechanical force was unknown before Archimedes and was probably suggested by his device. He also contrived war engines for the defence of his native city against the Romans.

The writings of Archimedes show a generous appreciation of the achievements of others. He had friendly personal relationships with Eratosthenes (pp. 76 f.). His lofty intellect, his compelling lucidity, and his terseness of exposition, made a profound impression on his fellow mathematicians. His mechanical skill must have been of a high order, for we hear also of his 'planetarium', a sphere of the heavens or orrery with models of sun, moon, earth, and planets, the movements displayed with elaboration of detail that showed even eclipses.

A well-worn story tells of his success in practical affairs. Hiero, on gaining power in Syracuse, vowed a golden crown to the gods. He contracted for its manufacture and weighed out the gold. The contractor duly delivered a crown of correct weight. But a charge was made that some gold had been abstracted and an equivalent weight of silver substituted. Hiero invoked Archimedes to put the matter to the test. While it was on his mind, Archimedes happened to go to the bath. On getting in, he observed that the more of his body was immersed, the more water ran over the top. This suggested the solution. Transported with joy he rushed home shouting 'Eureka! Eureka!' ('I have found it, I have found it!') What he had found was, in effect, the conception of specific gravity.

He made two masses of the same weight as the crown, one of gold, the other of silver. Next he filled a vessel to the brim and dropped in a mass of silver. Water ran out equal in bulk to the silver. The measure of this overflow gave the bulk of silver. The same was done with an equal weight of gold. The smaller overflow corresponding to the gold was, of course, as much less as the gold was less in bulk than the silver, for gold is heavier than silver. The same operation was now done with the crown. More water ran over for the crown than for gold of like weight, less than for silver of like weight. Thus was revealed the admixture of silver with the gold. Archimedes had, in effect, obtained the relative *specific weights* of gold, silver, and of the mixture of the two, by comparing the relative amounts of water displaced by the same weight of the three. The scientific aspect of the subject is set forth in his work *On Floating Bodies*. This is the first record of the scientific employment of what we should now call 'specific weights', though, of course, long before Archimedes, men were well aware that some substances were relatively heavier than others.

This question of the scientific use or development of a piece of common knowledge is important for the history of science. Discussion of it throws some light on the nature of the scientific process. Thus to Archimedes the ancient world owed a general exposition of the doctrine of levers (Fig. 28). This must not be taken to mean that Archimedes invented the lever any more than that he had discovered some bodies to be heavier than others. Levers in various forms were used from remotest antiquity, and an intelligent ape will use a stick as a lever. But it is one thing to use or even to contrive a device, and another to lay bare its exact mathematical principles and to follow them to their theoretical applications and conclusions. Important in this connexion is the statement of Archimedes of the possibility of moving a weight, however large, by a force, however small—a valuable theoretical application of levers. His saying is often recalled, 'Give me but a place to stand, and I can move the world.' He demonstrated this with a compound lever by which, with only the slightest effort, he was able to move a laden ship. Archimedes no more invented levers than the Greeks invented science. But science owes to the Greeks its formal and conscious development as a discipline and a method, and the doctrine of levers owes to Archimedes its first

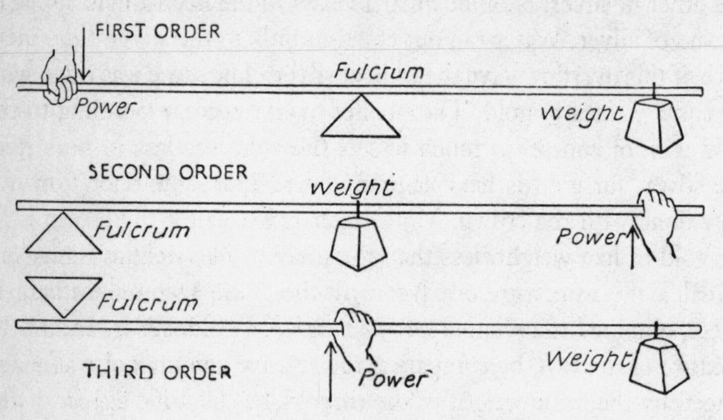

FIG. 28. The three orders of lever.

formal and systematic exposition as susceptible of exact analysis. Formal and systematic exposition is a main task of science and without it knowledge cannot rise into the realm of science.

Perhaps the earliest work of Archimedes that we have is that *On Plane Equilibrium*. In this some fundamental principles of mechanics are set forth as rigorous geometric propositions. The work opens with his famous 'postulate': 'Equal weights at equal distances are in equilibrium; equal weights at unequal distances are not in equilibrium but incline toward the weight at the greater distance.' This is, in effect, the principle of the steelyard. It led him in the end to the discovery of the centre of gravity in a variety of geometric figures.

Among the mathematical achievements of Archimedes a very high place must be given to his methods of measuring the areas of curved figures and surfaces. The simplest expression of this effort, 'squaring of the circle', had been broached by Hippocrates of Chios (p. 35). Eudoxus (p. 43), in estimating the volume of certain solid bodies, had propounded a method that involved in its essence the idea of 'limits'. This idea had been used by Euclid for a particular proposition of his twelfth book. Archimedes, however, employed limits systematically. This doctrine is of the utmost practical and historical importance, since it has formed a main foundation of modern mathematical development. It is essential to the 'calculus' as developed by Newton (p. 293) and Leibniz (p. 309). The calculus in its turn has been the starting-point for the development of many types of mathematical research.

The principle of the doctrine of limits can be expressed very simply. A square can be inscribed within a circle. Of such a figure two propositions are obvious:

(*a*) The sum of the sides of the inscribed figure is less than the circumference of the circle:

(*b*) The area of the inscribed figure is less than that of the circle.

It is quite easy to double the number of sides and make an eight-sided figure, still inscribed within the same circle. Propositions (*a*) and (*b*) remain true but the difference is smaller in each case. We can go on doubling the number of sides to 16, 32, 64, 128, 256 or to any higher number. The more we increase the number of sides the more nearly will the sum of the sides and the area of the inscribed figure approach the circle. 'In the limit', when its sides are so small as to be no more than points, the polygon may be conceived as becoming the circle. Archimedes realized that this limit can never be reached but that it can be approached as nearly as we wish (Fig. 29).

Archimedes proves that the area of a circle is equal to that of a triangle of base equal to the *circumference* of the circle and of height equal to the *radius* of the circle. To calculate this area it is necessary to find the ratio between circumference and diameter. In estimating this ratio Archimedes sought the limit approached by the sides of regular polygons both inscribed and circumscribed on the circle. The limits for their ratio to that of the diameter he found to lie between $3\frac{10}{71}$ and $3\frac{10}{70}$. The latter has, since his day, been generally accepted as a good approximate value of the quantity known as π.

In his *Quadrature of the Parabola* Archimedes relates that he had been led by the study of mechanics to the solution of the problem of finding the area of a segment of a parabola, and that he had then obtained geometric proof of the correctness of his solution. His method resembles that which he adopted for the circle, namely to take both an inscribed and a circumscribed figure in relation to the curve under investigation. The two rectilinear figures are, as it were, compressed one from within and the other from without until they coincide with the curvilinear figure.

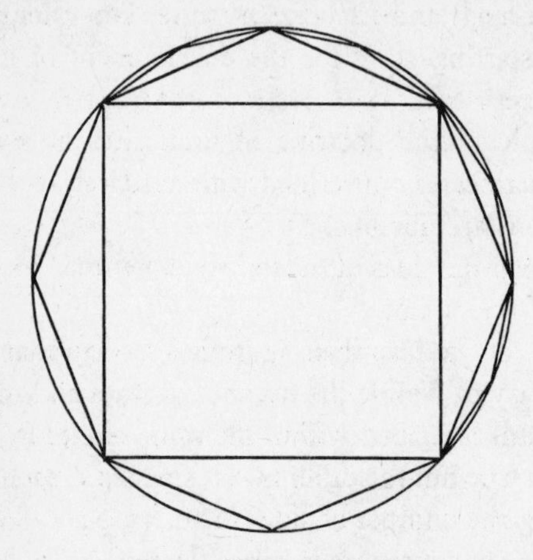

FIG. 29. Doctrine of limits.

This mode of procedure, as well as that of using mechanics for the solution of problems afterwards demonstrated by geometry, leads us to the consideration of an extremely interesting treatise by Archimedes, the nature of which is suggested by its title *On Method.*

For the most part, Archimedes, like other Greek men of science, gives us only his final results. He gives us his proofs, but does not tell us how he reached them. In the *Method*, however, Archimedes, addressing Eratosthenes (p. 76), recalls the mathematical discoveries which he had sent on a former occasion and proceeds to inform him that he is now sending a description of the way in which he elicited them.

In essence the 'Method' consists in the application of two principles. The first is that a plane figure may be regarded as an aggregate of an infinite number of parallel lines with certain common properties. The second is the consideration of the respective weights of the two plane figures as drawn on paper whose area has to be compared. The process is also applied to demonstrate relationship between the areas of solid figures considered as aggregates of an infinite number of parallel planes. It amounts to a practical solution of problems of the relation between areas or volumes of two figures by analysis, mechanical or other, after which the philosopher returns to a synthetical mathematical process. He thus gains by experiment some insight into the solution before he seeks its mathematical demonstration.

Finally we may mention the remarkable system used by Archimedes for expressing very large numbers. It is so efficient that it enables any number to be expressed, up to that which, in out notation, would require eighty thousand million million ciphers. Archimedes expressed the opinion that his system was adequate to express the number of grains of sand that it would take to fill the universe. He therefore called his work the *Sand Reckoner*. From his calculation of the size of the universe, we get our idea of the cosmic conceptions of Archimedes. He knew the view of Aristarchus (p. 65) that the universe was heliocentric, the earth revolving round the sun in a comparatively unimportant orbit.

The sum of the contributions to knowledge by Archimedes is enormous. With his character, his humanity, his width of interest, his simplicity of exposition, and his unity of purpose, no mathematician of any age has commanded such general sympathy and respect.

3. MIDDLE ALEXANDRIAN PERIOD (200–30 B.C.)

A worthy Alexandrian successor of Archimedes was APOLLONIUS (fl. 220 B.C.) of Perga in Asia Minor (not to be confused with Pergamum). He studied under successors of Euclid at Alexandria and also at Pergamum. Apollonius is specially remembered for his *Conic Sections*, a subject which he placed on a new footing.

Apollonius built on the work of Menaechmus (p. 44). That writer had derived the three types of conic section from three types of right cone. Apollonius showed, however, that all the three types of conic section can be derived from the same cone, whether right or scalene (Fig. 21). He established the terms *ellipse*, *parabola*, and *hyperbola* to denote the three types of section previously indicated by the angle of the cone of origin. The general geometric laws which give the properties of conic sections come to us, like the nomenclature of these figures, from Apollonius.

Archimedes and Apollonius between them originated two great problems which have ever since occupied geometers. The first is the quadrature of figures outlined by curves. This gave rise in due course to the infinitesimal calculus. The second is the theory of conic sections. This gave rise in due course to the theory of geometrical curves of all degrees so that, for example, a circle could be treated as a special case of the ellipse (Fig. 30).

The Ptolemies, in their zeal for learning, did not forget geography. Ptolemy III Euergetes (247–222 B.C.) rendered the greatest service to the science by his encouragement of ERATOSTHENES (*c.* 276–*c.* 194 B.C.), the librarian at Alexandria, and the most learned man of antiquity. His most important investigation, the measurement of the globe of the earth, was performed by an operation of beautiful simplicity. Eratosthenes started from the three propositions (Fig. 31):

(*a*) That at Syene on the Nile (the modern Aswan) at noon on midsummer day an upright rod casts no shadow.

(*b*) That Syene is 5,000 stadia from Alexandria.

(*c*) That Syene is directly south of Alexandria.

Now, it is clear that, if we consider the earth as a sphere, then the ratio

$$\frac{\text{angle at centre subtended by 5,000 stadia}}{\text{four right angles}} = \frac{\text{5,000 stadia}}{\text{circumference}}.$$

The problem is, therefore, to determine the angle at the centre subtended by 5,000 stadia. But if on midsummer day the shadow cast by an upright rod at Alexandria is measured, then we shall be able to estimate the angle which the sun's ray makes with the rod. Since, however, the sun is so vastly distant from the earth, the sun's ray at Alexandria is in effect parallel to the sun's ray at Syene. Therefore the angle that the sun's ray makes with the rod is equal to the angle subtended by 5,000 stadia at the earth's centre. There is thus but one unknown—the earth's circumference—in our equation. The circumference of the earth thus obtained is a very fair estimate.

Having measured the earth, Eratosthenes proceeded to consider

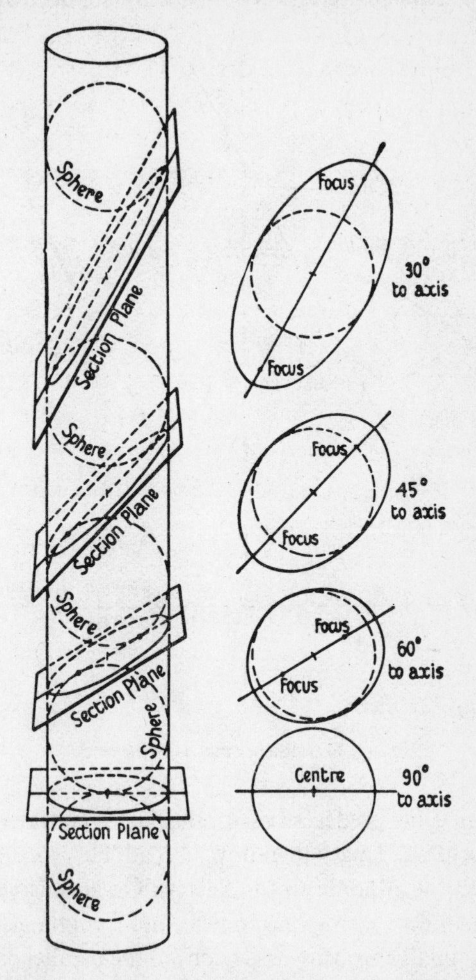

FIG. 30. The circle as special case of the ellipse, shown by series of sections through a cylinder. The cylinder of the diagram exactly contains a series of spheres; the points of contact of these with the section planes are the foci. The left figure is pictorial while the curves on the right give the true shape of the sections.

With a slightly more complex diagram the same relations may be shown in a series of sections through a cone, the cylinder being itself a special case of the cone (compare Fig. 21).

the known parts of it. Here, in common with almost all ancient geographers, he fell into an error, or rather a self-imposed limitation. Eratosthenes regarded the habitable world as placed wholly within the northern hemisphere and forming only about a third of that.

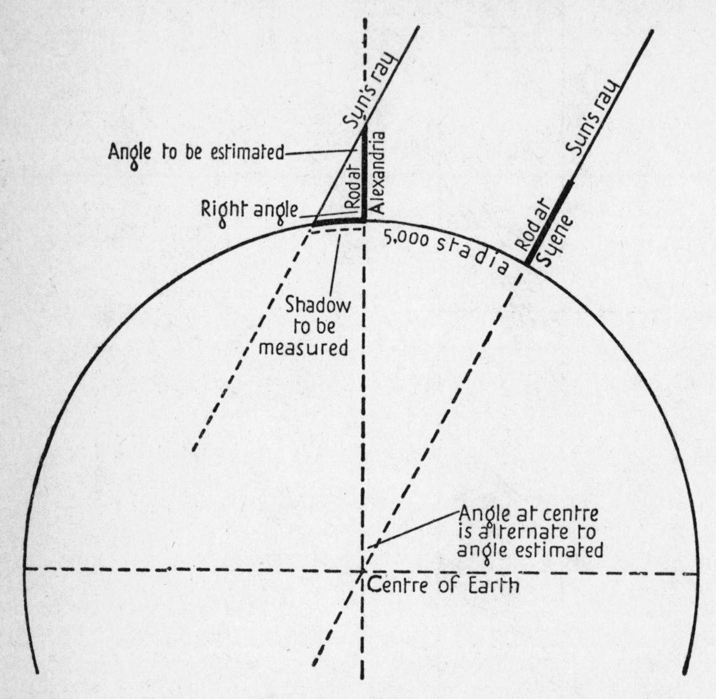

FIG. 31. Eratosthenes measures the earth.

Again following his predecessors, Eratosthenes considered that the habitable world was longer than it was broad. He estimated that the distance from the Atlantic to the Eastern Ocean was 78,000 stadia (that is, about 7,800 geographical miles), and from the parallel of the Cinnamon Land (Taprobane or Ceylon) to the parallel of Thule was 38,000 stadia. As Eratosthenes estimated the circumference or equator of the earth at 250,000 stadia, he was able to estimate the circumference at the parallel of the Pillars of Hercules which he knew was also that of Rhodes (latitude 36°) (Fig. 32).

This fundamental parallel passed, as he erroneously thought,

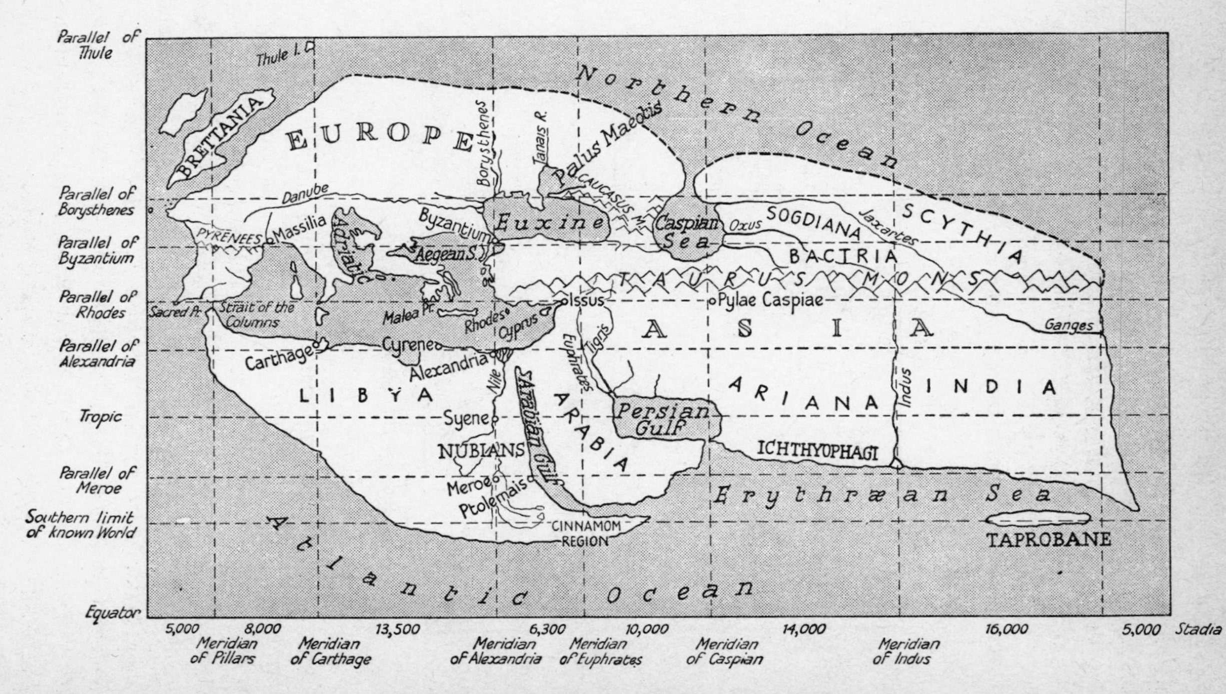

FIG. 32. The habitable world according to Eratosthenes.

through other important points—the westernmost point of Spain, for example, and the southern points of Italy and Greece and along the Taurus mountains. At this parallel the total circumference of the world he estimated at 200,000 stadia. The rest was sea, so that, as he observed, 'if it were not for the vast extent of the Atlantic one might sail from Spain to India along the same parallel'. This is the first suggestion for the circumnavigation of the globe.

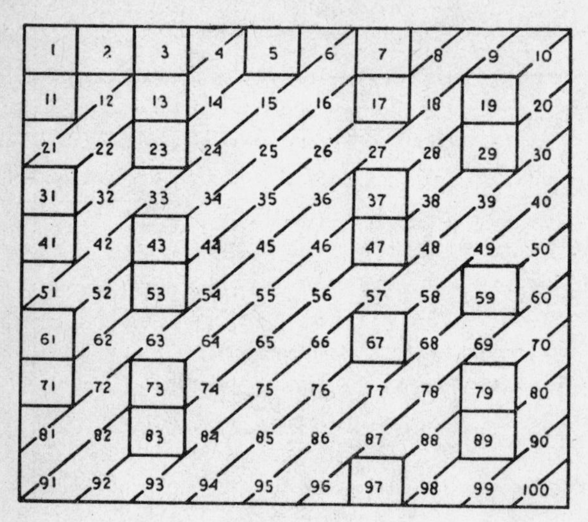

FIG. 33. The 'Sieve of Eratosthenes'.

At right angles to the important parallel of Rhodes, Eratosthenes determined a north–south line between Alexandria and Syene. This line, produced northward, he regarded as passing through Byzantium and, beyond, to the mouth of the river Borysthenes (now called the Dnieper). Southward, he considered that it passed to Meroë, and then along the Nile to the Sembritae.

Both these fundamental lines contain several errors of allocation. Their determinations, together with those on other parallels of latitude and lines of longitude, are, however, sufficiently accurate for the construction of a map of the Mediterranean area recognizably similar to one based on modern knowledge (Fig. 32).

Eratosthenes exhibited great ability as a mathematician. He advanced the knowledge of prime numbers, a subject to which Archimedes had paid much attention. The famous *Sieve of Eratosthenes* is a device for eliciting these numbers. Write down all integers in their natural succession. Then strike out all the multiples of 2, then the remaining multiples of 3, then those of 5, &c., through the other prime numbers (Fig. 33). The properties of prime numbers have attracted mathematicians in all ages, and it is astonishing how some simple rules concerning them have not been rationally explained to this day. Thus it is now well over a century since it was remarked that every even number is the sum of two primes. This has been verified up to 200,000,000, but no proof is yet forthcoming.

Mathematical advance in Alexandrian times made possible a great development of astronomical theory. The discussion of the supposed rotation of the celestial spheres and of the movements of the heavenly bodies gave rise to a nomenclature, parts of which have survived to our day, but parts of which have been modified by the Arabian and other authors through whose hands the Greek mathematical works have passed (pp. 140–64).

The astronomical observer regarded himself as being in the centre of the vast heavenly sphere bearing the fixed stars. He considered the earth so small that his distance from its centre was as nothing to his distance from the celestial boundary. Of this celestial sphere he could only see half, for the other hemisphere was hidden from him by the opaque earth. The limiting circle thus imposed on his vision was the *horizon* (from a Greek word meaning 'to bound' or 'to limit'). This horizon formed a *great circle* on the heavenly sphere. He recognized, too, the celestial *poles* or points on the sphere pierced by the axis about which the heavens seem to turn. On the sphere he marked out the *meridian*, which passes through the *zenith* (a word of Arabic origin) and the poles. The great circle at right angles to the line joining the poles was the *equator*. Starting from these elementary conceptions the Alexandrian observers worked out their whole astronomical system (Fig. 34).

Besides measuring the size of the earth Eratosthenes also made a remarkably accurate measurement of the angle which the circle of zodiacal constellations makes with the celestial equator, in other words a measurement of the *obliquity of the ecliptic*. His estimate works out at 23 degrees 51 minutes. This was only seven minutes from the true measure of the obliquity at that date.

The greatest astronomer of antiquity was HIPPARCHUS of Nicaea (*c.* 190–120 B.C.). He worked at Rhodes, where he erected an observatory and made most important researches. He developed trigono-

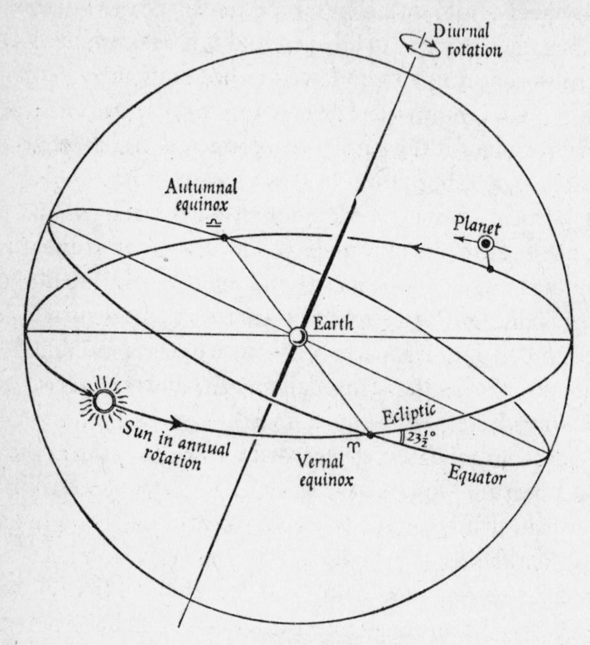

FIG. 34. The Astronomical Elements.

metry by which numerical calculations can be applied to figures drawn on either plane or spherical surfaces. The study is of great value to astronomy.

Hipparchus made numerous accurate astronomical observations. He also collected and collated the records of previous observers to see if astronomical changes had taken place in the course of the ages. There were available to him records of his Alexandrian and earlier Greek predecessors, and also those of the yet more ancient Babylonian astronomers. As a result of these comparisons he gave to the world two brilliant astronomical conceptions. (*a*) One of these, the precession of the equinoxes, was of permanent value. (*b*) The other, his theory of the movements of the planets and notably of the sun and moon, was of value to subsequent generations for the calculation of eclipses.

(*a*) *Precession of the equinoxes.* In 134 B.C. Hipparchus observed a new star in the constellation Scorpio. This suggested to him that he should prepare a catalogue of star positions. He therefore drew up a list of several hundred stars, each of which was given its celestial latitude and longitude. The constellations to which Hipparchus referred these stars are those which are today generally accepted. He showed great foresight in recording a number of cases in which three or more stars were in a line, so that astronomers of subsequent ages might detect changes in their relative positions.

Hipparchus proceeded to compare his observations with others of about 150 years earlier. He found that in this lapse of time there had been changes in the distances of the stars from certain fixed points in the heavens. The changes were of a kind that could only be explained by a rotation of the axis of the earth in the direction of the apparent daily motion of the stars. This causes the equinoxes to fall a little earlier each year. The knowledge of this *precession of the equinoxes* and of the rate at which it takes place was necessary for the progress of accurate astronomical observation. The complete cycle of precession takes 26,000 years.

(*b*) *Theory of motion of the planets.* When Hipparchus came to examine the apparent movement of the planets he had before him two theories, namely that of 'epicyclic motion' and that of 'eccentric motion'. Certain of his predecessors—notably Apollonius of Perga (p. 75)—had suggested the epicyclic view (Fig. 35). According to this each planet moves on a circle the centre of which moves on another circle, the centre of which is the centre of the earth. Others of his predecessors had set forth the view of eccentric motion. According to this the planet moves around the earth but in a circle whose centre is not at the centre of the earth. This secondary centre may also be represented as moving on a circle. Hipparchus explained the behaviour of the sun by a fixed and the moon by a moving eccentric. (The geometric results of moving eccentric and epicycle are identical.)

The epicyclic view largely prevailed through the mediation of the astronomer Ptolemy (pp. 89 f.). The theory of the eccentric motion of the moon and to a less extent of the sun, as enunciated by Hipparchus, was, however, of great service in that calculations based on it accorded much more closely with actual observations than did calculations based on any older doctrine of their movements. From the time of Hipparchus onward eclipses of the moon could be predicted

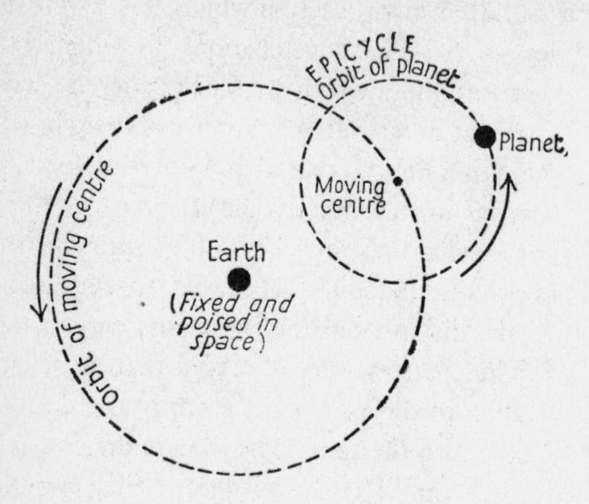

FIG. 35. To illustrate epicyclic motion.

within an hour or two. Eclipses of the sun could be predicted less accurately.

The Middle Alexandrian period, so brilliant in its development of the mathematical sciences, is disappointing when we come to consider its biological achievement. Of true scientific biology, apart from medicine, there was very little. The tradition almost died with Theophrastus (p. 57). There was, however, one real contribution, that of realistic plant painting. Of this we have considerable evidence, including copies of a number of figures by the herbalist CRATEUAS (*c.* 80 B.C.). This method, important still, was doubly valuable in the absence of a system of botanical nomenclature. The plants figured by Crateuas were all of medical application. They are of very special interest as the earliest specimens of biological draughtsmanship (Fig. 36), and the tradition that they created can be traced through the ages to our own time.

In more purely medical matters illustration is perhaps also the main contribution of the middle Alexandrian period. The medical writings of the time were mainly commentaries on the works of the *Hippocratic collection.* Copies of the sketches of operations and

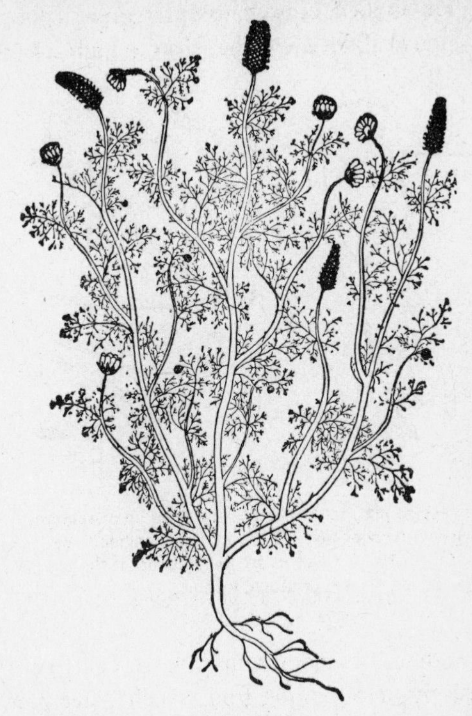

FIG. 36. 'Pheasant's eye', *Adonis aestivalis*, as represented by Crateuas about 80 B.C. and preserved in a derivative MS. of about A.D. 500.

bandaging by APOLLONIUS of Citium (*c.* 100 B.C.) have survived, and give some indication of the conditions under which ancient surgical practice was conducted.

4. LATE ALEXANDRIAN PERIOD TO A.D. 200

(i) *Physical and Mathematical Knowledge*

Egypt became a province of the Roman Empire in 50 B.C. Alexandria's achievement had now become an episode in her history. There remained little native power of initiative, but some scientific curiosity and considerable compilatory capacity. Creative efforts—as those of Strabo (pp. 110 f.), of Ptolemy (pp. 89 f.), and of Galen (pp. 98 f.)—were forthcoming only in response to definite imperial needs.

An ingenious writer of the age was one HERO of Alexandria (fl. A.D. 62). He applied himself to entertaining contrivances and sometimes to practical devices rather than to high scientific themes.

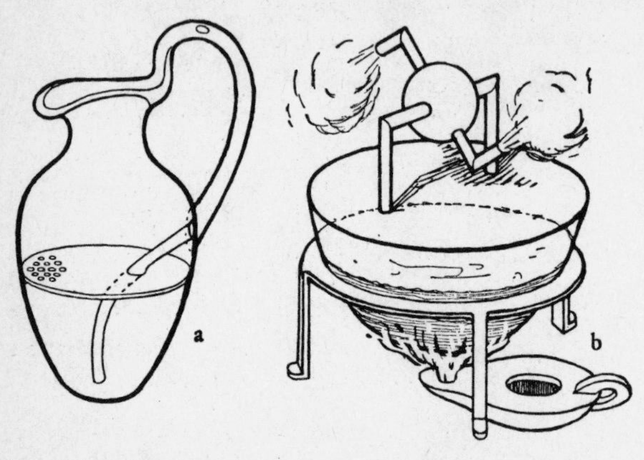

FIG. 37. (*a*) Hero's magic jug. As the thumb is pressed on or released from the hole in the handle, the jug will pour or not. (*b*) Hero's steam-engine. The globe is pivoted on tubes rising from the boiler. It revolves by the reaction of the issuing steam and is a primitive turbine.

His *Pneumatica* describes many conjuring tricks. Thus the principle of the siphon is applied to a jug from which water pours or not at will (Fig. 37 (*a*)). Most famous of his toys was a globe which whirls by force of steam—the first suggestion of a steam-engine (Fig. 37 (*b*)). In his *Mechanica* he shows understanding of the cogwheel, of rack and pinion, of multiple pulleys, of transmission of force from a rotating screw to an axis at right angles to it, and to the combination of all these devices with levers (Fig. 38).

Hero records advances in optics. The oldest treatise on the mathematical aspect of that subject is by Euclid (p. 64), who considered that light moves in straight lines and believed vision to be something that goes forth from the eye. Hero showed that when light is reflected from a surface, it is at an angle equal to the angle of incidence. One of his surveying instruments depended for its working on the equality of these angles. His Dioptra (Fig. 39) served many purposes for which the theodolite is now used. Hero was also particularly ingenious in his use of water-levels in surveying.

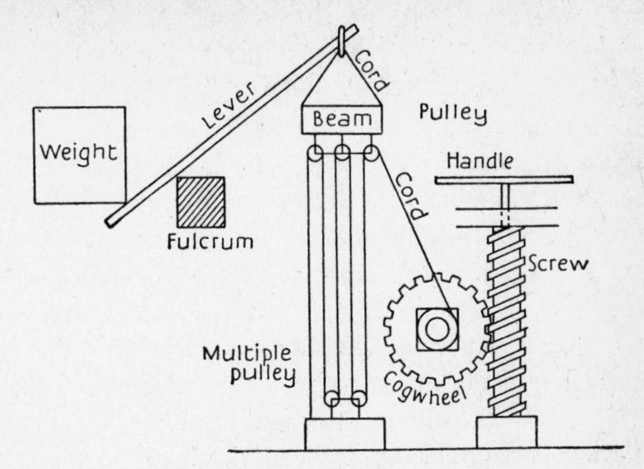

FIG. 38. Hero's mechanical repertory.

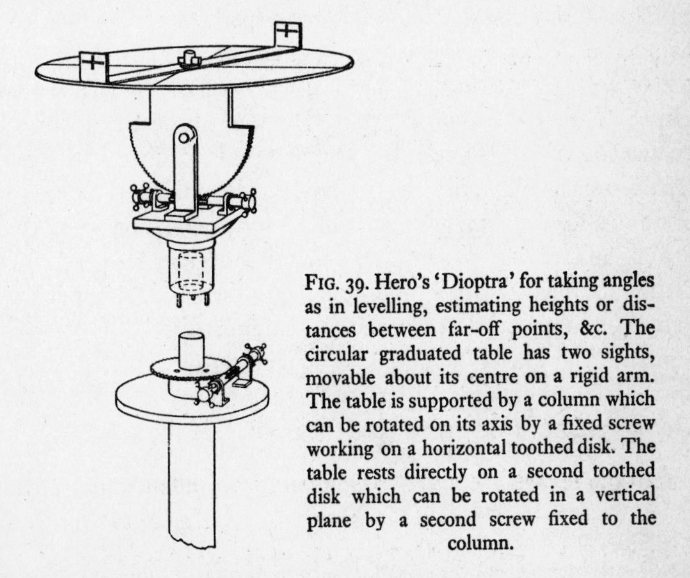

FIG. 39. Hero's 'Dioptra' for taking angles as in levelling, estimating heights or distances between far-off points, &c. The circular graduated table has two sights, movable about its centre on a rigid arm. The table is supported by a column which can be rotated on its axis by a fixed screw working on a horizontal toothed disk. The table rests directly on a second toothed disk which can be rotated in a vertical plane by a second screw fixed to the column.

Attempts were made to study refraction, that is, the behaviour of light in passing from one medium into another of different density, as from air into glass or water. The bent appearance of oars or rods dipped in water must have been observed very early. CLEOMEDES (first century A.D.) referred to the same principle the fact that an object, lying in an opaque basin and just obscured by the brim, could be rendered visible by pouring in water. He applied this principle to the atmosphere and suggested that the sun, even when below the horizon, might be visible under certain circumstances (see Fig. 40). It is remarkable that he failed to give a practical application to this view of atmospheric refraction, for he disbelieved statements of his predecessors that in certain lunar eclipses the sun seems to be still above the horizon while the eclipsed moon rises in the east.

That some beginning had already been made of the science which deals with the eye as an optical instrument we learn from a work by a medical writer, RUFUS of Ephesus (*c.* A.D. 100). He had a fairly accurate conception of the structure of the eye. Some of the names which he applied to parts of this organ have survived in modern scientific nomenclature. Rufus is the first to describe the eye as possessing a *lens*; he speaks of it as 'lentil-shaped'.

A late Alexandrian writer, DIOPHANTUS (perhaps of about A.D. 180), is important as the best ancient exponent of algebra. His work on that subject was commented on by HYPATIA of Alexandria, the only woman mathematician of antiquity. She was murdered by Christian fanatics in A.D. 415. The work of Diophantus is the first that employs signs systematically. He gives symbols for the unknown, for powers, for minus, for equality, and so forth. He solves equations of the first, second, and, in one instance, of the third degree. He sets forth a method for finding two or more square numbers the sum of which is a given number, while each of the two approximates to the same number. The device he adopts is a method of approximation to limits. Thus in dividing 13 into two square numbers each of which is to be greater than 6 he reaches the result that the sides of the required squares are $\frac{258}{101}$ and $\frac{257}{101}$. Diophantus solved other comparable problems.

Not only was Greek algebra far behind Greek geometry but it was also far less influential on later mathematical development. Thus the work of Diophantus did not appear in print until 1575 and then only in a Latin translation. It had, therefore, little effect on the revival of mathematics in the sixteenth century. With Diophantus creative Greek mathematics comes to an end.

(ii) *Ptolemy. Astronomy and Geography*

PTOLEMY of Alexandria (fl. A.D. 140),[1] who provided the great astronomical and geographical syntheses of antiquity, contributed also the most remarkable ancient experimental research on

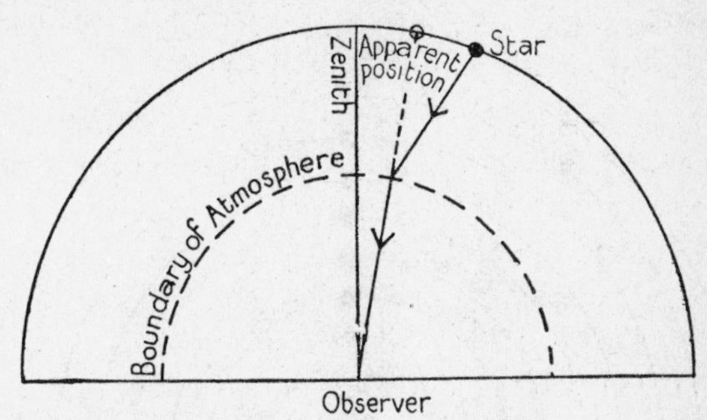

FIG. 40. Refraction of ray by atmosphere makes the apparent position of a star nearer the zenith than the real position.

optics. He knew that rays of light are deflected in passing from one medium to another and suggested the simple rule that the angles of incidence and refraction are proportional. This is not so, but for small angles it is approximately true (p. 229). Applying these ideas, Ptolemy pointed out that the light of a star on entering the atmosphere must be bent or refracted.[2] Thus it will appear to be nearer the zenith than is actually the case (Fig. 40).

The great work of Ptolemy, later known as the *Almagest*, is one of the most influential scientific writings of all ages and one of the most important surviving from antiquity. It provided a mathematical representation of the physical world and thus set the stage for many developments. In the later Middle Ages and the early Renaissance, it was the apex of scientific attainment. Because of it, the medieval science of mathematical astronomy outstripped all other kinds of explanation of the universe.

[1] Claudius Ptolemaeus, not to be confused with any of the Ptolemies, kings of Egypt.

[2] There is no sharp upper limit to the atmosphere and it gets denser the nearer to earth. This was, of course, unknown to Ptolemy.

The very name *Almagest* has a history. The Greeks called it the *megalē mathematikē syntaxis*, i.e. 'great mathematical composition', but used also the superlative form *megistē syntaxis*. This was transliterated by the Arabic writers as *al-magistē* which passed into medieval Latin as *Almagestum*.

The *Almagest* gives a full exposition of the descriptive and less mathematical parts of 'astronomy' (Fig. 41), but its main burden

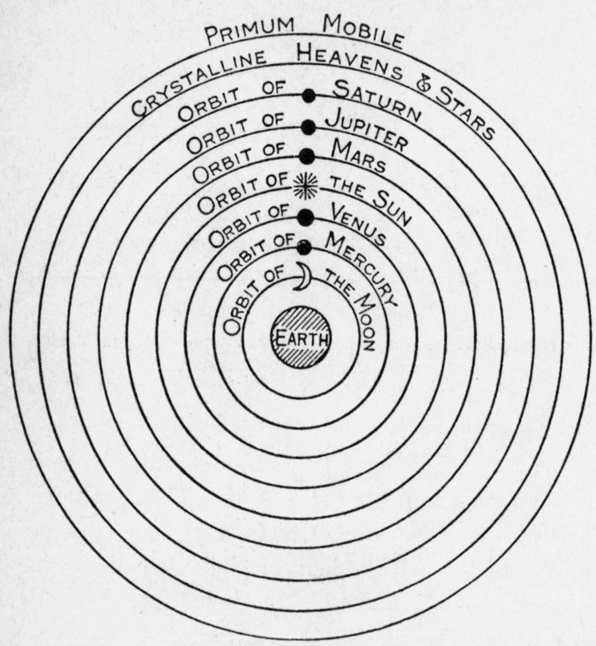

Fig. 41. The Ptolemaic World-System.

and message is the mathematical presentation of the paths along which the planets appear to move in the heavens. Ptolemy's success in this was very satisfactory to those few who could understand it. Nevertheless, his theory was perhaps as difficult, relative to his own time and subsequent centuries till the sixteenth, as Einstein's Relativity is to ours. It accounted for all observed planetary movements, in general to an accuracy a little better than the limitations of ordinary naked-eye observations would allow, i.e. within about five seconds of arc. Its theory remained virtually unchanged, so far as its mathematical details were concerned, even after Copernicus had vastly altered the philosophical standpoint (1543). Not until Kepler in the seventeenth century (1608, pp. 240–1) were fundamental changes made in Ptolemy's details to allow for the increased accuracy of observation of Tycho Brahe (died 1601, pp. 215 f.).

Because of its success, the *Almagest* displaced its predecessors. Thus it is hard to be sure how much of it is Ptolemy's own, and how much drawn, as he sometimes tells, from predecessors, notably Hipparchus (p. 82). Certainly the more elementary parts of his astronomy had long been common knowledge, and even to the mathematical planetary theory Ptolemy probably added only the final improvements. His great achievement, however, is his superb presentation of the scheme as a whole. His most effective contribution to it was in setting the earth eccentrically within the main circle representing the orbit of the planets and in making the rotation of each uniform about another point, the *equant*. This latter was a point an equal distance from the earth on the other side of the circle's centre (Fig. 42). This device gives so close an approximation to the result achieved fifteen centuries later by Kepler that it accounts for its long acceptance.

In addition to the eccentric circle, the *Almagest* uses epicycles (Figs. 35 and 42) to explain the second periodic variation of a planet's position (which is really caused by the motion of the earth around the sun). It should be remembered that it makes no difference to the mathematics whether we treat as stationary the earth with Ptolemy or the sun with Copernicus.

The *Almagest* contains a detailed and accurate star catalogue, partly adapted from Hipparchus. It has also the earliest extensive descriptions of astronomical instruments. Of these some were very simple. Such was a metal ring which could be set in the plane of the meridian and graduated with degrees. It had sights on an inner ring which could be turned, so that the elevation of a planet or star could be read (Fig. 43). Another was a block of wood sliding in a long frame with, at one end, an eyehole, by looking through which and sliding the block back and forth, small angles could be measured and the apparent width of the sun's disk estimated (Fig. 44); or again a simple ring set in the plane of the equator and used for equinox determinations.

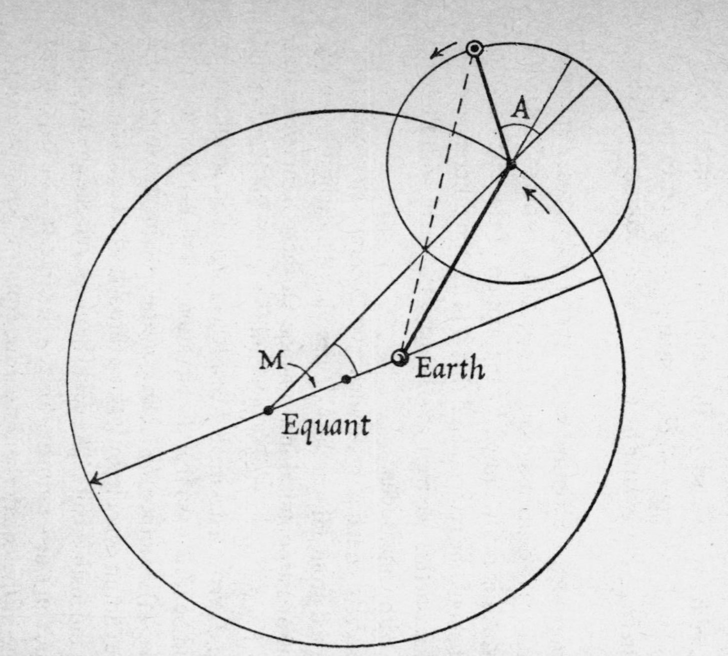

FIG. 42. Ptolemaic construction for movement of a planet. Angles M and A each increase uniformly with time at their own appropriate and different rates. For Mars, Jupiter, and Saturn, M follows the planetary period and A makes one revolution in a year. For the inner planets Venus and Mercury, the situation is reversed. The dotted line indicates the direction of the planet seen from the earth.

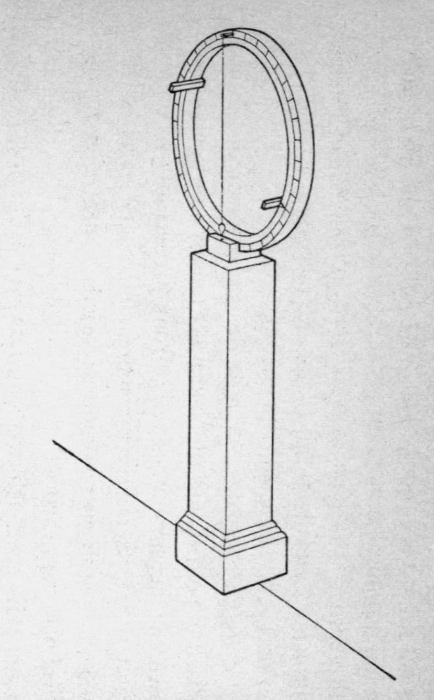

FIG. 43. Ptolemy's instrument for observing meridian altitude of sun. An inner ring with two sights is set within a graduated outer ring mounted vertically in the plane of the meridian.

More complicated instruments described by Ptolemy are of high historic interest. The *triquetrum* (Latin = triangular, Fig. 45) had a sighting rod with a large hole at the upper end which just framed

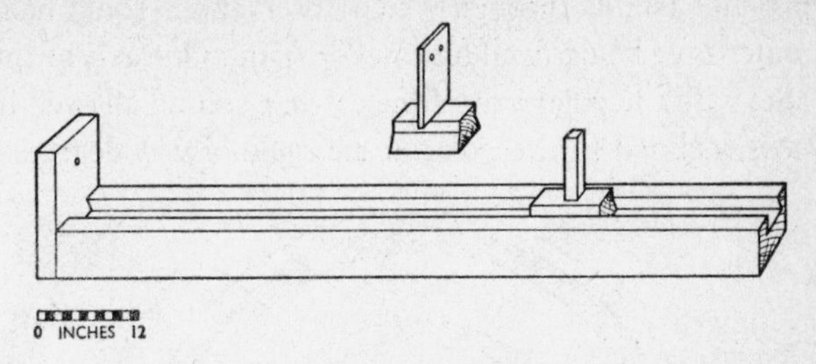

FIG. 44. Ptolemy's apparatus for observing apparent width of sun's disk and other small angles.

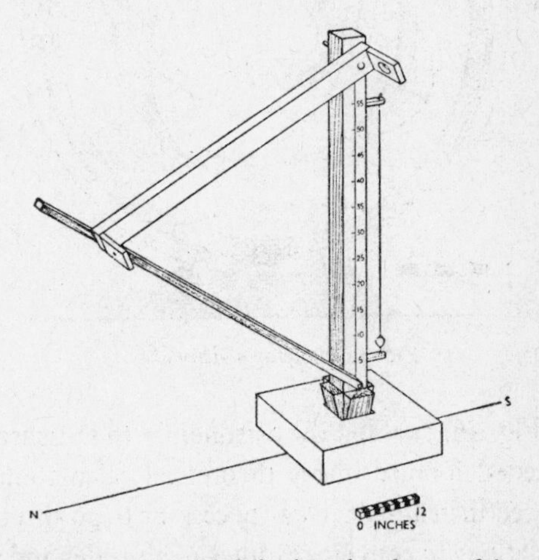

FIG. 45. Ptolemy's *Triquetrum*. An alidade is pivoted at the top of the vertical graduated post. The free end of the alidade is so attached to the movable lathe that its distance from the pivot is always equal to the distance between upper and lower pivots on the post. This distance along the lathe was a measure of the chord of the angle between alidade and vertical which could be read off from a table of chords.

the moon when one looked at it through the lower hole. With it the position of the moon's centre could be determined even when only a small crescent was visible. The *triquetrum* also had the advantage that the graduations were set along a straight rod and therefore easier to mark than along the circular scales of other instruments.

Most important of the instruments of Ptolemy was the armillary *astrolabon* (Fig. 46).[1] It consisted of a series of nesting metal rings, pivoted one within the other, the innermost having an alidade with a pair of pin-hole (sights through which observations could be made) and the outermost being fixed on a heavy stand. One axis of rotation was parallel with the polar axis of the earth, a second allowed for the ecliptic axis inclined to the polar at an angle of 23½ degrees. This

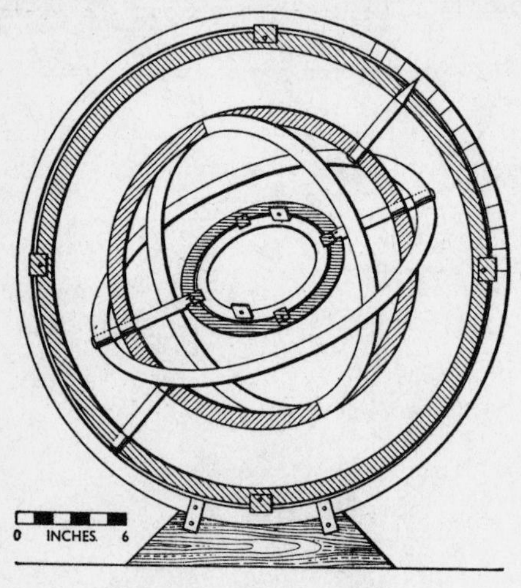

FIG. 46. Ptolemy's *Astrolabon.*

instrument (Fig. 46) enabled the astronomer to measure directly the angles he needed for use in his theoretical calculations. With any less complicated instrument it was necessary to go through long and tedious calculations to convert readings of altitudes and azimuths, or equatorial coordinates, into the ecliptic coordinates essential to the theory.

It is typical of the *Almagest* that Ptolemy's chief concern is to establish a mathematical theory for each of the planets separately. When it comes to the question of considering the planetary system as a whole, he makes the point that no observation known to him can

[1] Not to be confused with the *plane astrolabe*, or *astrolabium*, a calculating instrument with a sighting attachment. This also may have been known to Ptolemy but is not proved to have existed until later, and was not popular before the tenth century in Islam and hardly known until the thirteenth century in Latin Europe (pp. 149–50).

throw much light on this, and any assumption can make little difference to his theories. He therefore falls back on the old tradition of accepting rapidity of motion of the planets as the main test of nearness and agrees with the ancient order of having the earth at the centre and the Moon, Mercury, Venus, Sun, Mars, Jupiter, Saturn, and the Fixed Stars outwards in that order. Thus although the *Almagest* has very little to say about cosmology, Ptolemy's name has become attached to the main pre-Copernican world-system (Fig. 41).

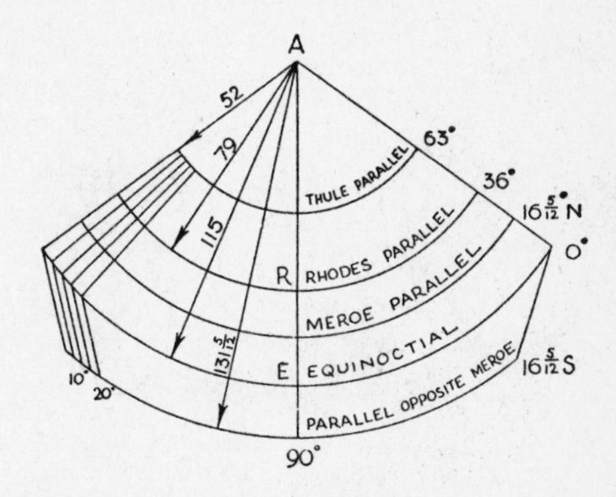

FIG. 47. Ptolemy's method of geographical projection.

Ptolemy's other great work was his *Geographical Outline*. This was essentially a product of the knowledge brought by the expansion of the Roman Empire. He studied itineraries of Roman officials and merchants. Thus he may be said to have preserved for us a summary of Roman knowledge of the earth's surface, presented, however, in a form probably quite beyond the capacity of any of his Latin contemporaries. Ptolemy may well have had access to the great map prepared by Vipsanius Agrippa at Rome (p. 112). He certainly developed his own manner of representing the curved surface of the earth on a plane surface. It was on a tangent cone of 36° with the equator as base. The standard parallel was through Rhodes at 36° N., the most northerly point being Thule at 63°, and the most southerly as far beyond the equator as Meroë was north of it, i.e. $16\frac{5}{12}$° (Fig. 47). It supposed the size of the habitable world to cover 180° in longitude and 80° in latitude. The meridians of longitude are represented by straight lines which converge to the Pole[1] (Fig. 48). He delineates in this manner the whole of the then known world. Its boundaries are: on the north, the ocean which surrounds the British Isles, the northern parts of Europe, and the unknown land in the northern region of Asia; on the south, the unknown land which encloses the Indian Sea, and the unknown land to the south of Libya and Ethiopia; on

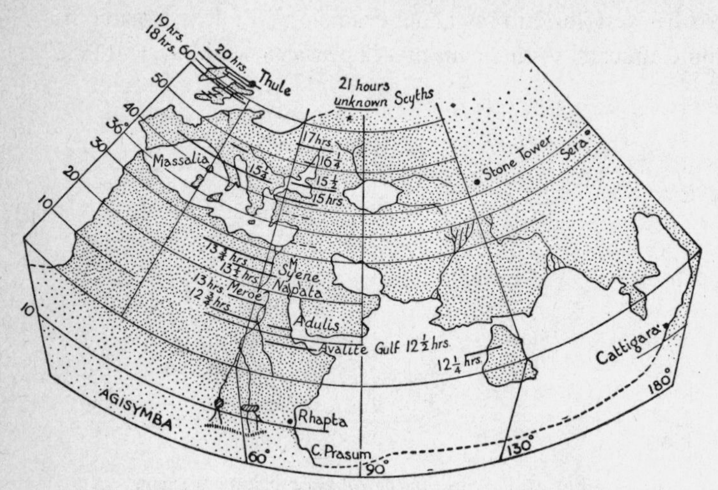

FIG. 48. Reconstruction of Ptolemy's map of the habitable world.

the east, the unknown land which adjoins these eastern nations of Asia, the Sinae (Chinese) and the people of Serica, the silk-producing land; on the west, the great Western Ocean and unknown parts of Libya.

As originally written Ptolemy's geography was furnished with maps. These have long since disappeared, but as Ptolemy gives the latitude and longitude of the places that he mentions his charts can be reconstructed. A peculiar interest attaches to the map of Britain, which can thus be put together (Fig. 48). Scotland is bent eastward with its axis at a right angle to that of England. This is an unusual degree of error for Ptolemy. It is probable that he was here working not on the records of travellers, but on maps of the island, and that he had made the error of fitting the map of Scotland on to that of England on the wrong side!

[1] He has another scheme of projection in which the meridians are curved.

Ptolemy exhibits the final extension of scientific geography in the Empire. How far the average educated citizen of the Empire was able or willing to appreciate science in general and geography in particular is another matter. It was the attitude of the Romans and especially of the Roman ruling class to things of the mind that determined the

FIG. 49. British Isles according to Ptolemy.

fate of science and with it, perhaps, the fate of the Empire. To estimate the attitude of the Roman to science we must turn to geographical works in Latin (pp. 108 f.).

The *Almagest* of Ptolemy was translated into Latin in the later twelfth century and his *Geography* in the fifteenth. Thus they could not directly influence the earlier Middle Ages. In place of the *Almagest* a simpler cosmic scheme based on Aristotle prevailed (Fig. 68). At the close of the Middle Ages conflict between the views of Aristotle and those of Ptolemy become of considerable importance for the history of science.

(iii) *Late Alexandrian Biology*

The picture presented by the exact sciences of the late Alexandrian period is that of a number of minor works followed by two great syntheses and then a steady decline. We have seen this for astronomy and geography. It is repeated for the biological and medical sciences. In those departments we need only note the figures of Dioscorides and Galen.

PEDANIUS DIOSCORIDES of Anazarba in Asia Minor was an army surgeon who served in his own country under Nero (A.D. 54–68). He wrote a work on drugs. It consists of short accounts of plants arranged on a system that has hardly any reference to the nature of the plants themselves. The descriptions given are often terse and striking, and sometimes include a few words on the habits and habitats of plants. This elaborate pharmacopoeia was early illustrated in the style of Crateuas (Fig. 36), and some fine copies of these figures have come down to us.

The history of the work of Dioscorides reveals it as one of the most influential botanical treatises ever penned, despite the absence from it of anything like general scientific ideas. It provided most of the little botanical knowledge that reached the Middle Ages and furnished the chief stimulus to botanical research at the time of the Renaissance. It decided the general form of every pharmacopoeia before the present century and determined much modern plant nomenclature, both popular and scientific.

The great biological and medical synthesis of antiquity was made by GALEN (A.D. 131–201) of Pergamum (p. 63). In his youth he visited Alexandria and other centres of learning, collecting all the knowledge of the day. Later he proceeded to Rome where almost all the rest of his very active life was passed.

In Galen's time the dissection of the human body had fallen into desuetude. The knowledge of anatomy had therefore declined. He made, however, accurate anatomical and physiological studies on many other animals. Among these were the Barbary and other apes, the structure of which is not very far removed from that of man. Galen also made numerous highly skilled experiments on living animals. He was thus able to evolve a complete and very ingenious physiological system. This was generally accepted by later antiquity and did not begin to be undermined until the work of Vesalius (p. 210) in the sixteenth century. Nevertheless the system generally accepted as Galen's cannot be entirely derived from his works.

The basic principle of life in this scheme was a spirit or pneuma (p. 59). This, as Erasistratus had long before held (p. 68) was

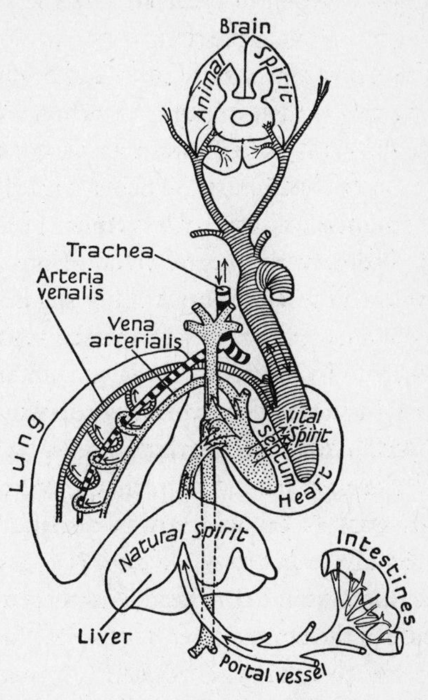

FIG. 50. Diagram illustrating Galen's physiological system.

drawn from the cosmic pneuma, that is the air. It entered the body through the wind-pipe and so passed to the lung and thence (through the 'vein-like artery', which we now call the *pulmonary vein*) to the left ventricle, where it encountered the blood (Fig. 50). But what was the origin of the blood? To this question his answer was most ingenious, and the errors that it involved remained till the time of Harvey. Galen believed that chyle, brought from the alimentary tract by the portal vessel, arrived at the liver. That organ, he considered, had the power of elaborating the chyle into venous blood, and of imbuing it with a second spirit, or pneuma, innate in all living substance so long as it remains alive. This pneuma was called the *natural spirit*. Charged with natural spirit derived from the liver, and with nutritive material derived from the intestines, the venous blood, Galen believed, was distributed by the liver throughout the venous system which arises from it, ebbing and flowing in the veins.

One great main branch of the venous system was the cavity that we now call the right ventricle of the heart. For the venous blood that entered this important branch, the right side of the heart, the Galenic scheme reserved two possible fates. The greater part remained awhile in the ventricle, parting with its impurities, which were carried off (by the 'artery-like vein'—now called the *pulmonary artery*) to the lung, and there exhaled. These impurities being discharged, the venous blood in the right ventricle ebbed back again into the general venous system. A small portion of this venous blood from the right side of the heart followed a different course. This small portion trickled through minute channels in the interventricular septum and entered the left ventricle drop by drop. There it encountered the pneuma brought thither from the outside world by the wind-pipe (through the 'vein-like artery'). These drops thus developed a higher type of pneuma, the *vital spirit*. Charged with this, the dark venous blood became bright arterial blood which was distributed through the arteries to all parts of the body.

Of the arteries, some went to the head, and thereby vital spirit was brought to the base of the brain. Here the arterial blood was minutely divided and became charged with yet a third pneuma, the *animal spirit*. This was distributed by the nerves, which were supposed to be hollow (Fig. 50).

Nearly all the biological conceptions, most of the anatomy, much of the botany, and all the ideas of the physical structure of living things from the third to the sixteenth century were contained in a small number of works of Galen or ascribed to him. The biological works of Aristotle and Theophrastus lingered precariously in a few rare manuscripts in the monasteries of the East; the output of hundreds of years of Alexandrian and Pergamene activities was utterly destroyed; forgotten were the Ionian biological works, of which fragments have marvellously survived; but the vast, windy, ill-arranged treatises of Galen lingered on. Translated into Latin, Syriac, Arabic, and Hebrew, they saturated the intellectual world of the Middle Ages.

Commented on by later Greek writers, who were in turn translated into the same list of languages, they were yet again served up, often under the names of other writers in the Middle Ages and later.

What is the secret of the vitality of these Galenic biological conceptions? The answer can be given in four words: *Galen was a teleologist*. He believed that everything is made by God to a particular and determinate end (*telos* = 'end', 'aim'). Moreover, Galen's teleology is of a kind which happened to fit in with the prevailing theological attitude of the Middle Ages, whether Christian, Moslem, or Jewish. According to Galen, everything which exists and displays activity in the human body was formed by an Intelligent Being on an intelligible plan, so that the organ in structure and function is the result of that plan. 'It was the Creator's infinite wisdom which selected the best means to attain His beneficent ends, and it is a proof of His omnipotence that He created every good thing according to His design, and thereby fulfilled His will.' To know man you must therefore know God's will. This attitude removes the foundation of scientific curiosity. After Galen there is a thousand years of darkness, and both medicine and biology almost cease to have a history. Men were interested rather in the will and purpose of God than in natural phenomena.

(iv) *Late Alexandrian Alchemy*

In leaving the Alexandrian period we may touch on one activity the influence of which has been peculiarly persistent. Antiquity had a very highly developed and exact technology, but the attempts to rationalize it are lost. That such there were can be inferred from the earliest traces of 'alchemy' that reach us from Alexandrian sources from about A.D. 100 onward. Alexandrian alchemy is very sophisticated and clearly the result of a long evolution.

The surviving texts are all in Greek and carry marks of Christian, Jewish, Neoplatonic, Gnostic, Greek, Egyptian, and perhaps Persian and Chinese elements. Of names associated with this strange literature, two are worth recording here. One is MARY THE JEWESS, who is still remembered in the steam bath of our laboratories, the *bain Marie* of French chemists. The other is ZOSIMUS, the first alchemist who can be treated as an historic figure. He flourished about A.D. 300 and some of his apparatus can be reconstructed from his descriptions. These Alexandrian alchemical texts contain ideas that persisted to the very dawn of modern chemistry. Some of them contain recipes which are eminently practical and are therefore hardly alchemical as that word is now commonly understood. The early Arabic alchemists (pp. 144–7 and 154) doubtless had access to far more Alexandrian material than now exists.

IV

THE FAILURE OF INSPIRATION

Science the Handmaid of Practice: Imperial Rome (50 B.C.–A.D. 400)

I. DEVELOPMENT OF THE ROMAN ATTITUDE TO NATURE

THE scientific idea, the conception of a reasonable universe, came to the peoples of central Italy much later than to the Greeks of the eastern Mediterranean and of southern Italy. Moreover, science with the Romans always remained somewhat of an exotic. Rome established her protectorate throughout the eastern Mediterranean soon after 200 B.C. The influence of Greek ideas on Roman civilization thenceforth grew rapidly. All educated men came to learn Greek and were inevitably affected by Hellenic philosophy. Yet despite the stimulus of Alexandrian thought, the Latins produced no great creative men of science.

The prevalent attitude towards nature among the Latin-speaking governing classes, whether Italian or provincial, was best expressed by the *Stoic* creed. The Epicurean philosophy gained fewer adherents among them. The Stoic system laid great stress on correct conduct and duty. It was based on a rigid conception of the interrelation of the different parts of the world. It provided little stimulus for the acquisition of new knowledge or for anything in the way of research. Thus, in place of knowledge accumulating progressively on a basis of a wide and far-reaching theory, we get, under Stoicism, either a type of exact but intellectually motiveless observation, or a rejection of all knowledge not of practical importance. The dogmatism of Epicurean teaching was even less favourable to scientific research than was the Stoic outlook.

There have been many attempts to explain why the Romans did not continue the scientific works of the Greeks. It has been said that the Roman mind could find no time from conquest and administration to attend to scientific matters. This will not explain the situation, for there were Romans who were able to answer the no less exacting claims of philosophy and literature. The matter, in fact, lies deep in the Roman character and tradition. It was related to the ethics of the favourite Roman philosophy, Stoicism, and is not unconnected with the Roman passion for Rhetoric. In general we may say that Roman science appears at its strongest in the department of the general study of nature and at its weakest in pure mathematics. The success or failure of the Romans in any scientific field may be roughly gauged by its nearness to one or other of these disciplines. But Roman culture is so large a source of our own civilization that it is desirable to consider the Roman influence on the course of science in greater detail than the direct Roman contribution would itself warrant.

We have several works by Latins which deal with the implications of science in general. None involves any expert knowledge of natural phenomena, and they are concerned rather with the philosophical relations of science than with science itself. Of such works the most striking and widely read is LUCRETIUS (*c.* 95–55 B.C.), *On the Nature of Things*. The book is magnificent as literature and important as our best representation of Epicurean views (p. 60). It is, however, too much a work of propaganda to be of high scientific value. Moreover, it neither records first-hand observations nor does it even present a typically Roman attitude of mind.

The attention of the scientific reader of Lucretius will naturally be drawn to his atomic views. Following his master Epicurus, Lucretius explains the origin of the entire world as due to the interaction of atoms. This interaction, he believes, is without the intervention of any creative intelligence. Even mental phenomena are of atomic origin and there is no reality save 'atoms' and 'the void' (*inane*, p. 60). 'Nothing is ever begotten of nothing by divine will.' Everything springs from determinate units (*semina certa*). The genesis of all things is typified by the generation of organic beings. The species of plants and animals give us models for all processes and natural laws. This conception of generation has its converse. 'Things cannot then ever be turned to naught.' Such an attitude involves that 'indestructibility of matter' which is the historical foundation on which nineteenth century chemical and physical knowledge was built.

The resemblance of the Lucretian theory to modern atomic views is, however, more apparent than real. Not only are the atoms of Lucretius of different sizes and shapes, but also he knows nothing of definite laws by which they hold together as molecules. He has no inkling of chemical combination. He is without that 'doctrine of energy' that is so characteristic a feature in nineteenth-century physical theory. His work indeed had little direct influence on the development of the modern doctrine and probably was not widely read even in its own day. Epicurean thought was not favourable to scientific development. Moreover, the atomic view of matter was practically lost during the Middle Ages, and Aristotelian philosophy, which implied continuity of matter, was paramount for centuries.

Some have seen in Lucretius the beginnings of a theory of evolution. He certainly exhibits a 'ladder of nature' not unlike that of Aristotle (p. 47). The earth produces first plants and then animals of ever higher type. 'Even as down and hair and bristles are first formed on the limbs of beasts . . . so the newborn earth raised up herbage and shrubs first, and thereafter the races of mortal things.' This idea of 'spontaneous generation' was inevitable until the realm of minute microscopic life could be explored (pp. 282 f.). It is thus no wonder that Lucretius follows Aristotle and all antiquity in assuring us that 'even now many animals spring forth from the earth, formed by rains and the heat of the sun'.

Did Lucretius take the matter further and did he have any conception of lower forms passing into higher forms? In a sense he did. Moreover, he invoked a process of 'survival of the fittest' for the formal exposition of which the world had to await the arrival of Darwin. But the Lucretian presentation of the manner in which the more perfect creatures reached their present state has no relation whatever to the historic geological record.

When we turn to the phenomena which Lucretius has chosen for special description we note that they are drawn from the magnificent, dramatic, or cataclysmic. His temper is far from the impartial spirit of science and there is nothing of the quietly scrupulous careful observer about him. Thunder and lightning, water-spout, volcano and thunderbolt, suffocating vapours and devastating pestilences—these are the themes he selects. There is no reason to give to Lucretius an important place among those who have helped or inspired the study of nature.

More characteristic of the Roman mind are the works of Varro (116–27 B.C.), of Pliny the elder (A.D. 23–79), and of Seneca (3 B.C.–A.D. 65).

VARRO, a country gentleman of the old Roman school, visited Athens and was influenced by Platonism, but developed definite Stoic leanings. He wrote an encyclopaedia of the sciences, and his works were the prototype of the numerous medieval treatises on the 'liberal arts'. He distinguished nine such disciplines, namely grammar, dialectic, rhetoric, geometry, arithmetic, astronomy, music, medicine, and architecture. Of these the last two were not recognized by the later Latin writers who handed down the tradition to the Middle Ages. The number of 'liberal arts' was thus reduced to seven (p. 107).

Varro tried to collect Latin learning and set it over against the Greek. He was in a good position to do this for he possessed the old Roman tradition and he had also received a good Greek education. He was employed by Caesar to arrange the great stores of Greek and Latin literature for the vast library which he intended to found. His work *On Farming* (*Res rusticae*) was written in his eightieth year. In it he records his own rich experience, but he has collected his material mainly from the writings of others. He thus exhibits the derivative tendency which is so disastrous a feature of Latin writers on scientific topics. He uses every opportunity to bring in etymology, rejoicing in artificial separations and divisions, so that the work gives much the impression conveyed by many treatises of medieval origin.

In PLINY (A.D. 23–79) the Greek leaven has worked further than in Varro. Pliny had a literary education in Rome, where he took to studying plants. Coming under the influence of Seneca (p. 107) he turned to philosophy and rhetoric, and practised as an advocate. After military service in Germany, and having visited Gaul and Spain, he returned to Rome. There he completed his *Natural History*, dedicating it to the Emperor Titus. As prefect of the fleet he was stationed in the bay of Naples at the time of the eruption which overwhelmed Pompeii and Herculaneum in A.D. 79. He owed his death to his desire to observe that phenomenon more closely. His education, career, opinions, and character are all typical of the Italian tradition of his day.

The *Natural History* of Pliny was drawn from about 2,000 works —most of them now lost—by 146 Roman and 326 Greek authors. Its erudite, travelled, and industrious author exhibits an interest in natural phenomena that is quite uncontrolled by scientific or critical standards. The main thought that runs through the book is that nature serves man. Natural objects are hardly described as such but only in relation to man. All things have their 'uses'. 'Nature and the Earth', he says, 'fill us with admiration . . . as we contemplate the great variety of plants and find that they are created for the wants or enjoyments of mankind.' This world of wonder is, however, effectively without a God and works by rule—though it is a crazy rule which these disordered, credulous, wonder-loving volumes set before us.

Many of the matters on which Pliny expresses a judgement would have been impressed on him in the manifold life of Imperial Rome. Many of the animals he discusses were brought to the capital for the arena or for the kitchen from the farthest ends of the earth. So too with plants. He describes a botanic garden kept by a Roman for the purpose of ascertaining the medical and allied properties of herbs. In descriptions of living creatures Pliny goes back to Aristotle and Theophrastus, but there is no systematic building of the subject and he is scientifically far inferior to his sources. Medical plants are treated in greatest detail, and he holds that all plants have their own special medical powers. The thought that nature exists for man constantly recurs. His philosophy, which accords in general with the Stoic scheme, is largely drowned and lost in his love of detail, and is often submerged in rhetoric. He presents a confused cosmology.

SENECA (3 B.C.–A.D. 65) is nearer to the Greek than either Varro or Pliny. A Spaniard by birth, he moved to Rome at an early age. There he came under Stoic influence and made his mark as an advocate and public servant. A member of one of the new provincial families, a brilliant rhetorician with a passion for philosophy, of which he was an eloquent but unsystematic exponent, a man whose undoubted balance and judgement had been earned in affairs rather than in action, with an interest in nature rather in its cosmical than in its detailed aspects, Seneca provides an interesting contrast to his contemporary Pliny.

Seneca's work is more philosophical and far more critical than that of Pliny. Yet his *Natural Questions*, even more than the *Natural History* of Pliny, is borrowed material. He, too, is a Stoic, but does not hesitate to criticize the opinions of that school. His subject is a general account of natural phenomena, but it is ill-arranged and imperfect. It deals chiefly with astronomy, meteorology, and physical geography. He shows, like Lucretius and Pliny, a special interest in the convulsions of nature. Moreover, Seneca was absorbed, like many Romans, by ethics, a moralist first and physicist afterwards. Thus physics—which for him meant a general description of the universe—led to a knowledge of man's destiny and through that to a consideration of man's duty.

Seneca repeatedly tells of the moral to be derived from the phenomena investigated. The relation is often of the most distant and strained character. Thus, terminating his discussion of the phenomena of light, he asks, 'What were nature's purposes in providing material capable of receiving and reflecting images?' And he answers, 'To show us the Sun with his glare dulled, for eyes are too weak to gaze at him direct. Secondly, that we might investigate eclipses reflected in basins. Thirdly, mirrors were discovered in order that man might know himself.' [Abbreviated.]

Such a point of view appealed greatly to the medieval Church, by which Seneca was mistaken for a Christian. He was included by St. Jerome among the 'ecclesiastical writers' and is frequently quoted by later Christian authors. But the ethical attitude to phenomena is inconsistent with the effective advancement of knowledge and has been one of the great enemies of science. In spite of the nobility of his sentiments, in spite of his lip-service to the advancement of learning, in spite of his faith in human destiny, Seneca could do nothing to stay the downfall of ancient science.

2. GEOGRAPHY AND IMPERIALISM

Just as the conquests of Alexander had opened up the East to science, so did the advance of Rome open up the West. Unfortunately it did not carry with it the spirit of scientific inquiry.

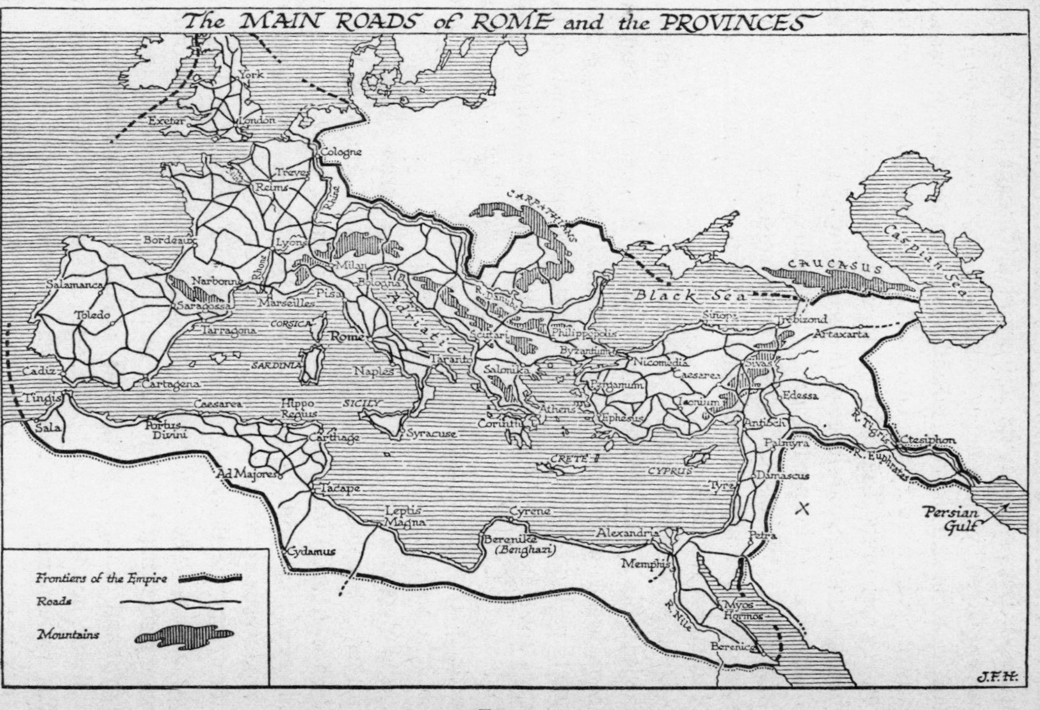
The MAIN ROADS of ROME and the PROVINCES
York
Exeter
London
Cologne
Treves
Reims
Rhine
Bordeaux
Lyons
Rhone
Milan
Narbonne
Salamanca
Saragossa
Marseilles
Pisa
Bologna
Toledo
Tarragona
CORSICA
Rome
Adriatic
R. Danube
CARPATHIANS
Black Sea
CAUCASUS
Caspian Sea
Cadiz
SARDINIA
Naples
Taranto
Scutari
Philippopolis
Byzantium
Salonika
Sinope
Trebizond
Artaxarta
Nicomedia
Caesarea
Sivas
Pergamum
Iconium
Edessa
Tingis
Sala
Cartagena
Caesarea
Portus Divini
Hippo Regius
SICILY
Carthage
Syracuse
Corinth
Athens
Ephesus
Antioch
R. Tigris
Ctesiphon
Palmyra
R. Euphrates
CRETE
CYPRUS
Tyre
Damascus
Ad Majores
Tacape
Leptis Magna
Cyrene
Berenike (Benghazi)
Alexandria
Petra
Persian Gulf
Cydamus
Memphis
R. Nile
Myos Hormos
Berenice
Frontiers of the Empire
Roads
Mountains
J.F.H.

FIG. 51.

A link between the Alexandrian and the Roman geographical standpoints is provided by the Arcadian POLYBIUS (204–122 B.C.), who had resided at Alexandria and later took service with the Roman army. He was present at the destruction of Carthage in 146 B.C., and was employed by the younger Scipio (185–129) to explore the coasts of Africa. He also visited Gaul and Spain. His descriptions, particularly of Spain, are very accurate, and he even attempts an estimation of the length of the Tagus. He has much valuable information about the Alps, and his knowledge of the geography of Italy was superior to that of any of his predecessors. Though an historian rather than a geographer, Polybius understood the necessity of constructing a correct map, and therefore gives much attention to the determination of distances and positions.

During the second and first centuries B.C., improved accounts of the Red, Black, and Mediterranean Seas, and the countries bounding them, began to be available for students. Determinations, even of points in India, were attempted. Mention should be made of the navigator EUDOXUS of Cyzicus (*c.* 150 B.C. not to be confused with Eudoxus of Cnidus, p. 43). After exploring the Red Sea, Eudoxus made at least two voyages southward along the African coast and brought back considerable new information.

The wars and military expeditions of the Romans yielded much further geographical knowledge. Thus STRABO of Amasia in Pontus (born *c.* 63 B.C.) had plenty of material when he began his general survey of the world. He was a considerable traveller and had journeyed westward to the part of Etruria opposite Sardinia and southward from the Black Sea to the borders of Ethiopia. 'Perhaps not one of those who have written geographies', he says, 'has visited more places than I within these limits.' He travelled right through Egypt and made a prolonged stay at Alexandria. Working for long at Rome, he was in a good position to receive authentic information. His mathematical qualifications were, however, inadequate and inferior to those of Eratosthenes (p. 76) on whom his work is based, though his circumstances gave him a greater knowledge of detail, especially for Europe.

Strabo opens by indicating the vast extension of knowledge as a result of the expansion of the Empire of Rome and that of her enemies on the east, the Parthians. Yet he is struck by the comparative smallness of the inhabited world. He makes the suggestion that there might be other continents still unknown. The length of the inhabited world from the Islands of the Blessed (that is the Canaries) to the Silk Land (that is China) was not more than about a third of the total circumference of the globe in the temperate zone. It was therefore possible that within the vacant space might be other lands inhabited by different races of men. In describing the inhabited world Strabo reduces its width from north to south to 30,000 stadia, an estimate below the 38,000 of Eratosthenes. The abbreviation is due to his scepticism as regards the northern regions. He rejects Thule, and disbelieves in any habitable land as far north as the Arctic Circle. Ireland, the most northerly of known territories, is 'barely habitable on account of the cold'. Southward, he considers the habitable world extends about 3,000 stadia beyond Meroë.

A feature of Strabo's work is his account of how a map of the world should be made. This, he points out, would not be difficult upon an actual globe, but such a globe would need to be very large for the insertion of details. He therefore considers the countries as though represented on a flat surface. Many of the distortions in Strabo's account are due to erroneous projection. His best accounts are of the countries bordering on the Mediterranean, where his map is distorted least. As he gets farther from the Mediterranean, his errors become greater. Even in the Mediterranean, however, he makes unexpected blunders. Thus the Pyrenees are represented as running north and south instead of east and west. Naturally with regard to the Caspian, Strabo shared the opinion of most ancient geographers except Herodotus that it was an inlet of the Northern Ocean (Figs. 10, 33, and 52). The north of Asia and the region east of Sogdiana was, he tells us, a mere blank to him. A vast chain of mountains extended, he thought, from east to west across Asia, bounding India on the north. From this range the Tigris and Euphrates took their rise in the west, the Indus and Ganges in the east. Thus the Himalayas are confused with the mountains of Asia Minor and with the Caucasus (cf. Fig. 52).

Among the very few native Romans who had a true conception of the nature of scientific inquiry was JULIUS CAESAR (102–44 B.C.). He formulated the splendid scheme of a complete survey of the Empire. The government of the provinces, the demands of trade, and the distribution of the fleet all made the need evident. The death of Julius left the execution of this plan to his successor, Augustus. The survey was superintended by his son-in-law, VIPSANIUS AGRIPPA (died 12 B.C.) and finally completed after nearly thirty years' work in 20 B.C. It was rendered possible by the fact that the Empire was well furnished with roads, marked with milestones. There was a regular service of skilled surveyors, whose work, incorporated in the reports of provincial governors, was available at headquarters. The vast chart prepared from these details was exhibited in a building especially erected for the purpose at Rome. In this map all other geographical elements were subordinated to indications for the marching of armies.

Geography in the limited sense, as distinct from cosmography, was a topic that might be expected to appeal to the practical and imperialistically minded Roman. He was, however, hardly in an intellectual position to appreciate geography, save in the form of a road-book or rough strategic chart. To general geography the Roman paid little attention. The only significant Latin writer on the subject is the Spaniard POMPONIUS MELA (*c.* A.D. 40), who refers to Britain as about to be more fully explored by an expedition then in progress. This was the visit of the Emperor Claudius in A.D. 43.

Pomponius Mela clearly meant his work as an easy account of his subject. He avoids mathematical topics in the true Roman manner. The world is a sphere, and the land upon it is surrounded on all sides by sea. Five zones may be distinguished. Of these the middle zone is as uninhabitable by reason of its heat as are the two extreme zones by reason of cold. We live in one of the two intermediate temperate zones while in the other dwell the 'Antichthones'. The land in our own hemisphere is completely surrounded by ocean, from which it receives four seas or gulfs, one at the north, the Caspian, two in the south, the Persian Gulf and the Red Sea, and the fourth to the west, the Mediterranean. He is a mere borrower from Greek sources.

Mela gives a general description of the three continents, Europe, Asia, and Africa. Between the three is the Mediterranean, which he speaks of as 'our sea'. He takes the river Tanis (*Don*), Lake Maeotis (*Sea of Azov*), and the Euxine Sea (*Black Sea*) as frontiers between

Europe and Asia, while it is the Nile that divides Asia from Africa. Asia is as large as Europe and Africa together. These ideas were passed on to the earlier Middle Ages and are expressed in the world-maps of which the earliest is in a seventh-century codex of St. Isidore of Seville (560–636). The so-called O.T. map of the Middle Ages is well known (p. 176).

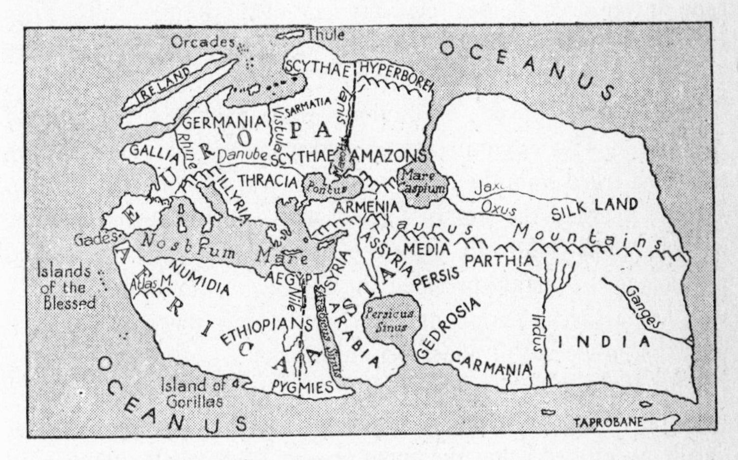

FIG. 52. Northern inhabited world according to Pomponius Mela.

The haziness of the geographical ideas even of an intelligent Roman of Imperial times may be gathered from TACITUS (*c.* A.D. 55–120). He tells how, under Agricola, the Roman fleet rounded Britain and proved it to be an island, discovering at the same time the Orcades (*Orkney Islands*) and coming in sight of 'Thule' (? *Shetlands*). Yet Tacitus, like Caesar and the elder Pliny, believes that Spain lies to the west of Britain. Like Strabo he describes the Pyrenees as running north and south (p. 111). He goes on to explain the phenomenon of the Midnight Sun—which he brings as far south as the north of Scotland—by telling us that 'the flat extremities of the Earth, casting a low shadow, do not throw the darkness up high, and the night does not reach to the sky and stars'. The statement implies the view that the earth is a disk with flattened edges. This from a Roman gentleman who had access to the ideas of Aristotle, Hipparchus, Archimedes, and Eratosthenes and at a time when Roman roads were covering much of Western Europe, the Near East, and North Africa (Fig. 51).

3. IMPERIAL ORGANIZATION OF MEDICINE AND HYGIENE

The original native Roman medical system was that of a people of the lower culture and devoid of scientific elements. Interwoven with ideas that trespass on the domain of religion, it possessed that multitude of 'specialist deities' characteristic of the Roman cults. Thus Fever had three temples in Rome, and was supplicated as the goddess *Febris* and flatteringly addressed as 'Divine Fever', 'Holy Fever', 'Great Goddess Fever'. Foul odours were invoked in the name of *Mephitis*, to whom a temple was erected at a place where asphyxiating fumes emerged from the earth. Lassitude was implored as *Fessonia*. *Uterina* guarded the womb. *Lucina*, with her assistant goddesses, had charge of childbirth. Over the entire pantheon of disease and physiological function presided the *Dea Salus*, 'Goddess Health', who had a special temple on the Quirinal. She was the deity who took the public health under her supervision.

The entire external aspect of Roman medicine was gradually transformed by the advent of Greek science. The change, however, hardly penetrated below the upper classes. Thus many references in the *City of God* of St. Augustine (354–430) show the ancient beliefs still current in the Italy of his day. After the fall of the Empire, they lingered among the barbaric peoples that entered into its heritage. Nor are they yet extinct. Prescriptions and practices of Pliny (p. 106) and of his even more gullible successors may still be traced in European and in American folk-customs and folk-beliefs.

During the Republic, medical education had been a private matter. The direct relation of pupil and master exhibited by the magnificent so-called *Hippocratic Oath* was evidently that which prevailed under the early Empire. The initiate declared:

> I will reckon him who taught me this Art as dear to me as those who bore me. I will look upon his offspring as my own brethren and will teach them this Art, if they would learn it, without fee or stipulation. By precept, lecture, and every other mode of instruction, I will impart a knowledge of this art to my own sons, and to those of my teacher, and to disciples bound by a stipulation and an oath, according to the Law of Medicine, but to none other. (See pp. 32 f.)

This passage, though traditionally associated with the name of Hippocrates, is more probably of Pythagorean origin and certainly embodies Pythagorean teaching (pp. 22 f.). Moreover it gives some hint of the arrangements for medical instruction that gradually came into use in the main centres of population in the Roman Empire.

The first important medical teacher at Rome was the Greek ASCLEPIADES of Bithynia (died *c.* 40 B.C.), a contemporary of Lucretius and like him an Epicurean (p. 60). He influenced deeply later medical thought, ridiculed, and perhaps we should add misunderstood, the Hippocratic attitude of relying on the *vis medicatrix naturae*, 'the healing power of nature', which he regarded as a mere 'meditation on death', and urged that active measures were needed for the process of cure. He founded a regular school at Rome which continued after him.

At first the school was the mere personal following of the physician, who took his pupils and apprentices round with him on his visits. Later, such groups met to discuss questions of their art. Towards the end of the reign of Augustus (died A.D. 14) these societies constructed for themselves a meeting-place with a regular organization. Finally the emperors built colleges for the teaching of medicine. At first the professors received only the fees of pupils, but before the end of the first century they were given a salary at the public expense. The system was extended in the second and third centuries. Thus Rome became a centre of medical instruction. Moreover, subsidiary centres were established in other Italian towns. These provincial schools were largely training places for army surgeons.

A very weak point in the Roman medical curriculum was the absence of any practical study of anatomy. Considering the indifference to human life which the Romans exhibited, considering their brutality to slaves and the opportunities offered by gladiatorial combats, considering the value—obvious to us—of anatomical knowledge for surgical practice, and considering the organization of the military medical service of the Empire, it is highly significant that the knowledge of antiquity was thus allowed to lapse.

Had a great Roman military leader been questioned on this point he would probably have replied, 'Of course doctors want anatomy, but isn't Galen's anatomy good enough? Cannot they read that?'

But he would have been wrong. It is not by reading that science is sustained. It is by contact with the object—by systematic observation and experiment. From these the Roman army doctor was cut off, and we see the result of his deprivation in the poverty of Roman science.

As regards the literature of medicine, the earliest scientific work in Latin bears the name of CELSUS and was prepared in Rome about A.D. 30. It is in many ways the most readable and well arranged of all ancient medical works. The ethical tone is high and the general line of treatment sensible and humane. The most interesting section is perhaps that on surgery, which gives an excellent account of what might be thought to be the modern operation for removing the tonsils. The dental practice includes the wiring of loose teeth and the use of the dental mirror. In view of the attractive character of the work it is disappointing to find that it is but a compilation from Greek sources. This fact also is significant of the status of science in Rome.

The remaining Latin medical writings of Imperial times are not of high scientific value. In this connexion we must recall Pliny (p. 106). A large section of his *Natural History* is devoted to medical matters. Yet he scorned medical science and the Greeks who practised it.

> Medicine, in spite of its lucrativeness [he says] is the one Greek art that the serious Roman has so far refused to cultivate. Few of our fellow-citizens have been willing even to touch it, and if they do so they desert at once to the Greeks. . . . Unfortunately there is no law to punish ignorant physicians, nor is the capital punishment inflicted upon them. Yet they learn by our suffering, and experiment by putting us to death!

The collection of Pliny that was to displace the works of the despised Greeks is a vast series of remedies chosen on the supposedly firm ground of 'experience'. Their selection is based on no theory, supported by no doctrine, founded on no experiment. Yet this drug book is the prototype of the medical output of the next fifteen hundred years. The cry of Pliny for 'experience' as against 'theory' has been plaintively echoed by the 'practical' man down the ages. Yet there are subjects and there are conditions in which the man without a theory may be the most unpractical of all. Medicine is such a subject; disease is such a condition.

When 'experience' is invoked by Pliny and by later writers, especially of the Middle Ages, we must beware against confusing it with the 'experience' of science. In scientific matters the essence of experience is that it be under control. Such experience is normally capable of repetition at will, as a chemical reaction, for instance, may be repeated. All true scientific experience, in fact, approaches the character of 'experiment'. Scientific experience is thus the result of a series of *observations provoquées*.[1]

A single example from Pliny will suffice to illustrate this distinction. 'The herb dittany', he says, 'has power to extract arrows. This was *proved* [note the word; it really means *tested*] by stags who had been struck by these missiles, which were loosened when they fed on this plant.' Had Pliny exhibited any desire to verify such a statement? Could he have verified it even if he had desired? The answer is not difficult. He had, in fact, taken his 'experience' from an interpolated and spurious passage of a work by Theophrastus (pp. 56 f.) and he omits to mention his source! Prepossession with the idea of the value of such experience led Pliny and the ages which followed him —as it leads men to this day—into innumerable absurdities. 'General experience' whether first-hand or second-hand is no substitute for exact scientific knowledge.

If in medicine itself the Roman achieved but little, in organization of medical service, and especially in the department of public health, his position is far more honourable. Thus Vitruvius (fl. *c.* 10 B.C.), a writer on architecture, gives much attention to the orientation, position, and drainage of buildings. Sanitation and public health early drew the attention of Roman statesmen. Considering the dread of the neighbourhood of marshes on the part of these practical sanitarians and in view of modern knowledge of the mosquito-borne character of malaria, it is entertaining to find the use of the mosquito net (*conopeum*) ridiculed as effeminate by poets such as Horace and Juvenal.

Sanitation was a feature of Roman life. Rome was already provided

[1] There are scientific experiences in which the mind comes to rest with conviction, even when not repeated. Thus an astronomical prediction, involving exact and detailed calculation, if confirmed in an exact and detailed way, may carry conviction as to the soundness of its principle even though verified by but a single observation.

with *cloacae* or subterranean sewers in the age of the Tarquins (sixth century B.C.). The first construction of the *Cloaca maxima*, the main drain of Rome, parts of which are in use to this day, is referable to that period.

The growth of hygienic ideas is seen in an interdict of as early as 450 B.C. against burial in the city. There is in this edict no reference

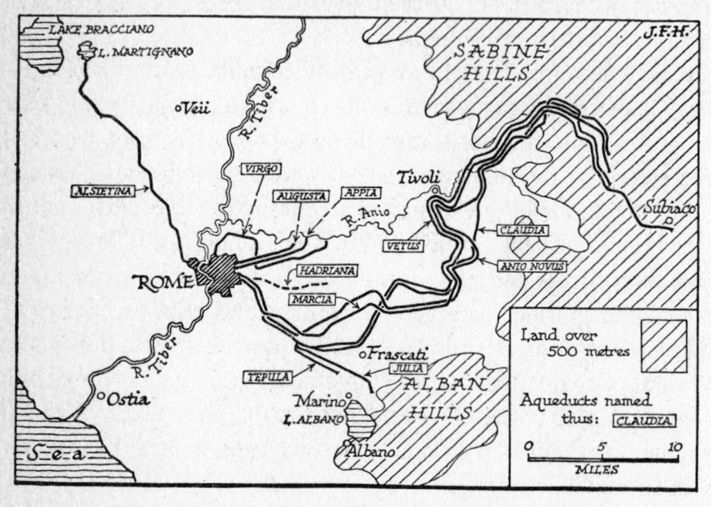

FIG. 53. Main aqueducts which supplied Ancient Rome.

to any physician. The same absence of professional intervention may be noted in the instructions issued to the city officers for cleansing the streets and for the distribution of water. Nor is any medical help or opinion invoked by the ancient law, attributed to Numa the first king of Rome, which directed the opening of the body of a woman who had died pregnant in the hope of extracting a live child. This is the so-called *Caesarean section* by which Caesar himself is said to have been brought into the world. The expression still has a surgical meaning.

The finest monument to the Roman care for the public health stands yet for all to see in the remains of the great system of aqueducts which supplied the city with 300,000,000 gallons of water daily (Fig. 53). Few modern cities are comparably equipped. The distribution of water to individual houses was also well organized, and excellent specimens of Roman plumbing have survived (Fig. 54).

Under the early Empire a definite public medical service was constituted. Public physicians were appointed to the various towns and institutions. A statute of the Emperor Antoninus of about the year A.D. 160 regulates the appointment of these physicians, whose main duty was to attend the needs of the poor. In the code of the great law-

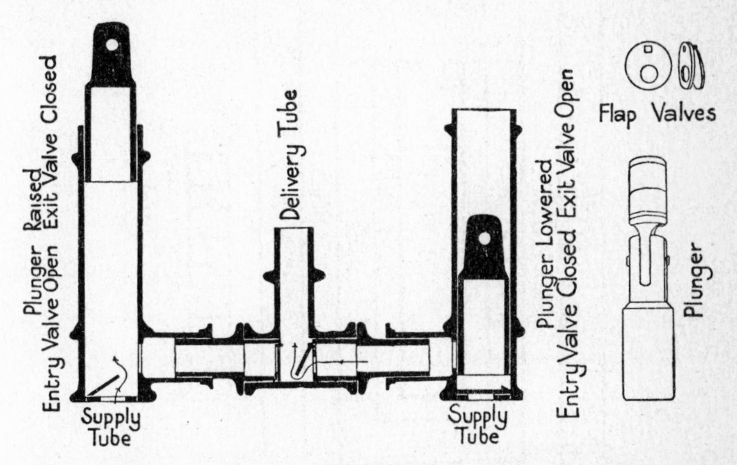

FIG. 54. Mechanism of Roman double action pump for domestic supply or for a fire-engine.

giving Emperor Justinian (A.D. 533) there is an article urging such men to give this service cheerfully and to prefer it to the more subservient attendance on the wealthy. Their salaries were fixed but they were encouraged to undertake the training of pupils.

Linked with the public medical service is the hospital system. It arose out of the Roman genius for organization and is connected with the Roman military system. Among the Greeks private surgeries were well known. Larger institutions were connected with the temples to Aesculapius, the god of healing, but there is no evidence of scientific medical treatment in these places. Such a temple had been established on an island in the Tiber in Republican times. On this island of Aesculapius writes the historian Suetonius (*c.* A.D. 120) 'certain men exposed their sick and worn-out slaves because of the trouble of treating them. The Emperor Claudius (41–54), however, decreed that such slaves were free, and that if they recovered they should not return to the control of their masters.' Thus the island became a place of refuge for the sick poor. It was an early form of public hos-

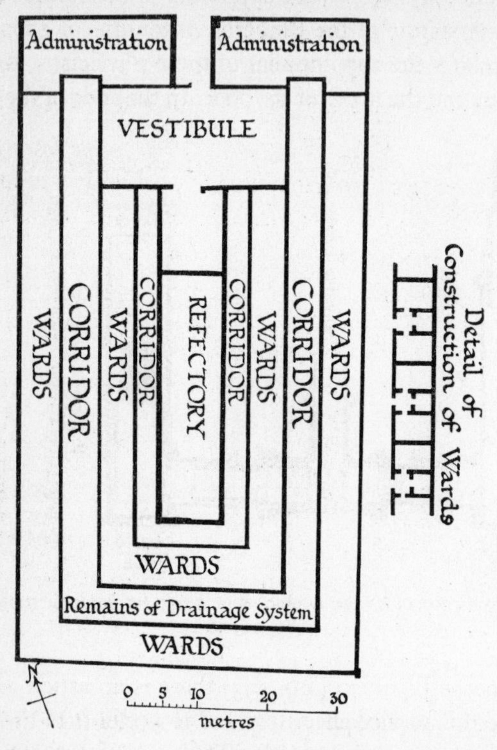

FIG. 55. Ground plan of Roman military hospital at Novaesium (Neuss) near Dusseldorf.

pital. The example was imitated, the facilities improved, and the service extended to free men.

The development of public hospitals naturally early affected military life. As the Roman frontiers spread ever wider, military hospitals were founded at important strategic points. Ground plans of several Roman military hospitals have been excavated, notably of one at Neuss (*Novaesium*) on the Rhine near Dusseldorf of the first century A.D. (Fig. 55). Later there were constructed similar institutions for the numerous imperial officials and their families in the provincial towns. Motives of benevolence, too, gradually acquired weight, and finally public hospitals were founded in many localities. The idea naturally passed on to Christian times, and the pious foundation of hospitals for the sick and outcast in the Middle Ages is to be traced back to these Roman institutions.

4. ROMAN MATHEMATICAL, MECHANICAL, AND CALENDARIAL SCIENCE

As with all peoples, the first system of numeration adopted by the Romans was finger counting. From it developed methods of mechani-

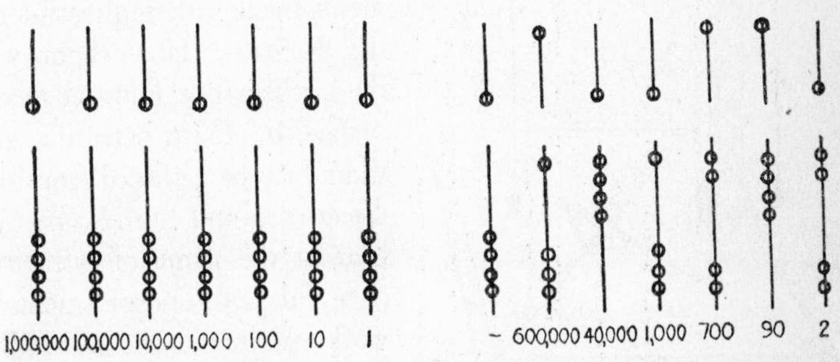

FIG. 56. Essentials of the Roman abacus, consisting of beads running on wires. On the left it is set for reckoning. On the right a sum of 641,792 is represented. Without an abacal representation and in Roman figures this would need twenty-one elements, namely CCCCCCXLIMVIICLXXXXII

cal reckoning. The simplest was a board covered with sand, divided into columns by the finger, counters being used in calculation. Such counters had graven upon them figures of the hand in various positions to represent different numbers. These symbols are identical with those which remained in vogue till late medieval times.

A more complicated apparatus was the true *abacus*. This began as a board with a series of grooves in which pebbles or *calculi* would be moved up and down, hence the verb *calculo* and the modern use of 'calculate'. In its more developed form the abacus consisted of an upper row of short rods and a longer row of long rods (Fig. 56). Each short rod had a single perforated bead running on it; each of the longer ones four such beads. The first rod on the right was marked for units, the next on its left for tens, and so on up to a million. The mode of application of the abacus was more complicated for multiplication and division than might be imagined.

The whole mathematical system of antiquity was handicapped by its inadequate notation. The scheme with which we are nowadays familiar, with nine separate integers and a zero, each of which has a local value, did not reach Europe until the Middle Ages. The Greeks used mostly geometrical methods where we should invoke the aid of algebra (pp. 26 f.), and their mathematical developments made little impression on the Romans. How slight was the mathematical knowledge absorbed by Latin scientific authors may be gathered from the *Geometrica* and the *Arithmetica* bearing the name of BOETHIUS (A.D. 480–524). Those elementary works ascribed to the 'last of the ancients' represent the mathematical legacy of antiquity to the earlier Middle Ages. It is interesting to note that Boethius divides mathematics into four sections, Arithmetic, Music, Geometry, and Astronomy, and that he is the first to describe these four disciplines as the *quadrivium* ('four pathways'). Even when Rome had world dominion, Cicero bemoaned that 'Greek mathematicians lead the field in pure geometry while we limit ourselves to reckoning and measuring.'

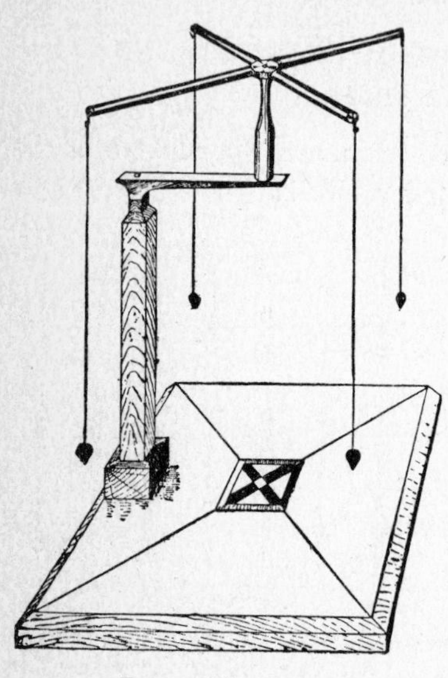

FIG. 57. The Groma.

The Romans held that the art of surveying was at least as old as their city, and had been practised from the first by the priests. In Imperial times a regular school for surveyors was established. The chief instrument in general use was known as the *groma* (Fig. 57). It consisted of two sets of plumb-lines fixed at right angles and arranged to turn about a vertical pivot. One set was used for sighting and the other to determine the direction at right angles to the first. As both agricultural and town-planning were mainly on rectangular lines this instrument was of wide application. A *dioptra* (p. 87) was in use and also a crude water-level. The latter was a mere open groove on a board.

Compasses and other instruments employed in mensuration recovered from Pompeii are well made, and the excellence of Roman

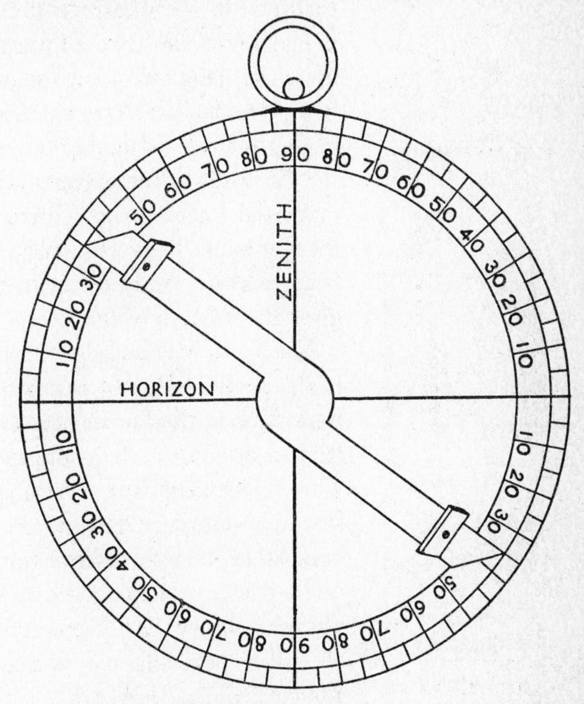

FIG. 58. A simple form of angular-distance measurer. It consists of a disk graduated in degrees around whose centre turns a limb, called later by Arabs and Latins *alidade* (Arabic = the humerus bone), provided with two sights. When the instrument is suspended the adjustment of the sights on a heavenly body gives its elevation; when it is laid horizontally it can give the angular-distance of two distant terrestrial objects.

masonry is a household word. Thus the inaccuracy of some Roman measurements is strange. For instance, $3\frac{1}{8}$ is given as the value of π by VITRUVIUS (*c.* A.D. 10), a competent architect who must often have had occasion to examine the drums of columns. A better result had already been suggested by Archimedes (p. 73).

Vitruvius gives a method of estimating the distance from an observer of an inaccessible point on the same level as himself, e.g. on the opposite bank of a river. A line is traced along the near bank, and is measured by rolling along it a *hodometer*, an instrument consisting of a wheel the length of the circumference of which is known and whose revolutions can be counted. This is in principle a 'taxicab'. From each end of the measured line a sight is taken by some simple form of angle measurer (Fig. 58). Angles and base being thus available a triangle congruent to that formed by joining the point on the far bank to the extremities of the measured line, can be constructed on the near bank. The vertical height of this triangle as measured by the hodometer gives the breadth of the river.

FIG. 59. Crane worked by treadmill and system of pulleys. Roman, *c.* A.D. 400.

Mechanical knowledge among the Romans always had a practical direction. Among the few devices of native Roman origin is perhaps the steelyard. This instrument is a device of considerable antiquity and may be traced back at least as far as the third century B.C. The principle of the pulley, too, was well known. An elaborate system of pulleys was adapted to cranes by Roman builders (Fig. 59).

The inadequate theoretical basis of the physical conceptions of Latin writers is shown in various directions. Thus Pliny recounts a fable of the Remora, a fish of the Mediterranean which has a sucker on its head. 'This tiny fish can restrain all the forces of ocean. Winds may rage and storms may roar, yet the fish withstands their might and holds ships still by simply adhering to them!' Three centuries before, Archimedes had said that with a place on which to stand (i.e. a fixed place), he could move the world (p. 72). The full understanding of the works of Archimedes failed for the next millennium and a half. Yet his simpler practical devices, such as the water-screw (Fig. 27), were familiar enough to the Romans.

Applied mathematics underwent some development in early Imperial times. JULIUS CAESAR (102–44) himself was an astronomical student and sought to improve the Roman calendar which had fallen into great confusion.

The early history of the Roman calendar is obscure. At an early date there emerged a lunar year of 355 days, which is almost exactly twelve lunations. Of this calendar Martius (the month of Mars) was the first month, Aprilis (probably for *aperilis* from *aperire*, 'to open'), Maius (perhaps related to *major*), and Junius (which may be related to *junior*) were named possibly in connexion with the opening, growth, and ripening of vegetation. The following six months, Quinctilis, Sextilis, September, October, November, and December were given merely the numerical names from fifth to tenth which the last four still bear. Januarius was named from the god *Janus*, and Februarius, the last month, was the season of ritual purification (*februare*, 'to purify' or 'expiate').

To obtain some relation of this lunar reckoning to the solar year a cycle of four years had been invented of which the first year contained 355 days, the second 377, the third 355, and the fourth 378. The cycle thus covered 1,465 days, and the average year was of $\frac{1465}{4} = 366\frac{1}{4}$ days. So variable a year had little value for agricultural purposes. The farmer had thus still to rely on the rising and setting of certain constellations for timing his operations. The year was variously modified at different periods, but until the reforms of Julius Caesar no adequate correspondence to solar events was attained.

In place of this system Julius Caesar, on the advice of the Alexandrian mathematician Sosigenes, substituted a solar year of $365\frac{1}{4}$ days and abandoned any attempt to adapt the years or months to the lengths of the lunations. In every fourth year one day was interpolated, thus introducing the system of leap years. This reform was probably a reproduction of an Alexandrian calendar enacted in 238 B.C. and had perhaps been designed at a yet earlier date by the Greek astronomer Eudoxus (p. 43). In 44 B.C., the second year of the

Julian Calendar, one of the months, *Quinctilis*, was named *Iulius*—our July—in honour of its founder. In 8 B.C. the month, *Sextilis*, was called *Augustus* after his successor. The Julian Calendar, the year of which began in the month of March, remained in general use until reformed by Pope Gregory XIII in 1582.

5. ROMAN ASTRONOMY AND ASTROLOGY

The Romans did not deal with astronomical matters until late, and then only for practical purposes such as the calendar, seamanship, or agriculture. Popular astronomy is represented in Latin by certain metrical writings bearing the name of AVIENUS (*c.* A.D. 380). These, which were popular in the early Middle Ages, are adapted from various Greek works. To one of the Greek sources of Avienus, namely ARATUS of Soli (271–213 B.C.), peculiar interest is attached. St. Jerome tells us that when, in Acts, St. Paul says 'In Him we live, and move, and have our being; as certain even of your own poets have said, *For we are also his offspring*' (Acts xvii. 28), he is quoting the *Phaenomena* of Aratus. The words 'for we are also his offspring' are in fact to be found in the opening invocation to Zeus in Aratus, and in an expanded form in Avienus. Aratus was a native of Cilicia, St. Paul's native province, and was one of the Stoics, who, with the Epicureans, were opposing the apostle at Athens (Acts xvii. 18).

Though backward in astronomy, the Romans had early developed a good knowledge of such elementary developments as the sundial, which was known to them in the third century B.C., and the results of which were early applied to calendarial reckoning (Fig. 60). Directions for the construction of sundials are given by the architect VITRUVIUS (*c.* A.D. 10, pp. 123 f.) who tells of a number of different forms in use in his time. Some of these, he says, were invented by various Greeks, of whom Aristarchus (p. 65) and Eudoxus (p. 43) are the best known. The construction of these various forms implies command of considerable mechanical skill and some efficiency in the making and recording of elementary astronomical observations. Sundials suitable for use by travellers were also not uncommon. Vitruvius describes also a water-clock of an extremely simple and effective type.

The difference in the length of day in different latitudes was well known to the Romans. From the fact that the longest day in Alexandria was 14 hours, in Italy 15, and in Britain 17, Pliny deduces that lands close to the Pole must have a 24-hours' day in the summer and a 24-hours' night in winter.

Many passages in Pliny reflect a contest concerning the form of the earth, reminding us of earlier disputes of the same order (pp. 17,

FIG. 60. Sundial from Pompeii, *c.* A.D. 50. On its base is a donatory inscription in Oscan, an Italic language allied to Latin but written from right to left.

27). He opens his work with a description of the general structure of the universe and discusses the spherical form of the earth:

> Science and the opinion of the mob [says Pliny] are in direct opposition. According to the former the whole sphere of the Earth is inhabited by men whose feet point towards each other while all have the heavens above their heads. But the mob ask how men on the antipodes do not fall off; as though that did not present the opposite query why they should not wonder at *our* not falling off. Usually, however, the crowd objects if one urges that water also tends to be spherical. *Yet nothing is more obvious, since hanging drops always form little spheres.*

To the moon and fixed stars the Romans had already, in Pliny's time, begun to attribute an influence on human affairs. 'Who does not know', he asks, 'that when the Dog Star rises it exercises influence on the widest stretch of Earth?' The influence of the Dog

Star is an idea that may be traced back in Greek literature at least as far as Hesiod (eighth century B.C.) and has given us our modern superstition of the 'dog days'. The moon's influence on tides was recognized, and it was thought that besides influencing the outer world, the *macrocosm*, the moon had influence also on the body of man, the *microcosm* (Fig. 61). With the waxing of the moon it was believed that the muscles became bigger and blood increased. This theory gave rise to the practice of periodical blood-letting which took so prominent a place in early monastic life.

The supposed influence of the heavenly bodies on the earth and on the life of man is a topic that leads directly to judicial astrology. A knowledge of that subject became under the Empire a professional possession, illegal and prohibited, but often tolerated and invoked even by emperors. Astrology was beginning to spread in Rome in the first century of the Christian era.

> There are those [Pliny tells us] who assign [all human events] to the influence of the stars, and to the laws of their nativity. They suppose that God, once for all, issues his decrees and never after intervenes. This opinion begins to gain ground, and both the learned and the vulgar are accepting it.

The art was of foreign origin. The credit of its invention is always ascribed to 'Chaldeans', but the main channel of transmission was Greek.

> As for the branch of astronomy which concerns the influences of the twelve signs of the zodiac, the five Planets, and the Sun and Moon on man's life [says Vitruvius] we must leave it to the calculations of the Chaldeans to whom belongs the art of casting nativities, which enables them to declare the past and future.

The original meaning of the zodiacal figures is disputed, but they were certainly in very ancient use in Mesopotamia (Fig. 2) whence came the methods of dividing time and the divisions of the heavenly sphere based on them. Against these Chaldeans Cicero directed his dialogue *On Divination*. He misunderstood the basis of astrology and marshalled ancient and fallacious arguments against it. Yet even Cicero accepted some astrological doctrine, and in his *Dream of Scipio* he spoke of the planet Jupiter as helpful and Mars as harmful.

To the early Christian writers astrology was completely abhorrent, for it seemed to them to be the negation of that doctrine of free will that was so dear to them. The fathers, Tertullian (*c.* 155–*c.* 222), Lactantius (*c.* 260–*c.* 340), and Augustine (354–430), all inveigh against it. With the spread of Christianity and disappearance of the Stoic philosophy, astrology passed into the background, to return with the Arabian revival and the rise of the universities in the thirteenth century.

At an early date there arose a large literature on the subject. Nevertheless astrology seems on the whole to have been rather less cultivated in Rome itself than the general state of society and the wide spread of the Stoic philosophy might perhaps suggest. Lovers sought to learn of astrologers a lucky day for a wedding, travellers inquired what was the best day for starting on a journey, and builders asked the correct date for laying a foundation stone. All these may easily be paralleled by instances among the empty-headed in our own time. But Galen (130–200), who practised medicine among the well-to-do and educated, assures us that they only bothered about astrology for forecasting legacies—and again a parallel might be drawn.

But astrology must not be considered only as a superstition and an occupation for empty heads and idle hands. The astrological system of antiquity was, in essence, a formal presentation of those beliefs concerning the nature and working of our mundane sphere which had been fostered by a scientific astronomy and cosmology. Faith in it was part of the Stoic creed. In the mechanism of the world there was no room for those anthropomorphic gods, the belief in whom was still encouraged by the priests and held by the multitude. The spread of belief in that mechanism had led at last to a complete breach between the official faith and the opinions of the educated classes. The idea of the interdependence of all parts of the universe produced in time a new form of religion. The world itself must be divine. 'Deity', says Pliny, 'only means nature.' From such a view to the monotheism of Virgil, in which the world as a whole is regarded as the artistic product of an external god, is perhaps no great step. Roman Stoicism, however, failed to take that step, and assumed among later Latin writers a fatalistic and pessimistic mood. 'God, if God there be, is outside the world and could not be expected to care for it', says Pliny. The idea of immortality seems to him but the 'childish babble' of those who are possessed by the fear of death, as Lucretius had once maintained. After death, so Pliny would have us believe, man is as he was before he was born—and this he tells us as he plunges into his magic-ridden pages!

Once and once only in these Latin scientific writings have we a clear note of real hope. It is significant that that note is sounded in connexion with a statement of a belief in the *progress* of knowledge, an echo of the Greek thought of the fifth and fourth centuries B.C. It is significant, too, that the note is sounded by one who approached, nearer perhaps than any other pagan Latin philosopher, to the idea of the divine immanence. In his *Natural Questions* Seneca wrote:

> How many heavenly bodies revolve unseen by human eye! . . . How many discoveries are reserved for the ages to come when our memory shall be no more, for this world of ours contains matter for investigation for all generations. . . . God hath not revealed all things to man and hath entrusted us with but a fragment of His mighty work. But He who directeth all things, who hath established the foundation of the world, and clothed Himself with Creation, is greater and better than that which He hath wrought. Hidden from our eyes, He can be reached only by the spirit. . . On entering a temple we assume all signs of reverence. How much more reverent then should we be before the heavenly bodies, the stars, the very nature of God.

But the science of antiquity as exhibited elsewhere in Latin writings contains very little of this belief in man's destiny, this hope for human knowledge. The world in which the Imperial Roman lived was a finite world bounded by the firmament and limited by a flaming rampart. His fathers had thought that great space peopled by *numina*, 'divinities', that needed to be propitiated. The new scientific dispensation—the *lex naturae* of the world that had so many parallels with the *jus gentium* of the Empire—had now taken the place of those awesome beings.

In the inevitableness of the action of that law Lucretius the Epicurean might find comfort from the unknown terror. Yet for the Stoic it must have remained a limited, fixed, rigid, and cruel law. His vision, we must remember, was very different from that given by the spacious claim of modern science which explores into ever wider and wider regions of space and time and thought. It was an iron, nerveless, tyrannical universe which science had raised and in which the Roman thinker must have felt himself fettered, imprisoned, crushed. The Roman had forsaken his early gods, that crowd of strangely vague yet personal beings whose ceremonial propitiation in every event and circumstance had filled his fathers' lives. He had had before him an alternative of the oriental cults whose gods were but mad magicians—a religion unworthy of a philosopher—and the new religion of science whose god, he now saw, worked by a mechanical rule. He had abandoned the faith of his fathers, had flung himself into the arms of what he believed to be a lovelier goddess, and lo! he found himself embracing a machine. His soul recoiled and he fled into Christianity. A determinate view of the world induced that essential pessimism which clouds much of the thought of later antiquity. It was reaction against this pessimism which led to the great spiritual changes in the midst of which antiquity went up in flames and smoke.

6. THE PASSAGE FROM PAGAN TO CHRISTIAN THOUGHT

We have gained a general view of the course of ancient thought in relation to science. Four stages may be distinguished:

(*a*) During the rise of Greek thought, philosophy is based on natural science. It neglects ethics and ignores popular religion (Ch. I). Here was the emergence of Mental Coherence.

(*b*) Plato and Aristotle seek to adjust the rival claims of ethics and science, while giving preference to the former. Popular religion is repudiated (Ch. II). This is the Great Adventure.

(*c*) Alexandrian thought develops separate departments for science, ethics, and religion. The age of the universal philosopher is over. The age of the specialist has begun. The Alexandrian period terminates with scientific deterioration (Ch. III) and the Second Adventure is over.

(*d*) Under the Empire the prevalent schools of thought, Stoicism and Epicureanism, are indifferent to science, which deteriorates further (Ch. IV). Great emphasis is laid on Ethics. Scientific inspiration at last wanes to nothing.

We must now consider certain aspects in this final stage of ancient thought. Stoicism, in the first two Christian centuries, divided the thinking world with Epicureanism and certain less important philosophical sects. The Stoic philosophy assumed that man's life in all its details is controlled by an interplay of forces. Both the nature and the behaviour of these were, in theory, completely knowable. The same assumptions were made by Epicureanism, save that different forces were held to control man's fate. The Stoic invoked the action of the spheres and astrology. The Epicurean invoked the play of atoms. Both schemes were *determinate*. In this they differed from the new and rising school of Neoplatonism, the *indeterminacy* of which fitted better the doctrine of free will on which Christianity came to insist. Atomism being opposed by the authority of both Plato and Aristotle and by Stoicism and Neoplatonism alike, Epicureanism fell into the background. All philosophical sects became ultimately absorbed into Neoplatonism, the history of which it is necessary to trace.

Alexandria of the third century of the Christian era presented an extraordinary mixture of religions, philosophies, and sects. The old scientific school was in decay. Christian, Jewish, and pagan elements jostled each other. The cults of ancient Egypt, of Greece, of Rome, and of the Orient appealed to the devout and to the superstitious. The decayed schools of Plato and Aristotle had still conservative followers. There were also those who called themselves Stoics and Epicureans. A common factor among these various elements was contempt for science.

It must be remembered that the science of those days differed from that of ours in that it had introduced no obvious and extensive amelioration of man's earthly lot. Nature had not been harnessed as we have harnessed her. Science was a way of looking at the world rather than a way of dealing with the world. And as a way of looking at the world—a way of life—positive knowledge, that is science, was a failure. The world was a thing that men could neither enjoy nor master nor study. A new light was sought and found. In its glare the old wisdom became foolishness and the old foolishness wisdom. Weary of questioning, men embraced at last and gladly the promises of faith. The faith that was immediately most successful was that which included within itself the experiences of the largest number of educated men. This was the syncretic system known as *Neoplatonism*.

The syncretic tendency exhibited itself very early in Alexandria. Philo, who was about twenty years older than Christ, developed a system that used the Jewish scriptures in the light provided by Plato and Aristotle, and with some admixture of mysticism. He introduced the doctrine of the *logos*, and his tendency is away from observational science. Following Philo in the first, second, and third centuries were writers of 'Neopythagorean' and 'Hermetic' leanings whose views and tenets were as syncretic as Philo's. They need not delay us. It would be possible to consider the earliest Christian writers as members of this syncretic group.

Early in the third century there arose in Alexandria a school of philosophy which became known as the *Neoplatonic*. The greatest of its exponents was PLOTINUS (204–70), himself a Roman. He carried Neoplatonism to Rome and thence to the world at large.

We are not here concerned with any general consideration of Neoplatonism and but little with a further discussion of its numerous sources. These included Plato and Aristotle and their successors and various religious cults, together with the philosophical sects such as Stoicism. There is, however, a certain doctrine of great historic importance which demands some notice here. It is a doctrine shared by Neoplatonism and Stoicism. Both philosophies saw analogies in the universe, the *macrocosm*, to Man, the *microcosm* (p. 139). The one was a reflection of the other. Broadly speaking, the Neoplatonist would have said that the universe had been made for Man who is the essential reality; the Stoic that Man has been made for the universe. The Neoplatonic view was victorious. The view of the macrocosm and microcosm as elaborated by Neoplatonism was not unacceptable to Christianity (Fig. 61).

Neoplatonism developed a characteristic metaphysic derived mainly from Plato but also from Stoicism whence it drew some of its ethics. The Platonic 'Idea' was greatly emphasized and almost personified. The Idea, as expressed by form, governs matter just as the soul governs the body. But matter may at times break away from the Idea and then the world of matter becomes one of strife, discord, even chaos. Idea is in the end identifiable with form. Matter, destitute of form or idea, is evil; with form it is at best neutral. It must be the soul's aspiration to free itself from such dangers. Then and then only can it hope for ecstatic union with the Divine.

During the fourth century Neoplatonism flourished. Associating itself with the theologies of various sects, it was a serious rival to Christianity. Its hopes rose high when Julian the Apostate became Emperor (361–3), but they fell again even before the end of his short reign to sink still lower with the victory of Christianity in the age of Valentinian (364–75) and Theodosius (379–95). Christianity in its spread absorbed, with the masses, some of their superstitions, their magic, their theurgy. Neoplatonism, on the other hand, at first saturated with these elements, became at last purged of them, though passing thereby out of touch with the spirit of the age. Towards the end of the fourth century the head of the Neoplatonist school at Alexandria was Hypatia (379–415). Her murder ended the effectiveness of the Neoplatonic school as such. She influenced Christian thought directly through her pupils, the most famous of whom, Synesius of Cyrene (373–414), became a very free-thinking bishop.

The passage of Neoplatonic doctrine into Christianity was in the main the work of ST. AUGUSTINE (354–430). After a youth and young manhood spent in devotion to Manichean studies he turned, at last, to study the exact sciences. In 383 he came to Rome whence he moved in 384 to Milan. There he became acquainted with Neoplatonic teachers. In 386 he became converted to Christianity. His great literary activity, begun in 393, ended only with his life.

We have it from Augustine himself that his debt to Neoplatonism was very great. In all his cardinal doctrines—God, matter, the relation of God to the world, freedom, and evil—Augustine borrowed freely from Neoplatonism. Through him we may regard Neoplatonism, itself the final stage of Greek thought, as passing in its final stage into Christianity. Through St. Augustine, above all men, early Christianity acquired its distaste for a consideration of phenomena. 'Go not out of doors,' said the great Father of the Church. 'Return into thyself. In the inner man dwells truth.' For a thousand years men responsible for the thought of the Western world did not go out of doors.

It was through St. Augustine that certain Neoplatonic doctrines, notably that of the macrocosm and microcosm, passed to the Latin West, where they awaited the Arabist revival (p. 164) for their fuller development. In a somewhat similar way such traditions lingered for centuries in the Byzantine East until, with the great outburst of

FIG. 61. Epitome of the doctrine of macrocosm and microcosm. A series of concentric circles, recalling the concentric spheres of the older astronomy with World, Year, and Man as centre. An outer circle bears the names of the four elements, within them are the four seasons, and within them again the four humours. On either side of each element is the name of one of the 'qualities', caught in the strand that weaves together all the items of the theory. The outermost circle that is not caught in the web is the rampart between the world that we know and the heavenly world of which we can have no direct knowledge. The diagram is from a work of Saint Isidore of Seville (p. 139), *De responsione mundi et astrorum ordinatione* printed at Augsburg in 1472. It would have been acceptable and intelligible from somewhat before the Christian era to the seventeenth century.

Islam, they were caught up and elaborated by the Arabic culture (pp. 140f.). Stamped with specific Islamic characters the same doctrines were sent forth a second time to Christian Europe in the process of translation from the Arabic (p. 158).[1]

[1] The student will need to distinguish the *Arabic* culture, which is mainly expressed in that language, from the *Arabist* culture, which is mainly expressed in Latin.

Through Neoplatonism there came to medieval Europe a way of thought destructive of science, viz. *reasoning by analogy*. Scientists often find analogy suggestive and often they seek in it hints for research, but they do not regard it as a way of reasoning and still less of proof. It was far other with the Neoplatonists and their Christian successors. Thus the four supposed ages of man and his four supposed elements, humours, and organs were constantly invoked as analogies to the world with its four arbitrary seasons, its four cardinal astronomical events, four compass-points, and so on. Endless complexity, based upon such unreal entities, were introduced as extensions of the view of microcosm and macrocosm and yielded a very frenzy of analogies (Fig. 61).

Before leaving the civilization of ancient Rome it must be said that, though its science was lost, its technology survived. Thus, for example, the elements of the mechanical repertory of Hero (Fig. 38) were not forgotten when the Western Empire fell, for its devices had already passed to barbarian Europe.

V

THE FAILURE OF KNOWLEDGE

The Middle Ages (about A.D. 400–1500): Theology, Queen of the Sciences

I. THE DARK AGE IN THE WEST (400–1000)

WE now enter the last and longest phase of the Great Failure. With the decline and fall of the Empire the eclipse of ancient philosophy was as pronounced as that of science. Neoplatonism gives place to the great philosophical and religious movement, Christianity. The standpoint of the early Christian champions, the Church Fathers, Tertullian (155–222), Lactantius (260–340), and, above all, St. Jerome (340–420) and St. Augustine (354–430), is outside the department with which we deal, but it was assuredly not conducive to the exact study and record of phenomena. Nevertheless, the Middle Ages, under the influence of the Church, developed a characteristic attitude towards nature.

For our purposes we may place the limits of the medieval period between about the years 400 and 1500. This millennium is divided unequally by an event of the highest importance for the history of the human intellect. From about 900 to 1200 there was a remarkable development of intellectual activity in Islam. The movement reacted with great effect on Latin Europe through works which reached it, chiefly between about 1200 and about 1300, in Latin translations from the Arabic. This intellectual event divides the medieval period in the Latin West into two parts, an earlier *Dark Age* which terminates in the twelfth century, and a later *Age of Arabian Influence* which expressed itself characteristically in Scholasticism. As we pass from one period to the other, the general outline of beliefs as to the nature of the external world changes relatively little, but their presentation is vastly altered and the whole doctrinal scheme of the material world assumes a formal rationality.

During the closing centuries of the classical decline, the body of literature destined to pass down to subsequent ages had been delimited and translated into Latin, the only language common to the learned West. We must briefly discuss the legacy that antiquity passed on to this Dark Age.

Of Plato's works, the *Timaeus* fitted well the views of Neoplatonic thinkers of the late Empire and fitted not ill to Christian belief. Part of a Latin commentary on the *Timaeus*, prepared in the third century, presents a doctrine held throughout the entire Middle Ages as to the nature of the universe and of man. This part became one of the most influential of all the works of antiquity, and especially it conveyed the central dogma of medieval science, the doctrine of macrocosm and microcosm (Fig. 61). The conception, that the nature and structure of the universe foreshadows the nature and structure of man, is basic for the understanding of medieval science.

Of the writings of Aristotle there survived only the logical works translated in the sixth century by BOETHIUS (480–524). These determined the main extra-theological interest for many centuries. Boethius had purposed to translate all of Aristotle, and it is a world-misfortune that he did not live to prepare versions of those works that display Aristotle's powers of observation. Had a translation of his biological treatises reached the earlier Middle Ages, the whole history of thought might have been different. Boethius repaired the omission, to some small extent, by compiling elementary mathematical treatises based on Greek sources. Thanks to them we can say that during the long degradation of the human intellect, mathematics, the science last to sink with the fall of Greek thought, did at least not come quite so low as the other departments of knowledge.

A somewhat similar service to that of Boethius was rendered by MACROBIUS (395–423) and by MARTIANUS CAPELLA (*c.* 500). The latter, especially, provided the Dark Age with a very elementary encyclopaedia of the seven 'liberal arts', namely the 'trivium', grammar, dialectic, rhetoric, and the 'quadrivium' (p. 122), geometry, arithmetic, astronomy, and music. This classification of studies was retained throughout the Middle Ages.

In addition to the little cosmography, mathematics, and astronomy that could be gleaned from such writings as these, the Dark Age inherited a group of scientific and medical works from the period of classical decline. By far the most widely read was the *Natural History* of Pliny (pp. 106 f.). Very curious and characteristic is a group of medical pseudepigrapha bearing such names as Dioscorides, Hippocrates, Apuleius, and others. These extremely popular works were translated into Latin between the fourth and sixth centuries. They provided much of the medical equipment of the Dark Age.

Such material, then, was the basis of the medieval scientific heritage. Traces of it are encountered in works of CASSIODORUS (490–585), perhaps the earliest general writer who bears the authentic medieval stamp. The scientific heritage is, however, much more fully displayed by Bishop Isidore of Seville (560–636) who produced a cyclopaedia of all the sciences in the form of an '*Etymology*' or explanation of the terms proper to each. For many centuries this was very widely read. The works of the series of writers, the Englishmen BEDE (673–735) and ALCUIN (735–804), and the German RABANUS MAURUS (776–856), who borrow each from his predecessor and all from Pliny, contain, with those of Isidore of Seville, almost the entire natural knowledge of the Age.

It must be remembered that the Dark Age presented no coherent philosophical system, and men were capable of holding beliefs inconsistent with each other. The world was but God's footstool, and all its phenomena were far less worthy of study than were the things of religion. In the view of many patristic writers, the study of the stars was likely to lead to indifference to Him that sitteth above the heavens. This is the general attitude of the fourth and fifth centuries, set forth for instance by St. Augustine, who speaks of 'those imposters the mathematicians (i.e. astrologers) . . . who use no sacrifice, nor pray to any spirit for their divinations, which arts Christian and true piety consistently rejects and condemns'.

By the sixth and seventh centuries the Church had come to some sort of terms with astrology. Thus St. Isidore regards astrology as, in part at least, a legitimate science. He distinguishes, however, between *natural* and *superstitious* astrology. The latter is 'the science practised by the *mathematici* who read prophecies in the heavens, and place the twelve constellations (of the Zodiac) as rulers over the members of man's body and soul, and predict the nativities and dispositions of men by the courses of the stars.' Nevertheless, St. Isidore accepts many of the conclusions of astrology. He advises physicians to study it, and he ascribes to the moon an influence over plant and animal life and control over the humours of man, while he accepts without question the influence of the Dog Star and of the comets. Other Dark Age writers on natural knowledge, accept successively more and more astrological doctrine.

A certain 'revival of learning' under Charlemagne, centred round about the year 800, is very important for its literary activity and certainly did much to preserve such few scientific texts as were available. This movement is greatly emphasized by general historians, but it cannot be considered in the light of a scientific awakening. Perhaps only one figure in the Dark Age is worth our attention here. It is that of GERBERT who became Pope as Sylvester II (died 1003). One of his merits is to have re-introduced the abacus (p. 121) which seems to have disappeared with Roman decline in the West. Its use lingered among the Byzantines whence it reached Arabic-speaking Spain in Gerbert's time. He had perhaps studied it there. He also visited the court of Otto I (913–73) in Southern Italy (970). From wheresoever he derived his abacus there can be no doubt that the details of his arithmetic, like the forms of digits that he used, were from the works of Boethius (p. 138). The immediate future of learning lay not in the West but in the East.

2. SCIENCE IN THE ORIENT (750–1200)

(i) *Byzantines and Moslems*

During the Dark Age the intellectual level of the Greek world stood higher than that of the Latin. Science, it is true, was as dead in the one as in the other, but in the Byzantine Empire there was more activity in the preservation and multiplication of copies of the works of antiquity. The classical dialect was not wholly unknown to the educated class. Despite intense theological preoccupation, classical learning was still occasionally cultivated. A few scholars still glossed the works of Plato and Aristotle.

The Byzantine Empire included many Syriac-speaking subjects. The Syriac language had, from the third century, replaced Greek in

Western Asia. There in the fifth century the heretical Nestorian Church had been established. The Nestorians, bitterly persecuted by the Byzantines, emigrated to Mesopotamia. Yet later, they moved to south-west Persia where, from the sixth century onward, they long exhibited great activity especially at their capital Gondisapur. Literature in Syriac became very extensive. It included translations of the works of Plato, Aristotle, Euclid, Archimedes, Hero, Ptolemy, Galen, and Hippocrates.

It was in the seventh century that the Arabs first entered into the heritage of the ancient civilizations of Byzantium and Persia (Fig. 62). From their desert home they brought no contributions save their religion, their music, and their language. Moreover, in the Byzantine and Persian Empires, Greek science was at a low ebb save among the Syriac-speaking Nestorians. Thus the Nestorian metropolis, Gondisapur, became the scientific centre of the new Islamic Empire. From Gondisapur during the Umayyad period (661–749) learned men and especially physicians came to Damascus, the capital. They were mostly Nestorian Christians, or Jews bearing Arabic names.

The rise of the Abbasid Caliphs (from 750) inaugurated the epoch of greatest power, splendour, and prosperity of Islamic rule, but Islamic thought was still in the absorptive period. The most important agents in the transmission of Greek learning through Syriac into Arabic were members of the great family of Nestorian scholars that bore the name of BUKHT-YISHU ('Jesus hath delivered'). This family produced no less than seven generations of distinguished scholars, the last of whom lived into the second half of the eleventh century. It was the skill of the physicians of this family that instigated the Caliphs to propagate Greek medical knowledge in their realm.

During the century 750–850 the old Syriac versions were revised and others added. The translators, mostly Nestorians of the Bukht-Yishu family or their pupils, had a command of the Greek, Syriac, and Arabic languages and often also of Persian. Most of them wrote first in Syriac. The venerable Yuhanna Ibn Masawiah, the JOHN MESUE of the Latins (d. 857), one of the Bukht-Yishu, and medical adviser to Harun al-Rashid (i.e. Aaron the Upright), the fifth Abbasid Caliph, produced many works in Arabic. As time went

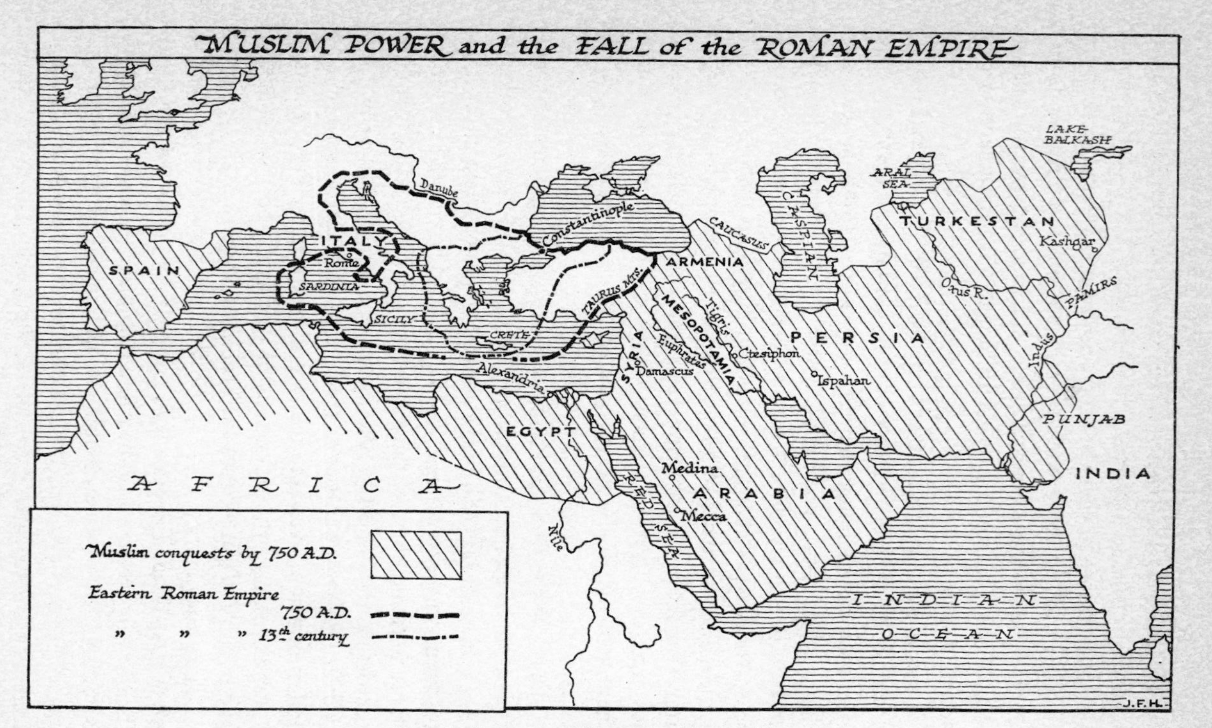

FIG. 62. Recession of Byzantium before Islam.

on Arabic began to replace Syriac for scientific and medical works. Just as 750 to 850 was the century of translation into Syriac, so 850 to 950 was the century of translation into Arabic.

The seventh Abbasid Caliph, Al-Mamun (813–33), son of Harun and a rationalist, created in Bagdad a regular school for translation. It had a library. HONAIN IBN ISHAQ (809–77), a particularly gifted philosophical and erudite Nestorian, was the dominating figure of this school. He passed his life in Bagdad, serving nine caliphs and exhibiting phenomenal intellectual activity. He translated into Arabic almost the whole immense corpus of Galenic writings. His predilection for the scholastic turn in Galen's theories contributed much to give Galen his supreme position in the Middle Ages in the Orient, and indirectly also in the Occident. He began the translation of Ptolemy's *Almagest* and of works of Aristotle. Honain and his pupils rendered also a number of astronomical and mathematical works into Arabic, as well as the Hippocratic writings. Many of these translations passed ultimately into medieval Latin.

Bagdad now rapidly replaced Gondisapur as the centre of learning. The Caliphs and their grandees furnished the necessary means to allow the Christian scholars to travel in search of Greek manuscripts and to bring them to Bagdad for translation. It was at Bagdad that most of the Aristotelian writings were first made accessible in Arabic, together with works on botany, mineralogy, and mechanics, as well as many Greek alchemical works. There was also an active intake of ideas and of texts from Indian and Persian sources. It seems likely that many alchemical methods were of Persian and some perhaps of Chinese origin, while there was a strong mathematical influence, expressed especially in the system of numeration, exercised by Indian civilization.

The general course of thought may be considered separately for Eastern and Western Islam. Of these the East is more important for the positive sciences, and we consider these under the headings (ii) *Alchemy* (p. 144), (iii) *Medicine* (p. 147), (iv) *Mathematics and Astronomy* (p. 148), and (v) *Physics* (p. 151).[1] Western Islamic science is considered under a separate heading (p. 154).

[1] There was also considerable geographical activity. As it contributed little to Western science, we omit it.

(ii) *Alchemy in Eastern Islam*

In discussing alchemy we must rid ourselves of the conception of it as a bundle of fantastic superstitions. The word 'alchemy' is usually said to be derived from the Egyptian *kem-it*, 'the black', or from the Greek *chyma* (molten metal), but in any event it comes to us through Arabic. The fundamental premises of alchemy are probably of basic Alexandrian origin (p. 101) and may be set forth thus:

(*a*) All matter consists of the same ingredients, the four elements, in various mixtures.
(*b*) Gold is the 'noblest' and 'purest' of all metals, silver next to it.
(*c*) Transmutation of one metal into another is possible, by an alteration in the admixture of the elements.
(*d*) Transmutation of 'base' into 'noble' metal can be achieved by means of a certain precious substance often called the fifth element (*quintessence*) or *elixir*.[1] (The earliest alchemical documents call the process 'tincturing' the base metal, and in fact describe an alloy.)

These conceptions, absurd though they seem to us, are no more so than those of many eminent chemists of as late as the eighteenth century. In fact they had the great merit of provoking experiment. It is a misfortune that at Alexandria, where alchemy specially flourished, mystical tendencies, largely of Neoplatonic origin, overlaid the experimental factor and thus tended to superstitious practice, passing into fraud.

JABIR (*c*. 760–*c*. 815) is the earliest alchemical writer in Arabic of whom we hear. He was known to the Latins as Geber though many Latin and Arabic works bearing his name cannot be by the Arab. Jabir came from South Arabia, belonged to the mystical brotherhood of the Sufis whose doctrines influenced the famous sect of the Assassins, became a friend of the Abbasid Caliph Harun al-Rashid (786–809) of *The Thousand and One Nights*, and associated with the Barmecides, his officers of state. Romance has thus cast her shimmering veil around him. Recent scholarship has begun to distinguish a

[1] *Elixir*. The original Greek word is *xerion* = dry thing, Byzantine parlance for a styptic drug. Some *xerion* acquired fabulous properties and the term passed into Arabic as *al-iksir* 'the sovereign remedy', as it were, for the baseness of metals. Thus it came into Latin and thence into Chaucer's English.

very few works of Jabir from many picturesque fables and a huge mass of occult works bearing his name.

Jabir had probably some knowledge of Greek and was well placed for obtaining alchemical information from the shrinking Byzantine Empire. He had a Pythagorean belief in numbers as real things and attached significance to the well-known magic square of the nine digits, which has been a source of wonder for centuries.

4	9	2
3	5	7
8	1	6

FIG. 63. Magic square of the digital numerals. Any column adds to 15.

He believed that numbers correspond to qualities or things and have specific relation to letters, to substances, and to the powers that change them all. We need not discuss this chaos of fruitless ideas nor must the history of science become a history of error, though errors often aid in tracing scientific and technical influences and stimulate further inquiry. By-passing such occult matters it remains that there are a few works, reasonably ascribed to Jabir, that contain genuine chemical knowledge. Thus in his *Book of Properties* he writes:

Take a pound of litharge, powder it well and heat it gently with four pounds of wine vinegar until the latter is reduced to half its original volume. Then take a pound of soda and heat it with four pounds of fresh water until the volume of the latter is halved. Filter the two solutions until they are quite clear and then gradually add the solution of soda to that of the litharge. A white substance is formed which settles to the bottom. Pour off the supernatant water and leave the residue to dry. It will become a salt as white as snow.

This is lead hydroxycarbonate (white lead), used for pottery glazes and as a paint.

Jabir classified minerals into: (*a*) *Spirits* which volatilize with heat.

These include sulphur, arsenic compounds, mercury, camphor, and sal ammoniac (ammonium chloride). The last was used industrially as a flux and was imported as a mineral product of burning coal. He knew how to make it from organic matter and gives the first Arabic reference to its preparation. (*b*) *Metals* of which he knew seven; gold, silver, lead, tin, copper, iron, and something that he calls 'Chinese iron' (? zinc), afterwards thought to correspond to the seven planets. (*c*) *Pulverizable substances* which are not malleable. There is a great variety of these and in another work of Jabir they are classed into intelligible groups.

Lastly, it is recorded that Jabir had a laboratory at Kufa on the Tigris. It was rediscovered during the demolition of houses there some two centuries after his death.[1]

RHAZES (865–925) was the greatest of the Arabic-writing alchemists. His name means 'man of Ray', a town in Persia. He gave the earlier part of his life to alchemy and produced works which relied on experience and experiment. Certain of them can justly be termed chemical and in one of them he makes the earliest suggestions for furnishing a laboratory, enumerating hearth, bellows, crucibles, refractory stills, ladles, tongs, shears, iron pans, scales and weights, flasks, phials, cauldrons, sand-baths, water-baths, ovens, haircloth and linen filters, stoves, kilns, funnels, dishes, &c. A working place thus equipped was no mere witch's kitchen.

The writings and views of Rhazes strongly influenced the Latin West. Despite his practical turn, being a man of his day he held to much alchemical theory including the transmutation of metals. He was, however, more interested in down-to-earth manipulation of substances. This emerges specially in his *Secret of Secrets* which has been said to foreshadow a laboratory manual. Of chemicals he had a great variety including probably nitric and possibly sulphuric acid. Many operations of a high-grade technology were open to him. The familiar division of substances into animal, vegetable, and mineral seems to have been first suggested by him. He also throws out the hint, developed centuries later by Paracelsus, that the three qualities of sulphur, salt, and mercury can be found in all things (p. 200).

[1] The account here given of Jabir is disputed in many respects by some authorities.

Rhazes distinguished six classes of minerals: (*a*) Four *Spirits*, mercury and sal ammoniac which sublime on heating, and sulphur and realgar (Arabic = powder of the mine) which burn on heating. (*b*) Seven *Bodies*, i.e. the seven metals (p. 146): (*c*) Six *Boraxes* including our borax and natron. The latter is a crude sodium carbonate occurring native in Egypt: (*d*) Eleven *Salts* including rock-salt, lime, salt of urine ('microcosmic salt'), potashes and *kali* or *al-kali* (Arabic = calcined wood ashes), the source both of our word alkali and chemical symbol K for Potassium: (*e*) Thirteen *Stones*, among them malachite (green basic copper oxide), haematite (Greek = blood stone, ferric oxide), gypsum (hydrous calcium sulphate), and alum which, being in both this class and the next, may be distinguished here as natural as against manufactured alum: (*f*) Six *Vitriols*, the Arabic names of some of these as given by Rhazes correspond to the Greek names given by Dioscorides (p. 98); thus *galgant* is Greek *chalkanthus*, copper pyrites, galgadis is Greek *chalkitis*, white vitriol, here alum. These and many other technical terms demonstrate Greek contacts.

AVICENNA, as he was known to the Latins, was the Persian Ali ibn-Sina (980–1037). A man of immense learning and versatility, he exercised great influence on philosophy and medicine both Eastern and Western. His influence on alchemy was less but is memorable for his denial of the possibility of transmutation. Though he did not maintain this consistently he did set out clearly in his *Book of the Remedy* (*c*. 1022) that, while it was possible to combine base metals so that they looked like gold or silver, this was very different from transmutation and that the metals remained separate in the alloys. All claims to transmutation were, he wrote, of this order. After Avicenna there were many Arabic alchemical writers whose works found their way into Latin. The quality of these is generally below those we have discussed.

(iii) *Medicine in Eastern Islam*

The first original Arabic writer on medicine was RHAZES (865–925). He studied in Bagdad under a disciple of Honain (p. 143) who was acquainted with Greek, Persian, and Indian medicine. The erudition of Rhazes was all-embracing, and his scientific output remarkable, more than half of it being medical.

The greatest medical work of Rhazes, and one of the most extensive ever written, is his 'Comprehensive Book' known to medieval Europe as the *Liber continens* (p. 163). It gathers into one huge corpus the whole of Greek, Syriac, and early Arabic medical knowledge and incorporates also the life experience of Rhazes himself. Rhazes was the first to describe smallpox and measles adequately. His account of them is a medical classic.

A prominent contemporary of Rhazes was the writer known to the Latins as ISAAC JUDAEUS (855–955). This Egyptian Jew became physician to the Fatimid rulers of Kairouan in Tunisia. His works were among the first to be translated from Arabic into Latin (p. 158). That *On Fevers* was one of the best medical works available in the Middle Ages.

AVICENNA (980–1037, p. 147) of Bokhara was one of the great thinkers of the Islamic world. He was less remarkable as physician than as philosopher, but his influence on medieval Europe was immense through his gigantic *Canon of Medicine*. It is the culmination and masterpiece of Arabic systematization and has been perhaps more studied than any medical work until modern times. The classification in it is excessively complex, and is in part responsible for the passion for subdivision which afflicted Western Scholasticism. The early Arabic literature of medicine is very extensive.

(iv) *Mathematics and Astronomy in Eastern Islam*

Of all the peoples of antiquity there was none except the Greeks that attained so high a standard in mathematics as the Hindus. Just as the Greeks developed geometry, the Hindus developed arithmetic and algebra. It is extremely difficult to fix the dates or even the chronological sequence of the Indian mathematical works. The Arabs, however, had much commerce with India and there can be no doubt that by the ninth century Hindu science was available in Arabic. Thus Arabic algebra and arithmetic are basically Indian.

The most influential mathematical work produced in Arabic was the *Arithmetic* of the Persian AL-KWARIZMI (*c.* 830). In it is used our so-called 'Arabic' numerical notation, in which the digits depend on their position for their value. The method is, in fact, of Indian origin. The *Algebra* of Al-Kwarizmi is the first work in which that word appears in the mathematical sense. 'Algebra' means in Arabic 'restoration', that is the transposing of negative terms of an equation to the opposite side. The mathematics of Al-Kwarizmi shows little originality. In general the achievement of the Arabs in pure mathematics is below the Greeks in geometry and below the Hindus in algebra. On the other hand, the Arabs exhibited great skill in applying their mathematics to physical and to a less extent to astronomical problems.

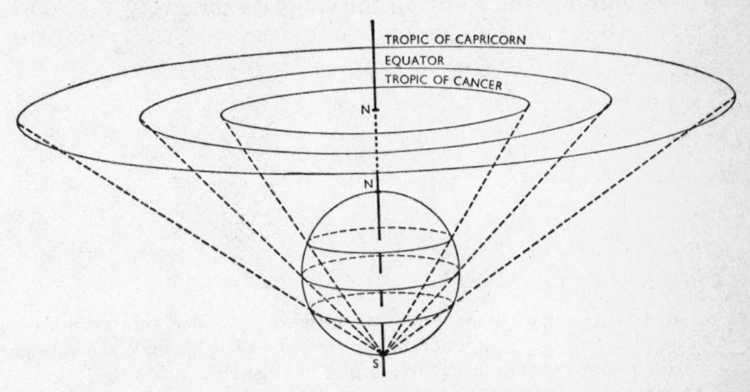

FIG. 64. Principle of stereographic projection

Astronomy and astrology were constant preoccupations of the Arabic-speaking world. Very early works were by the Syrian Nestorian Severus Sabokt (*c.* 650) and the Bagdad Jew MESSAHALA (770–820). Both wrote accounts of the plane astrolabe (not to be confused with the astrolabon, p. 94), and that of Messahala became very popular in the Latin West. The astrolabe is important for us as it shows the practical application of medieval mathematical skill at its best.

The earliest surviving astrolabes are Islamic and from the tenth century. There is one Byzantine astrolabe of A.D. 1062. Western astrolabes are known from *c.* 1275. In Islam it has remained popular and is still in use for determining times of rituals. Among the Latins the astrolabe reached its peak of popularity just before the invention of the telescope (*c.* 1610).

The principle of the plane astrolabe is stereographic projection (Fig. 64), that is a way of plotting the surface of a sphere on a plane. The instrument consists essentially of a circular star map which can be rotated about its north pole, resting on a plan of the sky as seen by an observer at some particular latitude. In practice the star map is made 'transparent' by an openwork metal plate, the *rete* or net (Fig. 65) with little pointers indicating the brightest stars. The plates for different latitudes, called *tympans*, are engraved with lines showing degrees of elevation from the horizon to the zenith, and another set of lines showing the points of the compass and every 5° or some

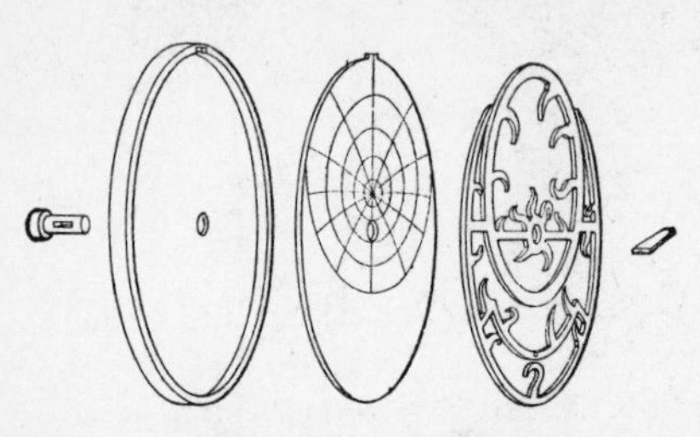

FIG. 65. Essentials of an astrolabe. From left to right: *pin*; *mater* with notch above; *tympan* engraved with lines of altitude and azimuth for a fixed latitude and with peg above to fit notch in mater; *rete* or map of important stars; *wedge*.

other suitable interval between. The tympan is enclosed in the *mater* or body of the instrument so that the appropriate plate may be set uppermost as required. The back of the astrolabe is usually marked out in degrees and fitted with a movable arm or alidade. This has sights to measure the angular heights of stars. These readings could be used in the calculations made with the main part of the instrument and, given adequate knowledge (which was rare), for telling the time at night. The alidade could also be used for surveying, for which a much simpler instrument sufficed (Fig. 58).

The astrolabe was relatively seldom used for actual observation but mainly for calculation and doubtless often to impress the clients of astrologers. Nevertheless, it was the most complex astronomical device available from the seventh to the sixteenth century. It was invaluable as a means of avoiding tedious routine calculations of spherical trigonometry such as always occur in astronomical and astrological work.

The Caliph Al-Mamum (813–33) built a fine observatory at Bagdad (829) where observations were long recorded. The greatest of all the Arabic astronomers, AL-BATTANI, Albategnius of the Latins (died

929), observed chiefly at his home Raqqa (Aracte) in Asia Minor, but also at Bagdad. He worked over the observations of Ptolemy in a searching and exact manner. He thus obtained more accurate values

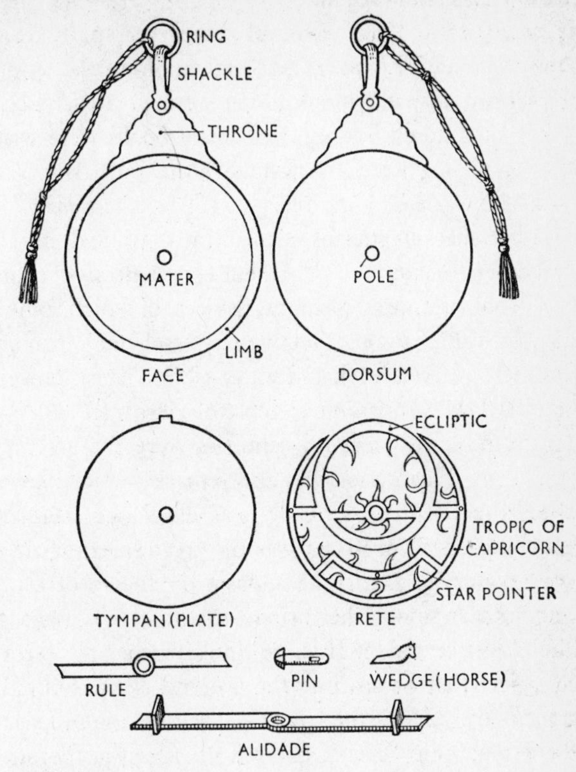

FIG. 66. Component parts of a plane astrolabe.

for the obliquity of the ecliptic and the precession of the equinoxes (pp. 82 f.). His improved tables of the sun and the moon contained his great discovery that the direction of the sun's eccentric (p. 83), as recorded by Ptolemy, was changing. Expressed in the terms of more modern astronomical conceptions, this is to say that the earth is moving in a varying ellipse (p. 310).

(v) *Physics in Eastern Islam*

Among the Arabic writers on physics ALKINDI (813–80) of Basra and Bagdad was the earliest. He is described as the 'first philosopher of the Arabs'. Of his works some are on meteorology, several are on specific weight, and others on tides. His best work is on optics, and deals with the reflection of light.

In the ninth century the technical arts were rapidly developing in Mesopotamia and Egypt, where irrigation works and canals for water-supply and communications were created. Theoretical mechanics roused much interest, and many books were written on such topics as raising water, on water-wheels, on balances, and on water-clocks. The earliest appeared about 860 as the *Book of Artifices* by the brother mathematicians Muhammed, Ahmed, and Hasan, sons of Musa ben Shakir, who were themselves patrons of translators. They describe one hundred technical devices, of which some twenty are of practical value, among them being vessels for warm and cold water, wells with a fixed level, and water-clocks. Most, however, are mere scientific toys comparable to most of Hero's (p. 86).

The tenth and early eleventh centuries were the golden age of Arabic literature. This period was also remarkable for its wealth of technical advance. Optics especially was developed. ALHAZEN (ibn al-Haitham 965–1038) of Basra was its greatest exponent. In his work, *The Treasury of Optics*, he opposes the theory of Euclid and Ptolemy and others among the ancients that the eye sends out visual rays to the object of vision. It is, he thinks, rather the form of the perceived object that passes into the eye and is transmuted by its 'transparent body', that is the lens. He discusses the propagation of light and colours, optic illusions and reflection, with experiments for testing the angles of reflection and of incidence. His name is still associated with the so-called 'Alhazen's problem'. 'In a convex mirror, spherical, conical, or cylindrical, to find the point at which a ray coming from one given position will be reflected to another given position.' It leads to an equation of the fourth degree which Alhazen solved by the use of an hyperbola. Alhazen examined also the refraction of light-rays through transparent media (air, water). In detailing his experiments with spherical segments he comes very near to the theoretical exposition of magnifying lenses which was made centuries later (p. 172).

Alhazen regards light as a kind of fire that is reflected at the spheric limit of the atmosphere. His calculation of the height of this atmosphere gives about ten English miles. He treats also of the rainbow, the halo, and the reflection from spherical and parabolic mirrors. He constructed such mirrors of metal on the basis of most elaborate calculations. His fundamental study *On the Burning-sphere* represents real scientific advance, and exhibits a profound and accurate conception of the nature of focusing, magnifying, and inversion of the image, and of formation of rings and colours by experiments. The work is far beyond anything of its kind produced by the Greeks. Alhazen records in it the semi-lunar shape of the image of the sun during eclipses on a wall opposite a fine hole made in the window-shutter. This is the first mention of the *camera obscura*.

Among the most characteristic products of Arabic thought is a group of writings on what we may call scientific theory and classification. An early exponent of these was the Turkish philosopher ALFARABI (d. *c.* 951), the author of the most important oriental work on the theory of music. His treatise on the classification of the sciences was very influential.

The Persian ALBIRUNI (973–1048), physician, astronomer, mathematician, physicist, geographer, and historian, is perhaps the most prominent of the versatile scholars of the Islamic Golden Age. His *Chronology of Ancient Nations* is an important historical document. Most of his mathematical work and many other of his writings await publication. In physics his greatest achievement is the very exact determination of the specific weight of eighteen precious stones and metals. His method was, in effect, that of the bath of Archimedes (pp. 70–71).

In the tenth and eleventh centuries several esoteric sects professing the doctrine of the atomic nature of matter opposed the orthodox Aristotelianism of Moslem theologians and established themselves in Mesopotamia. A struggle ensued comparable to that of later date in Europe. In the end the unorthodox atomists were vanquished. Among these societies, however, were the BRETHREN OF PURITY, a philosophic sect founded in Mesopotamia about 980. They combined to produce an encyclopaedia, parts of which deal with natural science, mainly on Greek lines. It contains discussions on the formation of minerals, on earthquakes, tides, meteorological phenomena, and on the elements, all brought into relation with the celestial spheres and bodies. The work of the Brethren, although burnt by the orthodox, spread as far as Spain where it influenced philosophic and scientific thought.

(vi) *Science in Western Islam*

In Western Islam the scientific tradition was established later than in Eastern. It first appears in Spain, during the glorious reigns of the Caliphs Abd ar-Rahman III and al-Hakam II of Cordova, in the person of HASDAI BEN SHAPRUT (d. *c.* 990), a Jew who was at once minister, court physician, and patron of science. He translated into Arabic, *c.* 950, with the help of a Byzantine monk, a splendid manuscript of Dioscorides (p. 98) sent, as a diplomatic present, to his sovereign from Constantine VI of Byzantium. (It is intriguing that the same text was turned into Anglo-Saxon about the same year.)

Alchemy also came to Moslem Spain from Eastern Islam in the tenth century. There were many Iberian alchemists during the next two centuries but none of comparable originality to Jabir (pp. 144 f.) or Rhazes (pp. 146 f.). Nevertheless, the Western school is important since it was mostly through it that texts on the subject reached the Latin West. Moreover the same group preserved many references to their Eastern brethren that would otherwise be lost.

A more progressive department of Western Islamic science was astronomy. A library and academy was founded at Cordova in 970, and similar establishments sprang up at Toledo and elsewhere. Astronomy was specially studied. The chief astronomer of Moslem Spain was known to the Latins as ARZACHEL. At Toledo he drew up so-called *Toledan tables* which attained a high degree of accuracy (1080). One of the last significant astronomers of Moslem Spain was Al-Bitrugi of Seville, known to the Latins as ALPETRAGIUS. He wrote a popular textbook of astronomy (*c.* 1180). The work attempts to replace the Ptolemaic by a strictly concentric planetary system and provided suggestions to Copernicus (pp. 212 ff.).

In the twelfth century a great change came over Islamic thought. Under the influence of the religious teacher Al-Ghazzali (d. 1111), tolerance gave place to persecution of studies thought to 'lead to loss of belief in the Creator and in the origin of the world'. Outstanding and independent works become rarer. Among the scientific writers an increasing proportion of Jews is to be observed, because they were relatively free from such restraints. The most eminent of these scientific Jews was the court physician, philosopher, and religious teacher, MAIMONIDES (1135–1204). Born in Spain, he spent most of his active life in Cairo under the great Saladin and his sons. In his medical works he even ventured to criticize the opinions of Galen. As a court official he wrote hygienic treatises for the Sultan which are typical specimens of the medical literature of Islam. His cosmological views influenced St. Thomas Aquinas and, through him, the whole thought of Catholic Europe. His *Guide for the Perplexed* is perhaps the most readable treatise on general philosophy produced by the Middle Ages. It has the crowning merit of relative brevity.

The latest and the greatest exponent of Islamic philosophy was the Spaniard AVERROËS (1126–98). He was born at Cordova, son and grandson of a legal officer. He himself held the office of judge, but also studied and practised medicine. His very voluminous philosophical writings earned the enmity of orthodox Moslem theologians, some of whom regarded him as having become a Jew. In fact, no writer exerted greater influence than Averroës on later medieval Jewish thought. His writings were burned by royal decree.

Averroës was certainly one of the most influential of medieval thinkers. He placed his thought in the form of a long series of commentaries on the works of Aristotle, though his teaching was basically modified by Neoplatonism, notably in his conception of the human soul as part of the Divine world-soul. His most discussed doctrine was that the world is eternal. This some of his interpreters represented as a denial of creation. Nevertheless, Averroës did accept the idea of creation, though not of an entire universe out of nothing as demanded by the current theology of Islam, Christianity, and Judaism alike.

Averroës believed, not in a single act of creation, but in a continuous creation, renewed every instant in a constantly changing world, always taking its new form from that which has existed previously. This is true philosophic evolutionism. For Averroës the world, though eternal, is subject to a *Mover* constantly producing it and, like it, eternal. This Mover can be realized by observation of the eternal celestial bodies whose perfected existence is conditioned by their movement. Thereby may be distinguished two forms of eternity, that with cause and that without cause. Only the *Prime Mover* is eternal and without cause. All the rest of the universe has a cause or, as we should say nowadays, is 'subject to evolution'.

Like all medieval thinkers, Averroës pictured the universe as finite in space. For any questioning of that doctrine we must wait for Nicholas of Cusa (p. 178) and Giordano Bruno (pp. 218 f.). With the thirteenth century there came a deterioration in Arabic philosophy and science. The future lay with the Latin West on which Arabic thought was now setting its stamp.

Perhaps the most significant of all Moslem influences on the West has been the philosophy transmitted through Averroës and chiefly by Jewish agents. By his doctrine of the eternity of the world, his denial of creation in time, and his conception of the unity of the soul or intellect, Averroës split Western thought from top to bottom. Orthodox Catholic philosophy of the Middle Ages may be regarded as an organized attempt to refute his views. That such organization seemed necessary tells of the gravity of the opposition. His influence may be traced in many medieval heresies, in the works of Nicholas of Cusa (p. 178), and of several Renaissance thinkers, in the standpoint of Copernicus (p. 212), in the thought of Giordano Bruno (p. 218), and beyond. It may seem strange that a professedly faithful exponent of Aristotle should have initiated that movement which led to the final overthrow of Aristotelian cosmology in the Insurgent Century (Ch. VII). It must be remembered, however, that Averroës, like the other Arabic philosophers, saw Aristotle through Neoplatonic spectacles. The Neoplatonic tinge became intensified in the Latin versions of his works.

3. ORIENTAL PENETRATION OF THE OCCIDENT TO *c.* 1100

The eleventh century and those that follow brought the West into relation with the wisdom of the East. In these centuries the relation of East and West was the reverse of that with which we are nowadays familiar.

In our time most Oriental peoples value Western civilization and accord it the sincerest form of flattery. The Oriental recognizes that with the Occident are science and learning, power and organization, and business enterprise. But the admitted superiority of the West does not extend to the sphere of religion. The Oriental nowadays gladly accepts Occidental standards of economics, technology, science, and medicine but repudiates, and perhaps despises, the religion, philosophy, and much of the social outlook of the West. In the Europe of the eleventh and twelfth centuries it was far other. The Westerner knew full well that Islam held the learning and science of antiquity. Moslem proficiency in arms and administration had been sufficiently proved—the Occidental belief in them is enshrined in our Semitic words 'arsenal' and 'admiral', 'tariff', 'douane', and 'average'. There was a longing, too, for the intellectual treasures of the East, but the same fear and repugnance to its religion that the East now feels for the West.

The first definitely Oriental influence that we can discern as affecting ideas about nature is of the character of infiltration rather than direct translation. Thus GERBERT, who died in 1003 as Pope Sylvester II had studied in north-east Spain, beyond the Moslem zone. He described an abacus (p. 121) that was almost certainly of Arabic origin though he used for it counters bearing numerals similar to those of Boethius (p. 122). He also instigated a translation from Arabic of a work on the astrolabe. He was clearly in touch with some sort of Arabic learning.

Similarly with HERMAN THE CRIPPLE (1013–54) who spent his life at the Benedictine Abbey of Reichenau in Switzerland. He wrote certain mathematical and astrological works which were extensively used in the following century. Herman was unable to read Arabic, and could not travel by reason of his infirmity. Yet his writings display oriental influence, which must have been conveyed to him by wandering scholars. His name is attached to the first account of the use of the plane astrolabe (p. 151) by a Latin. Similar evidence of Arabic infiltration is exhibited in lapidaries and herbals of the eleventh and twelfth centuries.

Arabic learning, thus beginning to trickle through to the West, was derived ultimately from Greek sources (pp. 141 f.). There was, however, just one channel by which the original Greek wisdom might still reach Europe, though in a much debased form. Communication between the West and the Byzantine East was very restricted in the Dark Age, but a Greek tradition still lingered in south Italy and

Sicily until late medieval times. But Saracens had begun their conquest of Sicily in the eighth century, and did not loosen their hold until the Norman attack of the eleventh. Thus between the tenth and thirteenth centuries the 'Sicilies' were a possible source of both Greek and Arabic science.

One seat of learning in the southern Italian area felt especially the influence of both Greek and Arabic culture. Salerno, on the Gulf of Naples, had been a medical centre as far back as the ninth century. There was a Greek-speaking element in the town and some traces of ancient Greek medicine lingered there, as in other parts of south Italy, after the downfall of the Western Empire. There was, moreover, a number of Jews in the town and many of these had affiliations with the Orient. From about 1050 onwards medical works were produced at Salerno. It is easy to understand why some of them contain Semitic words, and why others present unexpected and strangely altered Greek terms.

A very important carrying agent of Arabic learning was CONSTANTINE THE AFRICAN (1017–87), a native of Kairouan in Tunisia. He reached Salerno about 1070 and some years later acted as secretary to the Norman conqueror of that city. Later he retired to a monastery and spent the rest of his life turning current Arabic medical and scientific works into Latin. Constantine's sources are mainly Jewish writers of North African origin and Arabic language, among them Isaac Judaeus (p. 148). In the manner of his day Constantine often conceals the names of the authors from whom he borrows, or he gives them inaccurately. His knowledge of both the languages which he was treating was far from thorough. Yet his versions remained current in the West long after they had been replaced by the better workmanship of students of the type of Gerard of Cremona (p. 162). With Constantine is linked ALPHANUS, Archbishop of Salerno (d. 1085), who was himself a translator direct from the Greek, and turned a Neoplatonic physiological work of the fourth century into Latin.

4. MECHANISM OF TRANSLATION FROM ARABIC TO LATIN

The earliest oriental influences that reached the West had thus been brought mostly by foreign carriers, but the desire for knowledge could not be satisfied thus. The movement that was to give rise to the universities was shaping itself during the twelfth century. The Western student was beginning to become more curious and more desirous of going to the well-springs of Eastern wisdom.

Language was his main difficulty. The idiom of Arabic was utterly different from the speech of the peoples of Europe. Moreover, its grammar had not been reduced to rule in any Latin work, nor could teachers be easily procured. The only way to learn the language was to go to an Arabic-speaking country. This was a dangerous and difficult adventure, involving hardship, secrecy, and perhaps abjuration of faith. Moreover, a knowledge adequate for rendering scientific treatises into Latin meant a stay of years, since some understanding of the subject-matter as well as the technical vocabulary was needed. There is good evidence that such knowledge was very rarely attained by Western Christians, and probably never until the later twelfth century.

Until after A.D. 1000 the Iberian peninsula was Moslem save for Leon, Navarre, and Aragon, small kingdoms of the French march. In that northern area the grip of Islam had soonest relaxed, and this territory remained religiously and linguistically a part of the Latin West. The Moslem south was ruled from Cordova. At the more northern Toledo the townsfolk, while speaking an Arabic patois, were chiefly Christian, though with a large Jewish element. In 1085, Alphonso VI of Leon, aided by the Cid, conquered the town. A large Arabic-speaking population remained. It was at Toledo that most of the work of transmission took place (Fig. 67).

The question is often asked why in the Middle Ages the practice was to translate works from the Arabic rather than from the Greek, and why this tendency affected works originally written in Greek. The reasons may be set forth thus:

(*a*) Before about 1200 Moslem learning was better organized, more original, more vital than Byzantine.

(*b*) Byzantine Greek is far distant from the classical tongue. The language of Aristotle was incomprehensible to the monastic guardians of his manuscripts. But classical Arabic was intelligible to every well-educated man who spoke and wrote Arabic.

(*c*) The whole trend of Byzantine learning was to theology and away from philosophy and science.

(*d*) The channels of trade with the West were either direct with Islam or through Western enclaves within the Byzantine Empire.

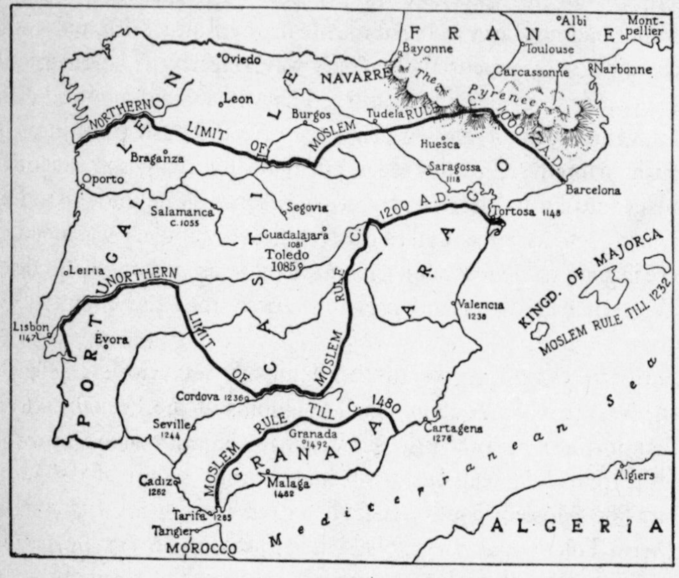

FIG. 67. Recession of Islam in the Iberian peninsula. Numbers after town names are dates of reconquest by Christendom.

(*e*) In the Middle Ages languages were learned by speaking and not from grammars. Spoken Arabic was more accessible than spoken Greek.

(*f*) Latin Christendom made little progress in occupying Byzantine territory. On the other hand, Islam was in retreat in the West.

(*g*) Jewish help could be obtained for Arabic, but seldom for Greek.

The process of translation from Arabic, especially in Spain, was frequently carried on by the intervention of Jewish students. Many of the translated works were themselves by Jews. The tenth, eleventh, and twelfth centuries, a time of low degradation of the Latin intellect, was the best period of Jewish learning in Spain. Arabic was the natural linguistic medium of these learned Jews, among whom were

Solomon ibn Gabirol (1021–58?) of Saragossa, who was disguised in scholastic writings as AVICEBRON, and Moses ben Maimon (1135–1204) of Cordova, more familiarly known as MAIMONIDES (p. 155). The writings of these two authors together with the Jewish versions of AVERROËS (p. 155) were the most philosophically influential of those rendered into Latin from Arabic during the Middle Ages. Their works helped to mould Western scholasticism.

Despite the activity of the translators, medieval Latin was not yet equipped with an adequate supply of technical terms. The meaning of some of these in the Arabic were imperfectly known to the translators themselves. Such words were therefore often simply carried over, transliterated from their Arabic or Hebrew form. The early versions are full of Semitic expressions. Thus of chemical substances we have *realgar* (red sulphide of arsenic), *tutia* (zinc oxide), *alkali*, *antimony*, *zircon*, and of chemical apparatus *alembic* for the upper, and *aludel* for the lower part of a distillation vessel. A chemical substance unknown to the Greeks which appears for the first time in the works of Geber (= Jabir, pp. 144 f.) is *sal-ammoniac*. The *ammoniacon* of the Greeks was rock-salt, and it seems that the transference of the old name to a new salt was effected by the Syrians. Of pharmaceutical terms, a number are Persian transmitted through Arabic, as *zedoary*, *alcohol*, *sherbert*, *camphor*, *lemon*, and *syrup*, while more purely Arabic are *alizarin*, *borax*, *amber*, *bezoar*, *talc*, and *tartar*.

In astronomy there are numerous Arabic star names as *Aldebaran*, *Altair*, *Betelgeuse*, *Rigel*, *Vega*, some technical terms as *nadir*, *zenith*, *azimuth*, *azure*, a few instrumental designations as *alidade* and *theodolite*, and several words which have passed into common language, *almanac* and *mattress*. To these may be added the mathematical terms *zero*, *cipher*, *sine*, *root*, *algebra* (pp. 148 f.), *algorism* (see below). Music was also deeply affected, as witness *lute*, *guitar*, *shawm*, *rebeck*. There was a complete Arabic-Latin anatomical vocabulary of which almost the sole remains is *nucha*, though the titles of the *basilic* and *saphenous* veins have passed through Arabic. The modern botanical vocabulary provides us with many plant-names of Arabic origin such as *artichoke*, *coffee*, *lilac*, *musk*, *ribes*, and *sumach* or names that have passed through Arabic as *jasmine*, *mezereon*, *saffron*, *sesame*, and *taraxacum*.

5. EARLY WESTERN TRANSLATORS FROM THE ARABIC (1100–1500)

Among the pioneer Western translators from Arabic to Latin was ADELARD OF BATH (*c.* 1090–*c.* 1150), who journeyed both to Spain and the Sicilies. His services to mathematics were very distinguished. He began early with a treatise on the abacus. Then he turned to translate into Latin the *Arithmetic* of Al-Kwarizmi involving the use of the 'Arabic', i.e. Indian, numerals (p. 148), which he thus introduced to the West. Al-Kwarizmi has, through him, left his name in *algorism*, the old word for arithmetic. Moreover, Adelard also rendered Euclid from the Arabic and so made the Alexandrian mathematician known for the first time to the Latins. He wrote a popular dialogue, *Natural Questions*, which is a sort of elementary compendium of Arabic science.

A generation later than Adelard was ROBERT OF CHESTER (*c.* 1110–*c.* 1160), who sojourned in northern Spain (1141–7). He was the first to translate the Koran (1143). Among his scientific renderings was the first alchemical text to appear in Latin (1144) (p. 173). His translation of the *Algebra* of Al-Kwarizmi (p. 148) introduced the subject to the Latins (1145). Later he returned to England (1147). There he produced astronomical tables for the longitude of London (1149–50) based on Albategnius (pp. 150 f.) and for the latitude of London based on Al-Kwarizmi (p. 148) and Adelard (see above).

Contemporary with Robert, and perhaps stimulated by him, were certain native translators who worked at Toledo. One, a Christian, rendered into Latin from Arabic the *Physics* and other works of Aristotle (*c.* 1140). Another, a converted Jew, was very active and translated (*c.* 1150) among many other works a pseudo-Aristotelian treatise which greatly influenced Roger Bacon, as well as various astronomical and astrological works.

The greatest of all the translators from the Arabic was GERARD OF CREMONA (1114–87), who spent many years at Toledo and obtained a thorough knowledge of Arabic from a native Christian teacher. He established a school of translators which turned into Latin no less than ninety-two complete Arabic works. Many of them are of very great length, among them the *Almagest* of Ptolemy (pp. 89 f.) and the *Canon* of Avicenna (p. 148).

Other achievements of Gerard's school are translations from the Arabic of Archimedes *On the Measurement of the Circle* (p. 73), of an optical work of Apollonius (p. 75), of many of the works of Aristotle both spurious and genuine, of Euclid's *Elements*, of many medical works of Galen, Hippocrates, Isaac Judaeus (p. 148), and Rhazes (pp. 147 f.), of alchemical works of Jabir (pp. 144 f.), of mathematical and astronomical works by Alkindi (pp. 151 f.), Alhazen (pp. 152 f.), Alfarabi (p. 153), Messahala (p. 149), and others. Gerard's school also translated certain important Neoplatonic works.

The Sicilian group was less active. Among its products was the *Optics* of Ptolemy (p. 89), translated about 1160 by the Sicilian admiral EUGENIUS OF PALERMO. He rendered it from the Arabic, though he had an effective knowledge of Greek. The great astronomical and mathematical system of Ptolemy, known to the Middle Ages as the *Almagest* (pp. 89 f.), was also first translated into Latin direct from Greek in Sicily in 1163, some twelve years before it was rendered from the Arabic by Gerard at Toledo (p. 162). This version from the Greek, however, gained no currency and only that from the Arabic was available until the fifteenth century.

The last important medieval translator from the Arabic was of Sicilian origin. He was the Jew MOSES FARACHI (d. 1285), a student at Salerno (p. 158). His works were among the latest of influence that issued from that ancient seat of learning. His great achievement was the translation for his master Charles of Anjou (1220–85), King of the Sicilies, of the enormous *Liber continens* of Rhazes (p. 148), a standard medical work of the Middle Ages. The Arabic to Latin literature is immense. The translators of some of its members are unknown. Such an orphan work is the Latin of Alhazen's *Optics* (p. 152) of about 1200. Another is the abbreviation from the Arabic of Galen's anatomy known as *De juvamentis membrorum* (*c.* 1270).

Special consideration among the translators may be given to MICHAEL THE SCOT (*c.* 1175–*c.* 1235) because we have more picturesque details of him than of the others. He had a career similar to Adelard. He visited Toledo and afterwards northern Italy, staying at Padua, Bologna (1220), and Rome (1224–7). He ended his days in the south in the service of the 'Stupor Mundi', Frederick II.

Michael's version of Alpetragius (p. 154) contained the first attack on traditional astronomy. His translations of Averroës (pp. 155 f.) were among the first works of that heresiarch available in Latin. His version of Aristotelian biology from the Arabic gave Aristotle's own scientific observations for the first time to the West. His work on astrology was the first major treatise on the subject accessible in Latin. Michael had Jewish and Moslem help and was long associated with that arch-enemy of the papacy, Frederick II. Thus in the popular imagination his name became associated with black magic. This was the fate of other translators from the Arabic. The vulgar attitude towards such men is faithfully reflected in Sir Walter Scott's *Lay of the Last Minstrel* where a traveller tells that

> Paynim countries I have trod,
> And fought beneath the Cross of God.
>
> . . . . . .
>
> In those far climes it was my lot
> To meet the wondrous Michael Scott;
> A wizard of such dreaded fame,
> That when, in Salamanca's cave,
> Him listed his magic wand to wave,
> The bells would ring in Notre Dame!
> Some of his skill he taught to me;
> And, warrior, I could say to thee,
> The words that cleft Eildon hills in three,
> And bridled the Tweed with a curb of stone:
> But to speak them were a deadly sin,
> And for having but thought them my heart within,
> A treble penance must be done.
>
> When Michael lay on his dying bed,
> His conscience was awakened;
> He bethought him of his sinful deed,
> And he gave me a sign to come with speed.
> I was in Spain when the morning rose,
> But I stood by his bed ere evening close.

6. SCHOLASTICISM AND SCIENCE (1200–1400)

The view of the material universe conveyed by Arabic science to Latin Christendom was new in tone and presentation rather than in kind. The thought of the Latins in their Dark Age on material things was Neoplatonic, with Aristotle's scheme and the theory of macrocosm and microcosm as keys (Fig. 61). With the advent of Arabic thought the outlines of this vision were sharpened, and details were

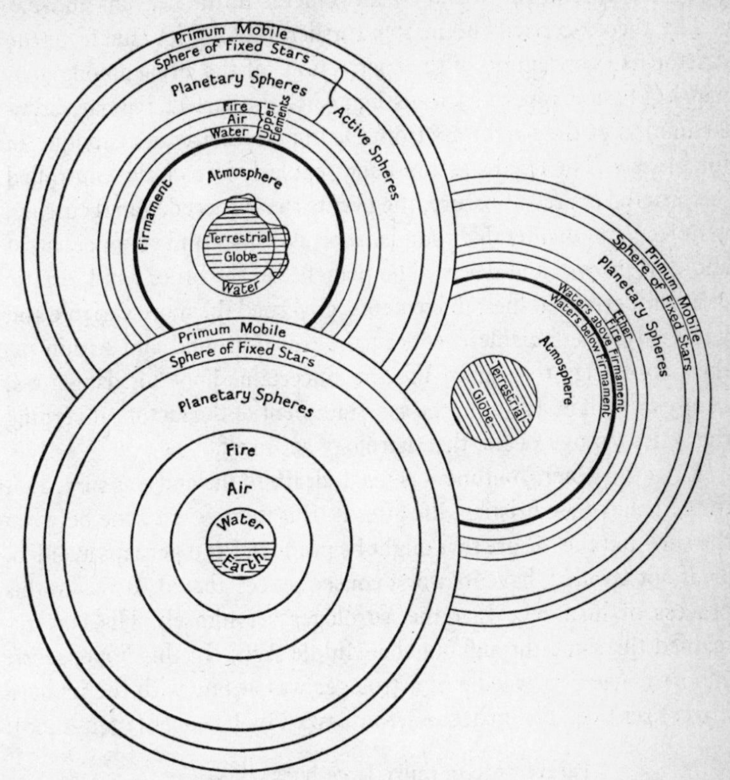

FIG. 68. Three typical medieval diagrams of world structure, all nominally derived from Aristotle. Above is the scheme of Maimonides, below that of Dante, to the right another favourite design. Seven separate spheres were held to contain the seven planets but these spheres are omitted in each case for simplicity.

elaborated from the Arabian commentators on the Aristotelian corpus (Figs. 68 and 72).

Thus Aristotle's views, or supposed views, as to the structure of the universe formed the framework on which the whole of medieval science, from the thirteenth century onward, came to be built. Aristotle conceived the stars as beings whose nature and substance were purer and nobler than that of aught in the spheres below. This was a point of departure from which the influence of the heavenly bodies over human destinies might be developed. Changes undergone by bodies on the earth—all the phenomena of our life—were held to be paralleled and controlled by movements in the heavens above.

The theory carried the matter further. Taking its clue from the Aristotelian conception of the 'perfection' of the circle among geometrical figures (pp. 52 f.), it distinguished the perfect, regular, circular motion of the fixed stars from the imperfect, irregular motion of the planets. The fixed stars, moving regularly in a circle, controlled the ordered course of nature, the events that proceeded in recurring, manifest, and unalterable rounds, such as winter and summer, night and day, growth and decay. The planets, on the other hand, erratic or at least errant in their movements, governed the more variable and less easily ascertainable events in the world around and within us, the happenings that make life the uncertain, hopeful, dangerous, happy thing it is. It was to the ascertainment of the factors governing this kaleidoscope of life that astrology set itself.

Thus the general outline was fixed, death in the end was sure, and, to the believing Christian, life after it. But there was a zone between the sure and the unsure that might be predicted and perhaps avoided, or, if not avoided, have its worst consequences abated. It was to this process of insurance that the astrologer set himself. His task remained the same throughout the Middle Ages. In this hope, *savoir afin de prévoir*, the medieval astrologer was at one with the modern man of science. The matter is summarized by Chaucer (1340–1400):

> Paraventure in thilke large book,
> Which that men clepe the heven, y-written was
> With sterres, whan that he his birthe took,
> That he for love sholde han his deeth, allas!
> For in the sterres, clerer than is glas,
> Is written, God wot, whoso coude it rede,
> The deeth of every man, withouten drede.
>
> . . . . .
>
> . . . But mennes wittes ben so dulle
> That no wight kan wel rede it atte fulle.
>
> (*The Man of Lawes Tale*, ll. 190–6 and 202–3.)

With the advent of the Arabian learning, astrology had become, in fact, the central intellectual interest. It retained this position until the triumph of the experimental method in the seventeenth century.

Especial attention had always been paid to the zodiacal signs (p. 18) and to the planets. Each zodiacal sign was held to govern some region of the body, and each planet to influence a special organ. The supposed relations of zodiacal signs and planets to bodily parts and organs, in relation to the advent of disease and calamity, had been set forth in many texts of late antiquity. This belief, conveyed to the Dark Age, but much corrupted and attenuated during its course, was reinforced and developed in the West by translations from the Arabic during the scholastic period which followed (pp. 127 f.).

Doctrine of this type, once received into Europe, was stamped with the special form of Western thought. Now it was characteristic of the scholastic thinker that he sought a complete scheme of things. He was not content to separate, as we do, one department of knowledge or one class of phenomena, and consider it in and by itself. Still less would he have held it a virtue to become a 'specialist', to limit his outlook to one department with the object of increasing the sum of knowledge in it, and in it alone. His universe, it must be remembered, so far as it was material, was limited. Its frontier was the sphere of the fixed stars. Of the structure and nature of all within this sphere he had been provided with a definite scheme. The task of medieval science was to elaborate that scheme in connexion with the moral world. This was first especially undertaken by mystical writers working under the stimulus of the new Arabian influence. Such authors produced most elaborate mystical schemes based on the docrine of the macrocosm and microcosm. These schemes took into account the form of the world and of man as derived from Arabian sources, and read into each relationship a spiritual meaning (Fig. 61).

For such an attitude of mind there could be no ultimate distinction between physical events, moral truths, and spiritual experiences. In their fusion of the internal and the external universe, these mystics have much in common with the mystics of all ages. The culmination of the process is with Dante (1265–1321) (Fig. 72).

There were other typical currents of medieval thought that were susceptible of more systematic development. It was the age of the foundation of universities and of religious orders. Among these new orders were two that specially influenced the universities, the Dominicans or Black Friars founded at Toulouse in 1215 by the austere Dominic (1170–1221), and the Franciscans or Grey Friars founded in 1209 by the gentle Francis of Assisi. The name of Dominic is associated with the extermination of the Albigenses. The Dominicans, whose title was paraphrased as *Domini canes*, 'hounds of the Lord', set themselves to strengthen true doctrine and extirpate error. The activity of the Inquisition was one of the less edifying interests of the 'hounds' of whom Torquemada was an unamiable representative. During the thirteenth century these two orders provided most of the great university teachers, who occupied themselves in marshalling the new knowledge and making it more accessible. Alexander of Hales (d. 1245), Robert Grosseteste (d. 1253), and Roger Bacon (d. 1294) were Franciscans; Albertus Magnus (1206–80) and St. Thomas Aquinas (1227–74) were Dominicans.

A foremost influence in the revival was the recovery of the writings of Aristotle. The interpretation of these works gave to Scholasticism its essential character. The first scholastic to be acquainted with the entire corpus of Aristotle was ALEXANDER OF HALES. ALBERTUS was the first who reduced the whole philosophy of Aristotle to systematic order with constant reference to the Arabian commentators, while ST. THOMAS AQUINAS remodelled the Aristotelian philosophy in accordance with the requirements of ecclesiastical doctrine. During the thirteenth century Aristotle, at first represented in translation from the Arabic, became partially accessible in renderings direct from the Greek. A very important translator from the Greek was the Dominican WILLIAM OF MOERBEKE (d. 1286), who was in close contact with St. Thomas.

It is remarkable that the process of codifying the new knowledge derived from the Arabic, involving as it did a rapid development in the whole mental life, did not early give rise to a more passionate and more conscious faith in the reality and value of progress in knowledge. The test of such faith, so far as nature is concerned, must be the direct appeal to nature. Yet there is little evidence of devotion to observation of nature recorded in the great physical encyclopaedias of the thirteenth century, such as those of the Dominican VINCENT OF BEAUVAIS (1190–1264), or of the Franciscan BARTHOLOMEW THE ENGLISHMAN (*c.* 1260). The explanation is that the medieval mind was obsessed with the idea of the world as mortal, destructible, finite, and completely knowable in both space and in time and at the same time as being hardly worth knowing. Hear St. Augustine:

Men seek out the hidden powers of nature, which to know profits not and wherein men desire nothing but knowledge. With the same perverted aim they seek after magic arts. . . . As for me, I care not to know the courses of the stars, while all sacrilegious mysteries I hate (*Confessions*, x. 35). Even if the causes of the movements of bodies were known to us, none would be important except such as influence our health. But since, being ignorant of these, we seek physicians, is it not clear that we should rest content to be ignorant of the mysteries of the heavens and the earth. (*De fide*, 16.)

Thus medicine—that is in effect pharmacy—is the one science that St. Augustine would allow. Is it wonder that medicine had deteriorated into mere traditional drug lists? In the Latin West during the Middle Ages the motive for detailed *research*, in our modern sense of the word, was almost absent and there were none who devoted their lives to the observation of nature.

One great Islamic philosopher there was, Averroës (pp. 155 f.), who took another view of the universe, denying it to be finite, at least in time. His works were available in Latin, but the great ecclesiastics set their faces against him, though he was widely and illicitly read. His theories were adopted mainly by Jews and by Latins with heretical leanings (pp. 155 f. and 161).

7. SOME MAIN PERSONALITIES OF SCHOLASTIC SCIENCE

The medieval world thus knew nothing of that infinite sea of experience on which the man of science nowadays launches his bark in adventurous exploration. Medieval science tended to the encyclopaedic form. The task of the writer of the encyclopaedia was to set forth such a survey of the universe as would be in accord with spiritual truth rather than to reveal new truths or new relations. The framework on which this scheme was built was Aristotle, largely as conveyed by commentaries upon his works. Yet it affords a reflection on the incompleteness of all philosophical systems that the great teacher and systematist, ALBERTUS MAGNUS (1206–80), who perhaps more than any other man was responsible for the scholastic world-system, was among the very few medieval writers who were real observers of nature. It is, after all, in the very essence of the human animal to love the world around it and to watch its creatures. 'Throw out nature with a pitchfork and back she comes again.' Albertus, scholastic of the scholastics, drowned in erudition, the most learned man of his time, has left evidence, in his works on minerals and on living things, that the scientific spirit was beginning to awake. As an independent observer he is by no means contemptible, and this element in him marks the new dawn which we trace more clearly in his successors.

Contemporary with the Dominicans, Albert (1206–80) and St. Thomas Aquinas (1227–74), were several Franciscan writers who were the first consciously forward-looking thinkers since antiquity. The most arresting of them was ROBERT GROSSETESTE (*c.* 1175–1253), Bishop of Lincoln. He determined the main direction of physical interests of the thirteenth and fourteenth centuries. He knew something of the action of mirrors and of the nature of lenses. He had actually experimented with lenses, and many of the optical ideas of Roger Bacon were taken from his master. The main optical source of Grosseteste was a Latin translation of the mathematical work of Alhazen (pp. 152 f. and 162). The great Bishop of Lincoln was an advocate of the study of Greek and Hebrew and an important forerunner of the Revival of Learning.

An influential writer was the Pole WITELO (fl. 1270), an acute mathematical investigator who worked in northern Italy and wrote a commentary on Alhazen (pp. 152 f. and 163). The Franciscan Roger Bacon was largely dependent on Witelo for his optical views. Another optical writer dependent on Alhazen was the English Franciscan JOHN OF PECKHAM (*c.* 1220–92), Archbishop of Canterbury. His works exhibit some mathematical skill, and one of them was printed as late as the seventeenth century after the appearance of the writings of Kepler and Galileo.

A much discussed figure in medieval scientific thought is ROGER BACON (1214–94). He was a Franciscan who taught at Paris and Oxford. He was essentially an encyclopaedist who realized better than most the urgent need for the enlargement of learning, especially in connexion with accurate knowledge of languages and the collection and collation of scientific data. He made an appeal, verbose, diffuse, yet definite, for the encouragement of the experimental spirit. He was not himself an experimenter or mathematician, but like Grosseteste he saw that without experimentation and without mathematics, natural philosophy is but verbiage.

Perhaps Bacon's greatest claim on our attention is that he recognized the usefulness of natural knowledge, forecasting man's control of nature set forth more clearly, three and a half centuries later, by his great namesake Francis (p. 264). Vaguely, too, he foresaw many modern scientific ventures; flying, the use of explosives, circumnavigation of the globe, mechanical propulsion, &c. A single anticipation of this kind would hardly deserve mention, but the convergence of so many in one head is impressive. Specially noteworthy—not so much for their originality as for their clarity—are Bacon's excursions into optics. He understood the nature of refraction and grasped its implications for curved surfaces. He thus attained to an approximately accurate view of the path of the rays in a burning-glass and he had more than an inkling of the mode of action of convex lenses (Fig. 69). He is the first to mention the use of lenses for spectacles and, perhaps, from hinting at the combination of lenses, can be regarded as the progenitor of optical apparatus.

Despite all this Bacon must not be considered as a man born out of his time. On the contrary, he was in many ways very typical of the scholastic movement and an important link in the chain of scholastic scientific development. He regarded the advancement of science as important for the support of religion. That he was in trouble with his superiors there can be no doubt, but to suggest that these differences were caused by his scientific views is not only to go beyond the facts but beyond probability.

During the century after Bacon, though his other works were still at times studied in the schools, it happened that, for a variety of reasons, mathematics and optics, in which sciences he was chiefly interested, made little progress. In this interval the chief advances were made by medical men of whom the last half of the thirteenth and the first half of the fourteenth century exhibit a specially able group. Bologna and Montpellier universities were the centres at which this progress was made.

At Bologna surgery may be said to have been born again with

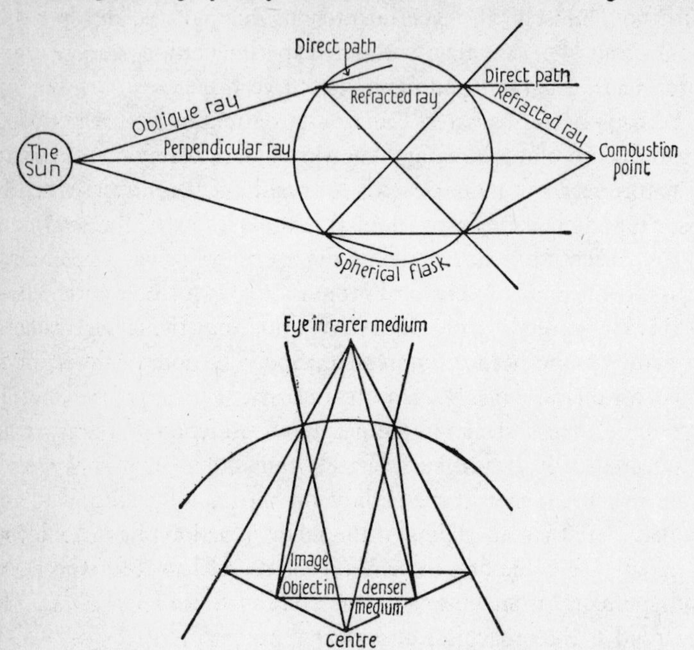

FIG. 69. Roger Bacon's diagrams of paths of light-rays through a glass sphere or flask filled with water and through a plane convex lens.

ROGER OF SALERNO (*c.* 1220) and his successor and faithful follower ROLAND OF PARMA (*c.* 1250), who link the new 'Arabic' medical movement with the old that had survived in southern Italy (pp. 157 f). Bologna at the beginning of the fourteenth century saw established a regular tradition of anatomization. This was expounded by MONDINO DA LUZZI (1276–1328), whose work was based on Latin translations of Arabic versions of the texts of Galen. The *Anatomy* of Mondino (1316) became the general textbook of the subject in Italy and Germany. During the fourteenth century the practice of dissection of the human body became recognized in several universities.

At the end of the thirteenth century the ancient foundation of the medical school of Montpellier was coming to the fore. The Catalan

ARNALD OF VILLANOVA (*c*. 1240–1311), one of the more remarkable personalities of medieval medicine, taught there. Arnald was not only the earliest modern exponent of the Hippocratic method of observing and carefully recording actual cases of disease, but he also influenced alchemy. That study was effectively of Arabic origin so far as the Western world is concerned (pp. 144 f.). It begins in 1144 with the translation into Latin by Robert of Chester (p. 162) of an alchemical work ostensibly by MORIENUS ROMANUS, supposedly a contemporary Christian of Jerusalem who derived it from an earlier Arabic source. Like other medieval studies, alchemy became linked with astrology. Thus the 'seven metals' were each controlled or influenced by one of the 'seven planets' much in the same way as were the organs of the human body (p. 146).

Of such ideas, Arnald was a prolific exponent. He had direct access to both Arabic and Hebrew and had personal relations with both Moslems and Jews. A student at Naples and Salerno, a traveller in Italy, Sicily, France, and Spain, he served as medical adviser to the Papal Curia both at Rome and Avignon, and was employed as ambassador on more than one special mission.

Astronomy—which cannot at this stage be distinguished from astrology—was certainly the main scientific interest of the scholastic age. The practical results of scholastic astronomical activity are, however, pitifully meagre. Western knowledge of astronomy was largely based on the activity of King ALFONSO THE WISE (1223–84) of Castile. He collected at Toledo a considerable body of scholars, mostly Jews, who calculated a set of astronomical tables (1252). These *Alfonsine tables* spread rapidly through Europe. They contain few new ideas, but several numerical data, notably the length of the year, were calculated with very remarkable accuracy. Alfonso is also responsible for a vast encyclopaedia of astronomical knowledge compiled by a similar group from Arabic sources.

The standard astronomical textbook of the scholastic period was by the Yorkshireman JOHN HOLYWOOD (Sacrobosco, died 1250) who was long a teacher at Paris. The work was universally popular, exists in numerous manuscripts, and was translated into most European vernaculars. It contains, however, no new or original element and is put together from translations of Arabic works. Holywood wrote also a book on arithmetic, or rather 'algorism' (p. 162). It was extremely popular and did more to introduce the Arabic notation than any other. Both the astronomy and arithmetic were very frequently printed.

One of the best medieval astronomical works is by the Provençal Jew, LEVI BEN GERSON (1288–1344). He did excellent mathematical work, perfected the camera obscura of Alhazen (p. 152), correcting the explanations of Witelo and Roger Bacon (p. 172), distinguishing between the clear images given by a pinhole and the blurred image

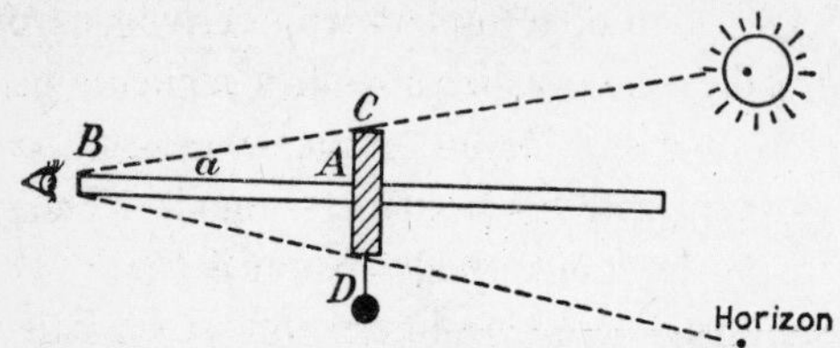

FIG. 70. Diagram of Levi ben Gerson's 'cross staff'. A graduated rod, *AB*, has a movable crosspiece *CD*, held vertical by keeping in line with a suspended plumb. To measure an altitude, e.g. of the sun, shift the crosspiece till *BC* is in line with the sun and *BD* with the horizon. The angle $2a$ = altitude, is given by graduations on the rod.

with a larger aperture. Levi's *Wars of the Lord* (1320, revised 1340) is a general philosophy, second in importance only to Maimonides among medieval Jewish writers. It includes astronomy and criticizes Ptolemy but, being in Hebrew, hardly influenced the Latins. Since Levi opposes those who hold that the earth moves round the sun, this view must have been held by some in his time. He invented the 'cross staff' (Fig. 70) and his account was turned into Latin for Pope Clement VI (1342–52) as *Revealer of Secrets* (Job 12. 22). The instrument is commonly called 'Jacob's staff', an erroneous ascription to another Jewish inventor.

After medicine, alchemy, and astronomy, the practical sciences in which the West exhibited activity in the Middle Ages were botany and optics. Botany was always studied in connexion with medicine. No advance was made in the use of drugs save what was borrowed from the Arabs. There is, however, some indication of a revived interest in nature in the graphic representation of plants. Numerous optical texts also show some advance in ideas. Nevertheless, none of the Latin texts is equal in value to the great work of Alhazen (p. 152) that itself became available in Latin about the beginning of the thirteenth century (pp. 152 and 163).

In pure mathematics the original achievement of the scholastic age was small. There was, however, a borrowed element that was to prove of high significance. At the end of the twelfth century a merchant LEONARDO OF PISA (*c.* 1170–*c.* 1245) resided for commercial purposes in North Africa. There he learnt of that use of Indian numerals in which the value of a digit depends on its place in a series. It is the ordinary method of numeration that we now employ. In 1202 Leonardo produced his famous *Book of the Abacus*, in which he advocates this system with great skill. It is the first book by a Latin Christian that popularized this system[1] and is the essential source of our modern system, which, however, was extremely slow of general adoption. Other works of Leonardo were much more original, but being before their time had less influence. He was undoubtedly a mathematician of extraordinary ability, but his positive contributions are as nothing compared to his importance as the carrier of the new method. A much more popular work than Leonardo's that employs the 'Arabic' numerals was the *Algorismus* of John Holywood (pp. 173 f.). It appeared about 1240 and was still being reprinted in the seventeenth century. It is one of the puzzles of history that this great improvement, represented by the 'Arabic' as against the Latin system, took three centuries to gain general acceptance. The scholastic age was over before the modern system came into general use.

8. STATUS OF MEDIEVAL SCIENCE

Characteristic of the medieval scene is the technical and scientific supremacy of the Orient, primarily of Islam and, to a less degree, of Byzantium. Technical ideas came to western Europe also from the Far East. Byzantium or East Rome, often used as a byword for static formalism, nevertheless survived her Western sister for a thousand years, not yielding her last stronghold till 1453. In this millennium her technology always equalled and usually far excelled Western, but for science the Byzantine contribution hardly extended beyond preservation of ancient Greek texts. We can therefore disregard it except in this sense.

Perhaps this is the place to mention the common misunderstanding that in the Middle Ages men believed that the earth was flat.

[1] The system was used in the West a century earlier for calendarial purposes.

True this was held by a sixth-century Byzantine eccentric, whose entertaining figures are often reproduced. Nevertheless, his opinion can be discounted. For the Middle Ages—as for our own—it would be hard to say what illiterates thought on the matter—if they thought at all—but educated medieval men assumed the earth to be a sphere. Indeed it almost became part of Western Christian doctrine, borrowed from Aristotle, that the heavens are a perfect series of perfect spheres with a spherical earth as centre.

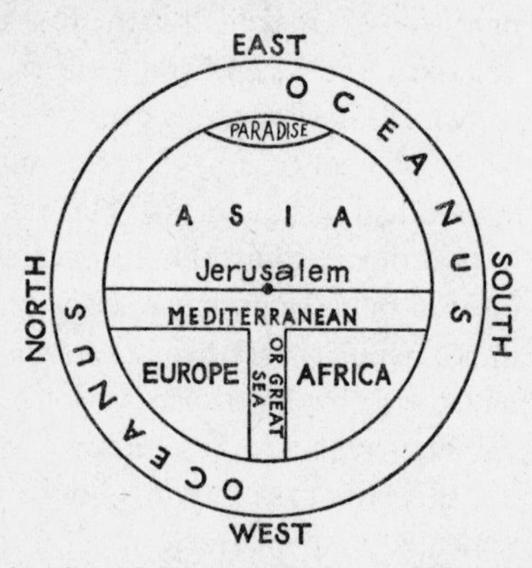

FIG. 71. Conventional medieval O.T. map as used by St. Isidore of Seville in the seventh century and by many writers till the seventeenth century.

Some confusion as to the medieval view has arisen from the medieval 'world-maps' (*mappae mundi*). These were not maps in our sense but plans with edificatory aim. They portrayed the *habitable* surface of the earth as a circle with Jerusalem as the centre to which all looked (Fig. 71). This idea provided one stimulus for the Crusades. Typical of the age was the conception of Dante (1265–1321). He took much interest in the structure of the world. It is easy to reconstruct his mental image in detail (Fig. 72).

Many have sought to raise the low estimate of the scientific status of the Middle Ages, and thus soften the abruptness with which the revival of science follows the completion in the recovery of the classics. They have succeeded only for technology. The Middle Ages improved many techniques of antiquity, borrowed others from the Near East and not a few from the Far East, invented many and transmitted a multitude to the following age. But for the disinterested and systematic study and record of phenomena, that is for science proper, the low estimate stands. It is true that there was much able discus-

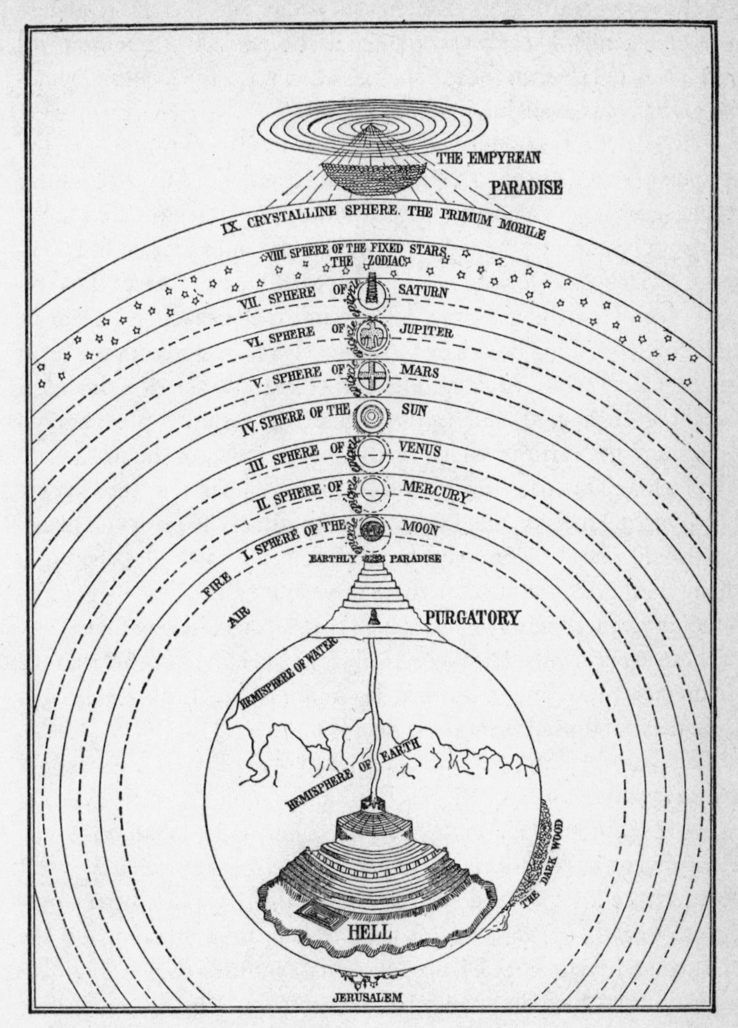

FIG. 72. Dante's view of the world-system as imagined by him about 1300.

sion by scholastics of how nature could best be studied and a few of them made remarkable forecasts of the results of such study. But neither discussions, nor prophecies, nor devices, nor all together, make science. What we fail to find before the fifteenth century is a man who devoted his life to the observation of nature, to the recording of his observations, to disinterested deductions therefrom, and/or to experiments. These were not scholastic ways. But science owes much to scholasticism for evolving a better way of presenting knowledge than was available in antiquity.

The passage from scholasticism to science is illustrated by the Rhinelander, Cardinal NICOLAS OF CUES (Cusa, 1401–64). The form of his writing was scholastic as was most of his thought. Like some of his predecessors, he sought for Greek texts and had a slight knowledge of Hebrew. He wrote a work (1450) which advocates the constant use of weighing by the balance in experiments, and suggests the timing of falling bodies by the water-clock, thus preceding Stevin (p. 223) and Galileo (p. 230). He further suggests the experiment of weighing earth, seed, and, in due course, the resulting plant and its ashes, and the earth in which it has grown, much in the manner of Van Helmont in 1644 (p. 269) and Stephen Hales in 1727 (pp. 143 f.). He thought, however, that the access in weight of the plant and plant-ash would reveal the amount of the 'element' earth absorbed into them without any idea of increase of weight from the air. It is disappointing that there is no evidence that he actually conducted any of these experiments. His theoretical views led him to a belief that the earth may move and the world be infinite though he reached no formal astronomical theory. He deeply influenced Bruno (pp. 218 f.).

If we seek the first real man of science perhaps the choice should fall on a contemporary of Nicolas, the Italian architect, LEONE BATTISTA ALBERTI (1404–72). He discovered experimentally, about 1450, the elementary mathematical laws of perspective without which the great Renaissance artists could not have done their work. Alberti's perspective methods were truly scientific and are still valid. On the science and technology of the following centuries they had an immense effect which deserves further discussion. Alberti's whole outlook makes a clear break with scholasticism.

9. MEDIEVAL SCIENCE AND THE USEFUL ARTS

Besides perspective certain other useful arts, developed or introduced during the Middle Ages, became so closely linked with science that they demand special discussion. Notable were (*a*) the Compass and Cartography, (*b*) Lens-making, (*c*) Printing, (*d*) Time-keeping, (*e*) Chemical and Alchemical processes and apparatus, (*f*) Fire-arms and Gunpowder.

(*a*) *Compass, &c.* The attraction of the lodestone for iron was known from antiquity. By the twelfth century its directive property

FIG. 73. Lines connecting windrose of 16 points with as many others similarly connected.

and the mode of its conveyance to a steel wire had become familiar in the West. Such a needle, floated in water or pivoted on a card, released the mariner from constant dependence on the stars for direction. The practice of indicating the cardinal points above or below the needle and of enclosing it in a box were introduced. The points were increased to 8, to 16, to 32 to form the 'windrose' (Fig. 73). The earliest work on the compass (1269) was by the Picard, PETER PEREGRINUS, a friend of Roger Bacon.

Windroses radiating from various points were inscribed on early sailing charts. The resulting criss-crossings are confusing to us (Fig. 74), but with a compass the mariner who knew his position could set his course along one of the lines. Such charts were probably made in the twelfth century but the earliest known is of about 1275.

(*b*) *Lenses* were unknown in antiquity but there are ancient references to the magnifying powers of glass globes. These have been recovered from medieval sites and were probably used as burning glasses for making fire.

Something was known of the properties of convex lenses from a

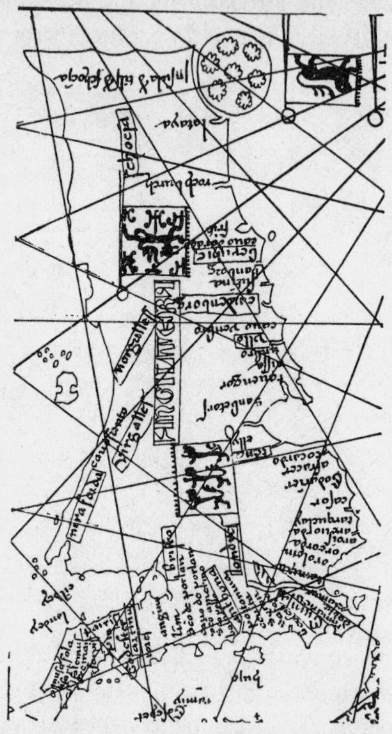

FIG. 74. Coasts of Britain, being a detail from a Catalan sailing-chart of 1375. The criss-crossing of the windrose lines can be seen. *Angletera* (England), *Schocja* (Scotia), *London*, and many other names may be read.

Latin version of Alhazen (*c.* 1170, pp. 170 and 152). Grosseteste (died 1253) developed from them a theory of the formation of the rainbow. Roger Bacon (died 1294) mentions their use for aiding the sight of old people (p. 171). Convex spectacles were invented probably before 1280 and perhaps in France. By 1300 they were being made at Venice, the centre of the glass industry. Concave lenses were not made until the mid-fifteenth century. By 1500 spectacles, both convex and concave, were common.

Making lenses was simple but laborious. Having secured a suitable disk of glass it was cemented to a handle so that it could be rotated on a metal mould, using a progressively finer abrasive. Lenses early affected science in two ways. The power of accommodation of the eye for near distance normally begins to fail in middle life. Reading then becomes more and more difficult. Thus spectacles effectively doubled reading life, no small contribution to the Revival of Learning. Secondly, lenses aroused curiosity concerning refraction. Thus optics became the first science to be experimentally developed.

(*c*) *Printing* is basically very ancient. What else is a seal and especially a rotating cylinder-seal? The taking of an inked impression from a stamp was a medieval practice, early adapted to decorative initial letters in manuscripts. Thence it is a short step to the cutting of pictures and short inscriptions and so to playing-cards and block-books. These were made at least as early as 1300. But the device that produced an intellectual revolution was the use of movable types of cast metal. For these parchment is less adapted than paper. Thus though the seminal idea of printing-types may have come to the West early from the East—as many think—its wide application in the West awaited the development of the paper trade.

Paper is a felt of vegetable fibres. It was known in China in the early Christian centuries and came West across Central Asia by the old Silk Route. It became known in Islam but did not reach Latin Europe till the thirteenth century; probably via Spain. By the mid-fifteenth century paper-making was common in Europe. The most suitable paper for printing was made, as it still is, from linen rags. Machines were devised for disintegrating these. Watermarks in paper may be as early as the end of the thirteenth century.

The date and place of invention and/or introduction of movable types in Europe is disputed, as is the question of whether the idea came from the Far East. It is, however, fairly certain that its effective application in Europe began in 1447 at Mainz with JOHANN GUTENBERG (*c.* 1400–*c.* 1466). The form of the early printing-press was identical with that of the linen-press long in use (Fig. 75).

Printing spread rapidly. The multiplication of books brought an immediate change in human outlook but not, be it noted, in scientific outlook. Effective interest in science before 1500 was very rare—Alberti and Leonardo da Vinci were lonely figures. Printing made little impact on science till the Greek scientific classics had been not only printed, but also translated and studied. This was not until the end of the first third of the sixteenth century. It is for this reason that

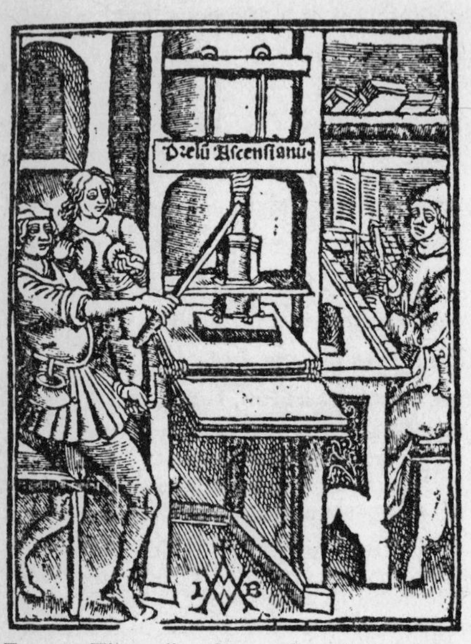

FIG. 75. The earliest figure of a printing press, that of Iodocus Badius Ascencius, Paris 1507.

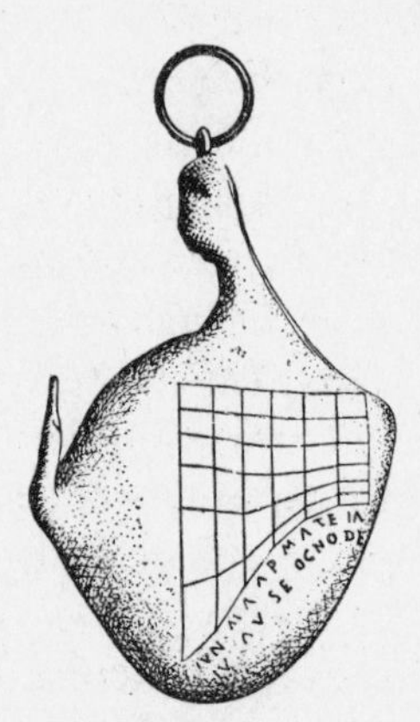

FIG. 76. Roman 'ham' dial with names of months.

the 'Revival of Science' is a laggard as against the 'Revival of Learning'.

(*d*) *Time-keeping*. In antiquity it was possible to keep time with the shadow-clock of which many varieties were devised (Figs. 4, 60, 76). There were also, from an early date, simple and ingenious forms of water-clock or *clepsydra* (Greek = water-thief). The need remained for a timekeeper that was portable and reliable.

The simplest portable time-measurer, other than the mere shadow-rod or *gnomon* and its elaboration the sundial (Fig. 60) was the 'ham dial' known in Roman times (Fig. 76). In this the gnomon projects from a suspended plate which is turned till the tip of the shadow falls on the scale for the appropriate month. The daylight was divided into 'tides'—as the Anglo-Saxons called the eight diurnal lapses of three hours. A beautiful development of the same device was made in England about the year 900 (Fig. 77). In this the gnomon is a detachable pin placed in a socket above the appropriate month-column. There are many variants and refinements on these devices. The astrolabe (pp. 149 f.) could also be used for telling the time and often had an alidade and scale for that purpose on the back.

FIG. 77. Anglo-Saxon dial from Canterbury.

A greater impact on science was made by mechanical time-pieces. Clocks were first devised in the thirteenth century. Their essential is the control of the force of a falling weight (and later of a spring or of a pendulum) so that it should act uniformly on an indicator. The earliest way was to interrupt the fall of a weight periodically so that the indicator should be moved in regular steps. This mechanism is the *escapement* of which several forms were devised (Fig. 78). Clocks had become effective and common by about 1500. The technical development of the clock and watch is susceptible of indefinite elaboration and adaptation to any regularly recurring natural event, as moon-phases, cycle of the seasons, length of day or of year, and to any required accuracy. The basic principle is simple but no department of technology has

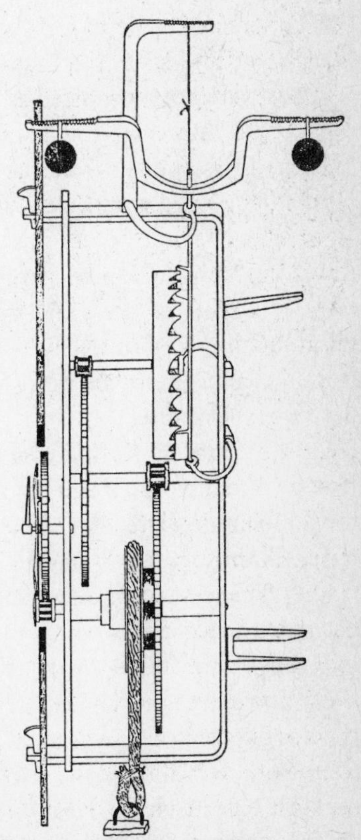

FIG. 78. Works of fifteenth-century mechanical clock.

developed independently such refinement and such complexity as horology.

(*e*) *Chemical and Alchemical Processes and Apparatus.* There was much medieval industrial production from native materials of such substances as metals, alkalis, soap, acids, pigments, mordants, alcohol, gunpowder, &c. The methods were mostly derived from those of both West and East Rome. Many medieval techniques were improvements on these but science had little or no hand in this.

Alkalis were available in some variety. Potashes were in wide use. They normally contain a high proportion of potassium carbonate which was purified by lixiviation and used for making glass and also for soap or other means of cleansing. Soap was probably invented by the Tartars and came to Europe in the eighth century. Both potash (hard) soap and soda (soft) soap were known.

Acids were not recognized as a class before the sixteenth century. Nitric acid, however, was known to the Arabs and thence to the West and was familiar from the fourteenth century. It was obtained by distilling nitre with a dehydrated sulphate at high temperature. It was used for the separation of silver and baser metals, which it dissolved, from gold. The separation could also be effected by *aqua regia*, a mixture of nitric and hydrochloric acid obtained by adding sal ammoniac to nitric acid. It dissolves gold. Nitre was obtained from the efflorescence on heaps of sheep-manure mixed with decaying vegetable refuse.

Of mordants by far the commonest was alum. It was mainly an Eastern product but from 1450 was prepared in Europe from the mineral alunite. This was roasted, weathered, and added to boiling water till the solution crystallized on cooling. The high purity of some medieval alum was not due to chemical knowledge but to its dependence for effectiveness on its whiteness, that is by its freedom from iron salts. Since these are coloured, since crystallization eliminates them, and since alum is easily crystallizable, the method was really a process of chemical purification. Alum was the only chemically pure compound available in the Middle Ages. From 1450 the Vatican sought ineffectively to make alum a papal monopoly, partly as a measure against the advancing Turks who controlled alum imports from the East (Fig. 62).

Alchemy presents a difficulty in a history of science. The word has come to suggest magic, obscurantism, futile symbolism, and fraud. Most of this is just, but the words alchemy and chemistry are from the same root, whose separate meanings were not clarified till the Middle

FIG. 79. Part of an alchemist's laboratory of about 1477 showing his scales in a glass case. In front are stills on furnaces.

Ages had closed. Many alchemical works have scientific elements. Moreover the alchemists contributed certain processes and apparatus. They claimed to use twelve processes: Calcination, Congelation, Fixation, Solution, Digestion, Distillation, Sublimation, Separation, Ceration, Fermentation, Multiplication, and Projection. All these are easily intelligible, except perhaps the last. 'Projection' is presentation of gold in the last stage of the process.

Many instruments and appliances of alchemy passed direct to the modern scientific chemist. Chief of these were the furnace and the still. Of methods for raising temperature by increasing draught, many were taken from industrial processes with adaptations of chimneys or stink-cupboards. Some alchemists were expert in the use of assay-scales (Fig. 79), sometimes enclosed in a glass case as in a modern laboratory. Other apparatus of the alchemist has passed into modern chemistry often carrying Arabic names. Substances with a comparable history have been discussed (p. 161). The alchemists apparatus *par excellence* were his furnace, which differed little from those in industrial use, his stills (Fig. 80), the cooling part of which was often greatly elaborated but also largely used in industry, and his balances (Fig. 79) which resembled those used by assayers.

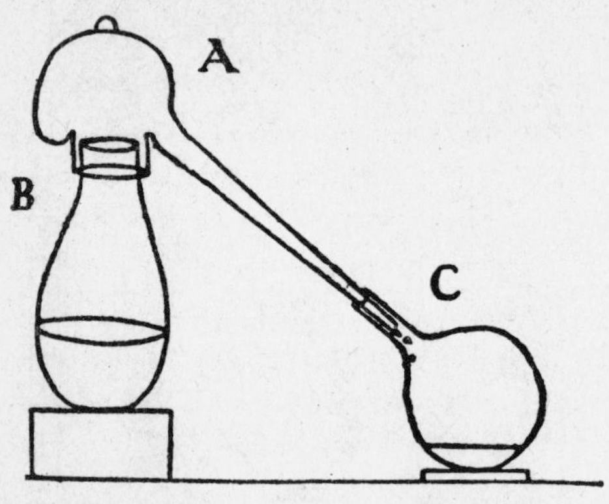

FIG. 80. Traditional form of still. *A* is the alembic (Arabic = the still), *B* the matrass (a word perhaps related to *mater*), *C* receiver. The alembic was elaborated into various forms of condenser.

Alcohol has a special history. The word is from the Arabic *kohl*, represented also in Hebrew (Ezekiel xiii. 40, translated 'paint'). This was a fine powder of antimony trisulphide, used by women to stain the eyelids. It came to signify any fine chemical product but how it was transferred to alcohol is obscure. Alcohol was first a product of distillation of wine (German *Brandtwein*, our *brandy*). The process was made possible by special condensers in the thirteenth century. Thus was produced *aqua ardens*, identified by alchemists with the 'quintessence' (p. 145). Methods were improved and by about 1300 what we would call 'absolute alcohol' was being prepared by Arnald of Villanova (p. 173).

A certain type of alchemy had become indistinguishable from chemistry by about 1600. It is a fair statement that the origin of chemistry is to be sought in attempts to understand industrial processes, with the adoption of the technical methods of the alchemists.

(*f*) *Gunpowder* and *Firearms* were great medieval contributions from an economic point of view, for ultimately they gave Europe ascendancy over the other continents. The history of these devices is obscure. The idea of propelling projectiles by explosives is doubtless of Chinese origin but in its development Byzantium and Islam had a share. Fireworks were known in Europe in the thirteenth century, and from a bamboo or parchment rocket-cylinder to a metal gun-barrel is no great step. Incendiaries were mechanically hurled for military purposes by both Byzantine and Moslem armies before 1300. A Byzantine compendium of about that date describes an explosive mixture of sulphur, nitre, and vegetable oil which would act like gunpowder, the oil providing the carbon. Sulphur was readily obtainable either native or by distilling pyrites, while nitre had long been an item of commerce (p. 184). The use of such a mixture as a propellant is probably a European re-invention of the later thirteenth century. The proportions and preparation were improved and, in the fourteenth century, we hear of the use of cannon. Hand-arms soon followed.

The passage from medieval to modern science is more abrupt in practice than theory. The nature of scientific method was apparent to such writers as Grosseteste and Bacon in the thirteenth century, yet its philosophical foundations were hardly appreciated by Kepler and Harvey in the seventeenth. The key to this discordance is the rapidity in the advance of technology. Not till the seventeenth century did science use technology as its tool. The sixteenth century was the transition period.

We think now of technology as a product of scientific knowledge and the press treats it as science itself. Perhaps the two may now be inseparable. Historically, however, systematic observation and experiment were made possible by technology. The matter comes out best perhaps in the passage from alchemy to chemistry within the limits of the sixteenth century. The first true textbook of chemistry was issued in 1597 by Libavius (1540–1616).

The sixteenth century marks a frontier for another reason. Until about 1500 the East was culturally as active as the West. Many devices came westward from both Near East and Far East. By 1500 the process had been reversed and the history of science can be told without reference to the East.

VI

REVIVAL OF LEARNING

Rise of Humanism: Attempted Return to Antiquity (1250–1600)

I. HUMANISM

THE advent of Catholic philosophy is one of the most impressive events in the history of Western thought. This great effort to rationalize Christianity is closely linked with the recovery of the Aristotelian texts.

Until the thirteenth century the only works of Aristotle available were those on logic. These had been turned into Latin from Greek by Boethius in later antiquity (p. 138). Early in the thirteenth century versions from the Arabic associated with the commentaries of the Moslem philosopher Averroës (pp. 155 f.) began to circulate. The centre of the intellectual world at that period was the University of Paris. There the reading of these Averroan interpretations of Aristotle met with ecclesiastical opposition. This was, however, lifted by the middle of the thirteenth century, perhaps because versions less coloured by the Averroan outlook had become available. The architect of Catholic philosophy, the Dominican St. Thomas Aquinas (p. 168), was able to work at Cologne and Paris, largely on versions of Aristotle prepared directly from the Greek by William of Moerbeke and others. These thirteenth century Latin versions both from Arabic and Greek remained in use after the end of the fifteenth century.

This summary of the knowledge of Aristotle in the thirteenth century requires some expansion. Before the days of printing, and indeed for long after, a work seldom replaced completely another that was actually in circulation. Manuscripts were too expensive to jettison. Libraries were small, scholars conservative and uncritical, catalogues inadequate. The better or newer versions did not commonly drive out the worse or older. The two generally continued in use, some at one centre, some at another, and often both at the same seat of learning.

So far as science is concerned the versions of Aristotle, and the Aristotelian commentaries and interpretations in most common use, continued to be those from the Arabic and *not* those from the Greek. The hold of the Arabic-Latin versions began to be shaken by the rise of humanism. But 'Arabist' versions retained their supremacy through the fifteenth and sixteenth centuries. Even in the seventeenth they were still in use in universities where the old Aristotelian philosophy flourished. The true and ultimate Arabist defeat was not the work of the Greek scholars who ranged themselves under the banner of 'Humanism'. It was rather the men of science, adherents of the new 'Experimental Way', who swept away the whole medieval philosophic approach—whether based on Greek or Arabic or Aristotle or Averroës. Their triumph was not fully apparent till the eighteenth century. There are backward centres where it is not complete even now.

In the nineteenth century scholarship itself was transformed by the experimental method. Adepts in that method came at last to prepare modern critical versions of the Aristotelian corpus. Then, and in the fullness of days, the scientific powers of the great teacher came to be properly appreciated. The beauty and symmetry of his mind appeared as never before. They will not again escape the historian of science, a study of which Aristotle is himself the progenitor.

We turn now to consider the small beginnings of a true appreciation of ancient science. The process is wrapped up with the advent of translations of scientific works prepared from the Greek. One of the first to appreciate these was the heretical PETER OF ABANO (1250–1318). He had a knowledge of Greek, acquired at Constantinople, and he translated works from that language. He professed medicine at Paris and later at Padua in the generation after that in which the newly won Aristotelian works on physics had entered the curriculum. He earned a reputation as a magician.

The best-known work of Peter, the *Conciliator*, expresses his mediation between the new Greek and the old Arabist school. It shows traces, too, of wider contacts, for from it we learn that he had met the great traveller Marco Polo (*c.* 1254–1324). He was much less conservative than most medieval writers on scientific themes. Among

Peter's views most worth record may be mentioned his statements that the air has weight, that the brain is the source of the nerves, and the heart the source of the blood-vessels—novel ideas in his time. He made a remarkably accurate measure of the length of the year as 365 days, 6 hours, 4 minutes.

With the fourteenth century appeared a great movement the hand of which is still heavy on our own day. The ancient classics began to be recovered and Greek began to be studied. Historians have linked the 'humanistic' movement too intimately with a knowledge of the Greek language. Instances of familiarity with that language in the West can be adduced far back into the Dark Age (e.g. John Scot Erigena, *c.* 840), while many of the greatest of the humanists, including PETRARCH himself (1304–74), were without any facility in Greek.

The backward-looking habit, strong in man from his nature and strengthened by Christian teaching, was yet further enforced by the humanists. From Petrarch onward the humanist was brooding on the past that had been Greece and Rome. Seeking to penetrate the dark shadows of what was now recognized as a 'Middle Age', the humanist tried hard to discern the antiquity that was beyond. And as he strained his eyes another vision, a reflection perhaps of himself, came sometimes to him. In the cloud-land of the past he caught or thought he caught a glimpse of what was to come—nay of what was in the act of becoming. And then again the vision would be clouded over by that terrible erudition which, in the absence of general ideas, has been and is one of the enemies of science.

In the thirteenth century Grosseteste, Bacon (pp. 170 f.), and a few isolated souls had something of this twofold vision, but for a whole school to possess it was something new. In his *Book of Memorable Things*, Petrarch says outright, 'Here stand I as though on a frontier between two peoples, looking both to the past and to the future.' While studying the classics some of these men were also forging new intellectual weapons by developing those national vernaculars that have made possible modern literature, modern philosophy, and modern science. It is no mere coincidence that Boccaccio (1313–75), friend and contemporary of Petrarch, should have been at once the first modern literary man to study Greek and the first great master of Italian prose.

Italy was the birthplace and nursery of humanism. Save for reference to the supreme poet in their own tongue, DANTE (1265–1321), the backward gaze of the Italian humanist is always fixed on the more distant classical past, not on the nearer period that came to be regarded as an abyss across which he sought to reach back to antiquity. To the humanist the abyss seemed real enough and dark enough. It stood for the period during which the 'sweet Greek literature' had been forgotten. But even in this new age it could be understood by few, except in Latin dress, and the work of translation and interpretation remained a specialist's occupation. An effective knowledge of Greek continued to be rare even among the learned. Some of the most important philosophical teachers of the sixteenth century were still quite without it.

The great influence of the masterpieces of Greece, therefore, was then as now something indirect, often conveyed through translators and special interpreters; something esoteric, the full intricacy of which was shared only by a few adepts; a subtle thing that influenced men's way of thinking rather than the actual content of their thought. The mere capacity for translation from the Greek goes back very far. It was not simply the discovery of the actual Greek language which brought about the revival of letters. How, then, can we account for the change of heart that came over the world when humanism was born? Or is that change of heart but an illusion, a difference of degree rather than of kind, in a world where everything is in a state of becoming?

Some answer to this absorbing question we may glean by comparing the earlier Greek works which came to the West with those of later advent. The general character of the earlier translations was determined by the outlook of a world becoming ever more deeply Arabicized. Islam, the inheritor of antiquity, entered into the enjoyment of its legacy with great spirit, but with a taste already fixed. The ancient literary and artistic works were debarred to the Moslem scholar. Homer and Hesiod, Sophocles and Euripides, Greek Art and Greek Architecture were chapters as closed and forbidden to Islam as to early Christian Europe. It was the philosophical and the scientific, mathematical, and medical works that appealed. The bulk and number of these gave sufficient material for thought and created an illusory impression of completeness with which Islam long rested content.

These very works, to which the world of Islam clung, were naturally the first to be rendered into Latin from the Arabic. The Latin taste being thus determined, such knowledge of Greek as was possessed in the Middle Ages made little change. It was works similar to those already accessible from the Arabic that were the first to be turned into Latin direct from Greek, for, so far as science and philosophy were concerned, Byzantine literary taste differed little from Arabic. The texts were merely improved by direct access to the tongue in which they had been written, but they were still the same philosophical, medical, mathematical texts.

Such material—and it is bulky and intricate enough—represents the Western access to Greek wisdom before the fifteenth century. It does not lack quantity—it lacks life. They err who think the discovery of the humanists was the Greek language—here the humanists were but followers where others had been pioneers. It is something much deeper and more fundamental which they have handed on, something the nature of which they hardly understood and the meaning of which they often missed—and perhaps still often miss.

The humanists discovered the literary works of antiquity. In them they became absorbed to the exclusion of all else. Their eagerness passed into a literary vogue, and cast the blight of a purely literary education on the modern world. The barren striving after form as distinct from substance, the miserable imitativeness that is an insult to its model, these features, exhibited typically in the literature of the late Empire, were repeated by the humanists as they have been often repeated in modern times. They still remain a curse to our educational system. The importance of the humanist is neither that he gave us the knowledge of a language, nor that he gave us an insight into the life of antiquity. What the humanist really gave was a something which, added to the heritage already there, made possible a completer reconstruction of the Greek spirit. That reconstruction, indeed, he was himself never able to make. It was the succeeding generations that made it for themselves. With that reconstruction Greece lived again, the modern world was ushered in, and modern science, art, literature, and philosophy were born. It is an illuminating reflection, not without bearing on our present state, that both the medieval heritage of Greek science and the Renaissance heritage of Greek literature proved barren by themselves. It was not until the one fertilized the other that there was vital growth.

Modern thought, modern science, modern art, modern letters are offspring of that union. Forget the timeworn fallacy that they are the virgin births of one of these elements alone. Men accomplished alike in the arts and in the sciences, Alberti, Leonardo, Dürer, Vesalius, Galileo, were more truly the heirs of Plato and Aristotle than were the men who spent their lives in editing the works of these giants of old. It is literature, art, and science, not classical scholarship, that enjoys the legacy of ancient wisdom. Plato and Aristotle could have talked to these and to our own scientists as man to man, but hardly to our textual critics.

2. PRINTING OF THE ANCIENT SCIENTIFIC CLASSICS

There are certain aspects of early printing in connexion with science to which attention must be directed.

(*a*) We now use the printed page to express our views on current matters. In science we mark a discovery by its first publication. Readers demand of a new book that it should contain something new. But in the early days of printing the press was not thus employed. The Bible and other sacred writings were the first to be printed, and then works of theological authority. Next medieval treatises on law, and especially ecclesiastical law, largely occupied the press, and were followed by medieval medical texts. The writings of classical antiquity mostly came later. Of early printed books hardly any are by contemporaries.

(*b*) In the process of recovery of the classical originals the attention of scholars was first directed to works of literary merit. Scientific treatises appealed to few. Thus the revival of classical science came later than that of other sections of classical literature.

(*c*) The great influence of the Revival of Learning on the subsequent history of thought and of education has distorted our view of fifteenth- and sixteenth-century cultural interests. Those who had real facility in Greek were extremely few. Right through the sixteenth and even into the seventeenth century, the overwhelming mass of published philosophical and scientific literature was still of the medieval type.

(*d*) The publication of Greek scientific writings had little influence until such works began to appear in Latin translations. The humanists seldom had adequate scientific equipment to read these documents at all and the men of science seldom had adequate linguistic equipment to read the originals in Greek.

Bearing these matters in mind, it is interesting to follow the chronological course of the appearance in print of the classical scientific works of antiquity. Ptolemy's *Geography* perhaps excepted (p. 95), the earliest scientific classics to be printed were those of the Latins. The first was the *Natural History* of Pliny, which appeared at Venice as early as 1469. But Pliny was in no sense 'recovered'. He had never ceased to be read throughout the Middle Ages (p. 139). The work was in fact so familiar that the Venetian printer did not think it worth while to attach the name of an editor to it. Throughout the sixteenth century Pliny was as popular as during the Middle Ages and was very frequently reprinted. In 1601 his *Natural History* was translated by Philemon Holland into English and was the second work of ancient science to appear in that language, the first being Euclid (1570, pp. 83 f.).

Following on Pliny were editions of Varro (p. 107, Rome, 1471), of a collection of agricultural writers (Venice, 1472), and of the poem of Manilius (Nuremberg, 1472). These were all of practical application. Manilius is interesting as the earliest classical scientific treatise to appear outside Italy. It was printed at the private press of Regiomontanus (p. 196). The interest in it is explained by its astrological content, for astrology had become part of the University curriculum. Lucretius followed in 1473 (Brescia). But Lucretius (p. 104), is not properly speaking a scientific writer. Celsus (p. 116), again of immediate practical value, followed some years later (Florence, 1478). The medical work of Celsus was thus the first technical scientific classical work to appear. It had been unknown in the Middle Ages and was a real discovery. The architectural writers, Vitruvius, Frontinus, and Vegetius (Rome, 1486–7)—again practical works—complete the short list of early printed ancient Latin science.

The Greek writings that deal with the true abstract sciences are both more numerous and have a more complex history. We may first note how backward was the treatment of the Aristotelian scientific corpus. For the most part the Renaissance reader was content with medieval Latin versions mainly from Arabic (pp. 158 f. and p. 188). The first 'modern' translation and first important scientific book to be printed was the Latin version by Theodore Gaza (1400–78) of the three great Aristotelian biological treatises (Venice, 1476).

Actual Greek type was hardly used before 1476, and it was near the end of the fifteenth century before the scholar-printer Aldo Manuzio (1449–1515) produced an adequate edition of the Greek text of Aristotle and Theophrastus (Venice, 1495–98). He added to his services by issuing the Greek text of Dioscorides 1499, (p. 98), of Pollux, a classical writer who influenced Renaissance anatomical nomenclature (1502), and of Strabo (1516, pp. 110 f.). Aldo's successors in the 'Aldine' firm were responsible for the first Greek editions of Galen (1525) and Hippocrates (1526).

Very influential for the whole course of Renaissance science were the editions of Euclid. He first appeared in Latin dress at Venice in 1482. Editions continued to flow from the press throughout the sixteenth century. The first edition in Greek appeared at Basel in 1533 and the first in English in London in 1570.

A work that had a large share in fixing the geographical ideas of the Renaissance was the *Geographia* of Ptolemy, which first appeared in Latin at Vicenza in 1475 and in Greek at Basel in 1533. The maps illustrating early editions are most interesting (p. 97). No less influential was the *Almagest* of Ptolemy, which was first printed in Latin at Basel in 1538 and very frequently at later dates. The works of Ptolemy in Renaissance versions are common in comparison to those of most Greek mathematicians and astronomers. Thus a collection of Archimedes was not made until 1544 (Basel) and was not reprinted till the seventeenth century.

Of medical works those of Hippocrates, Dioscorides, Galen and others appeared in scores of editions in Greek, Latin, and the vernacular throughout the sixteenth century. In conjunction with the Arabic medical writers Rhazes, Mesue, Avicenna, and Albucasis they came to provide the basis of the actual medical practice of the age.

3. SCIENTIFIC ATMOSPHERE OF THE EARLY RENAISSANCE

The humanists as a class exhibited little sympathy with the scientific outlook. Their interests were literary; their peculiar aversion was the Arabist tendency of the age that they were leaving behind. An exceptional position among the humanists was occupied by Cardinal JOHANNES BESSARION (1389–1472), a resident in Italy though a Greek by birth, who was equally anxious to aid the progress of astronomical knowledge and to diffuse Greek literature in the West. Bessarion's friendship, extended to two German students in Italy, Purbach and Regiomontanus, made possible their work which formed the foundation of that of Copernicus.

GEORG PURBACH (1423–61) of Vienna followed with great avidity the study of Ptolemy's *Almagest* (pp. 89 f.). He died prematurely and had only translations from the Arabic on which to work. He improved on his original, however, by calculating a table for every ten minutes, using sines instead of chords. His pupil Johannes Müller (1436–76) of Königsberg, known from his birthplace in Bavaria as REGIOMONTANUS, had the good fortune to have Greek originals on which he worked for years in Italy. Finally he moved to Nuremberg, then the centre of the trade in fine instruments. There he made the first scientific observations on a comet in 1572. Next year he published Purbach's work on planetary theory, said to be the first to indicate clearly the astronomical discrepancy between Aristotle and Ptolemy. He also produced the first systematic treatise on trigonometry[1] and a table of sines for every minute and of tangents for every degree. His astronomical tables were used by Columbus. Regiomontanus died at Rome, whither he had been summoned by the Pope to aid in the long-contemplated reform of the Calendar. This, in the event, was deferred for a century.

The Renaissance of Letters was contemporary with the Renaissance of Art, which had its reaction upon scientific thought. The great painters had begun to study nature more closely. Antonio Pollaiuolo (1428–98) and Andrea del Verrocchio (1435–99), among others, made careful investigations of surface anatomy, while the

[1] It was not printed till 1533, that is, 57 years after its author's death.

exquisite figures of plants in the pictures of Sandro Botticelli (1444–1510) mark him out as a very accurate observer. There was, however, one artist of the time who takes a quite peculiar place among students of nature. LEONARDO DA VINCI (1452–1519) stands for many as the turning-point of the Renaissance into modern times. The ingenuity of his ideas, the marvellous rapidity of his insight, the sureness of his intuitions, the exactness of his observations, the extreme versatility of his extraordinary genius, made earlier students place him in an isolated and almost superhuman position. His very limitations increase the apparent gulf which separates him from other men, and hamper us in our comprehension of him. To understand his scientific work and its fate we must recognize his defects.

Leonardo's great limitations were literary and linguistic. He obtained only late in life an adequate knowledge of Latin, and he exhibited no power of literary expression. The vernacular that he employs is that of a middle-class Florentine which seldom rises to lofty dignity. He created no great phrase or saying. His sentences are often ungrammatical and frequently unfinished. In a literary sense he was incoherent. The very rush of his ideas obstructed the channels for their expression. He might have said, with Petrarch,

> E l'amor di saper che m'ha sì acceso
> Che l'opera è retardata dal mio desio.
>
> My love of knowledge so inflamed me
> That my work is retarded by my very desire.

Among the great artists he was notorious for the smallness of his output and for the extreme slowness with which he worked. Did his art consume the major part of his energy and his thoughts? His private papers contain evidence not only of a unique scientific insight but of an industry which is almost incredible. He covers the whole field of science from mathematics to physiology, and there is nothing that he touches which he does not illuminate. Thus he presents us with a model of a flying-machine and suggestions for a helicopter and a parachute and, interested in the problem of flight, he analyses the nature of the flight of birds in a way that has been surpassed only during recent years. He designed a parabolic compass on a principle adopted only late in the seventeenth century (Fig. 81). He has drawings of quick-firing and breech-loading guns. He makes many ingenious suggestions for engineering apparatus. He has mastered the theoretical principles of perspective. He sets forth some of the homologies of the vertebrate skeleton. His anatomical, physiological,

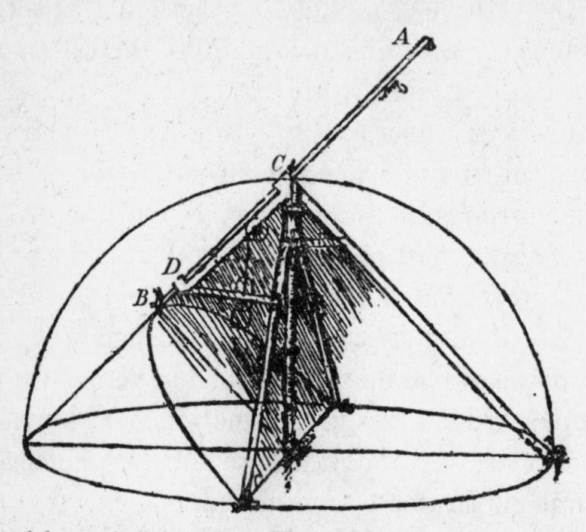

FIG. 81. Leonardo's parabolic compass. The pencil *AB* writes on an inclined surface with the point *B*. It slides freely through two casings at *C* and *D* on a rigid arm. If the pencil were further produced it would trace a circle on a horizontal surface. The lower casing is jointed to the vertical axis around which the arm and pointer are pivoted. The axis is steadied by two legs.

and embryological studies were not surpassed in certain respects for hundreds of years (Fig. 82).

Marvellous as were the attainments and achievement of Leonardo, he does not occupy a completely isolated position. Others of his age rival him both in versatility and penetration. Of Alberti we have spoken (p. 178). Leonardo's German contemporary ALBRECHT DÜRER of Nuremberg (1471–1528), apart from his achievement as an artist, made a profound and painstaking investigation of the proportions of the human body at different ages and in the two sexes, was an exceedingly close observer of the habits and growth of animals and plants, conducted experiments in optics, perspective, and the properties of sound, had a remarkable command of the mathematics of his day, and, in his great drawing, *Melancholia*, set forth in allegorical form the changes in thought and attitude with which the age was instinct. Dürer worked long in Italy. Though a German, all that he does and says is touched by the spirit of the Italian renaissance.

Very different was the Swiss Aureolus Philippus Theophrastus Bombastus von Hohenheim, more compendiously known as PARA-

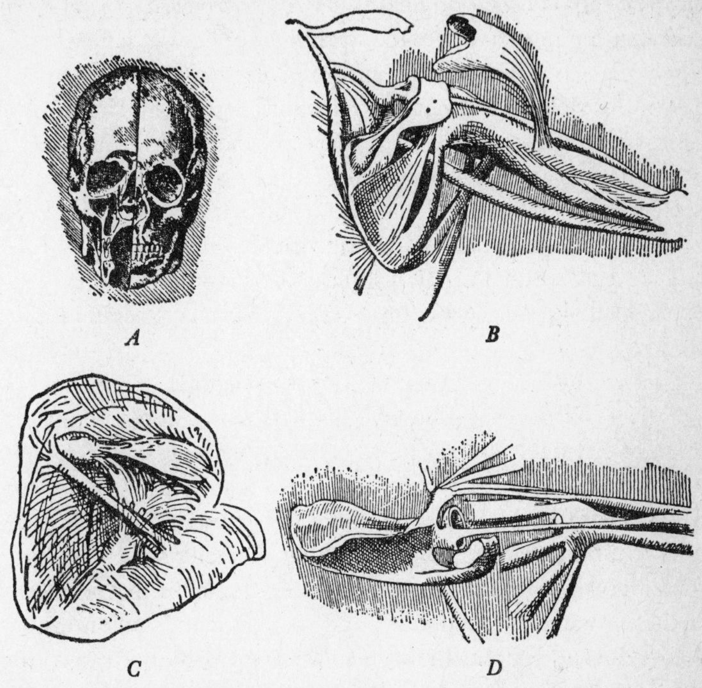

FIG. 82. Some anatomical sketches of Leonardo. A is a dissection of one-half of a skull showing cavities in the upper jaw and forehead not recognized for another 150 years. B and D are dissections of the parts around the human shoulder, the most complicated joint in the body. They exhibit muscles and tendons attached to the bones. C shows the right ventricle of the heart laid open to display the 'moderator band' passing obliquely between the outer wall and the septum. This structure was not recognized by anatomists till the nineteenth century and its physiological significance only in the twentieth. The moderator band is conspicuous in the hearts of ungulates and Leonardo may have made his drawing from one.

CELSUS (1493–1541). He was a person of violent, boastful, and repellent temper, whose iconoclasm, garrulous and often incoherent though it was, probably did something to deter men from the worship of the old idols. His symbolic act of burning the works both of the Greek Galen and of the Arab Avicenna, as an introduction to his lecture course at Basel, was meant to typify the position of the independent investigator. A writer of excessive obscurity, an obscurity of language and of form as well as of thought, very few claim the privilege of penetrating to his full meaning. It is unfortunate that most of these few have a vagueness of expression and an obscurity of style that rival those of their original. There is, however, a general agreement that his aim was to see the world in the 'light of nature'. That light of his is dimmed for us because of his extreme gullibility in some matters, his violence and self-contradiction in others, and his involved and mystical presentment in all. 'Nature' included for him the influence of the stars upon the lives of men and many other relationships then generally credited and now universally discredited. He believed still in a relation of microcosm and macrocosm (Fig. 61)—as in a residual sense we all do—but his free modification of that theory may have helped to pave the way for its rejection in the generation which followed.

It is not easy to ascribe any positive scientific contribution to Paracelsus. He did, however, give currency to one important modification of Aristotelian doctrine whereby alchemy was deflected into a direction which led to chemistry. He held that, apart from the 'four elements' of Greek philosophy, there were certain proximate principles that gave matter its distinguishing characteristics. The 'principles' were three in number. Unfortunately the names that he selected for these, 'mercury', 'sulphur', and 'salt', were already in use for definite substances and the principles seem to be borrowed from Arabic alchemists (p. 146). By *mercury* he means the nature, principles, or characteristics which are common to the metals; by *sulphur* he means the power of combustibility and the essence of changeability; and by *salt* the principle of fixity and of resistance to fire. This was an advance in the limited but important sense that these principles relate to experience and therefore do not demand that nature must of necessity be simple and accord to some rigid scheme.

4. PRACTICAL ARTS IN THE SIXTEENTH CENTURY

Whatever the influence of Paracelsus on the scientific ideas of his time, it is easier to trace that of the practical engineers and miners. Science must deal primarily with phenomena and only secondarily with generalizations from them—theories or, when secure and over a wide area, 'natural laws' as we have become wont to call certain of them. It is the conscious and constant search for generalizations based on things observed that marks the full change of technology—historically the earlier stage—into science proper. When, at length, science becomes recognized as a way to knowledge, a stage is reached

FIG. 83. Device for working bellows by cams on a shaft driven by an overshot waterwheel. From Biringuccio, 1540.

when it becomes itself the source of new technologies. All through the Middle Ages there were skilled and inventive men. Many of their devices are known but their authors seldom recorded their own experiences and none of them is known for his generalizations. But by the sixteenth century literacy had become more valued and more widespread. It was normal among directors of industry and engineering undertakings. There was a growing number of systematic technological writings. These provided a main basis for the great forward scientific move of the next century.

The earliest of the engineers whose writings were printed was VANOCCHIO BIRINGUCCIO (1460–1539) of Siena. He worked in the mines of northern Italy and central Europe. In 1540 appeared his *Pirotechnia*. This word, in the Italian usage of his time, included metallurgy with a certain amount of what we should now call industrial chemistry. The book is the product of long experience. Biringuccio sets forth repeatedly and clearly the economic advantage of chemical production on a large scale. His adaptation of water-power to bellows (Fig. 83) for furnaces illustrates this, while his presentation of a bell-scale (Fig. 84) is no bad alternative for a 'graph'.

The most influential technological writer of the sixteenth century, GEORGE AGRICOLA (1490–1555), studied medicine in Italy where he became interested in mining. In 1527 he settled at Joachimsthal in Bohemia. There he became familiar with the mechanisms, devices,

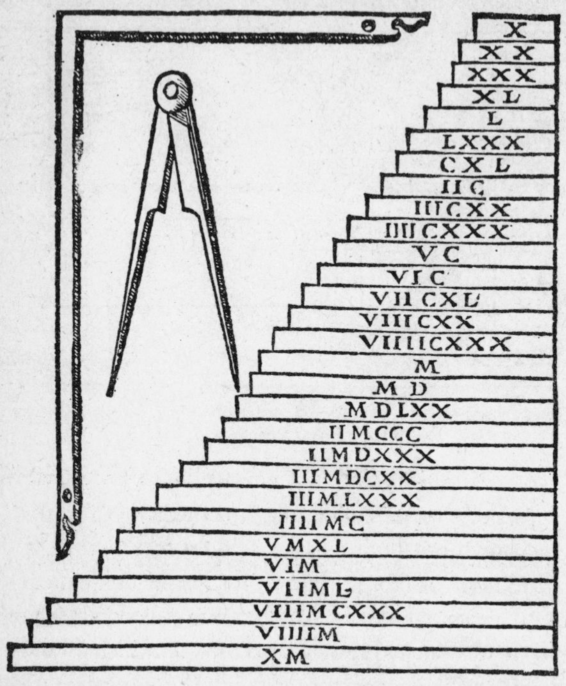

FIG. 84. Biringuccio's bell-scale, 1540. The length of each line represents the thickness of the rim of a bell of the weight in roman numbers.

and living-conditions of the miners. In 1546 he published the foundation work of modern mineralogy. Just after his death there appeared his better known *De re Metallica* (1556) which sometimes borrows from Biringuccio. It describes and illustrates the current practice of mining, of extraction, and of assaying. Because of its admirable illustrations it became standard but it is often hard to say how far back into the Middle Ages its devices may extend. Yet many of them are mentioned in this book for the first time (Fig. 85) and the best of his instruments were of delicate accuracy (Fig. 86).

Fig. 85. Agricola's pump for draining a mine. Water is sucked through a pipe by stuffed leather or fabric bolsters on an endless chain worked by gears from a hand-turned shaft, 1555.

Fig. 86. Agricola's assaying scales.

FIG. 87. Extracting gold. Ercker 1580. D and E are earthenware and iron stills for distilling mercury from the amalgam. At F is a wooden mortar for amalgamating gold ore. At G a man is squeezing a leather bag of amalgam to get rid of excess mercury.

A metallurgical writer whose book remained current for even longer than Agricola's was LAZARUS ERCKER. His German *Treatise on Ores and Assaying* was first published at Prague in 1574. Like Agricola, he worked in Bohemia and was a skilled chemist and assayer. His book is systematic, well arranged, and positive in its treatment. An idea of his ways of working may be formed from his

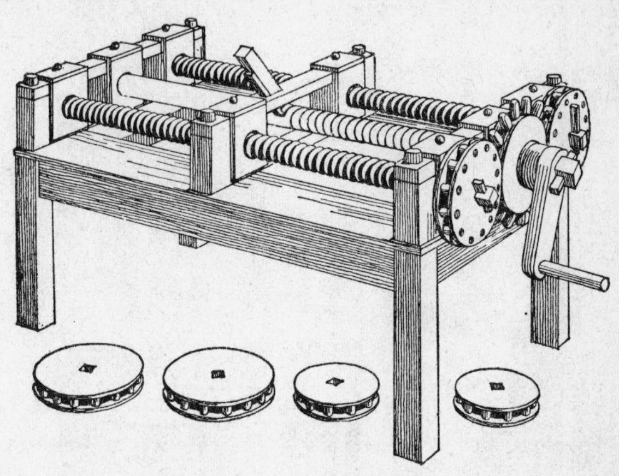

FIG. 88. Leonardo's screw-cutting machine. Restoration. The rod to be cut is in the centre and turned by the crank. A carriage bearing the cutting tool is moved by two control-screws rotated through gears by the same crank. Varying the gear-ratios varies the pitch of the thread cut on the rod.

laboratory for extracting gold from amalgam (Fig. 87). Yet, as it has little trace of theory, it is more representative of the pre-scientific than of the scientific stage of technology.

The basic change in technology during the sixteenth century was the increased control over metals and consequent replacement of wooden by metal machines. This came to affect the course of all the sciences largely through the improvement of screw-cutting by the lathe, as forecasted by Leonardo (Fig. 88). Many authors set forth this change but none more graphically than AGOSTINO RAMELLI (1531–90) of Milan. He worked mainly in France where he produced his beautiful book (1588). It paid special attention to pumps (Fig. 89). As with Leonardo it is doubtful to what degree his figures represent realities or creations of his imagination. Less picturesque but perhaps more truthful was Ramelli's contemporary JACQUES BESSON (fl. 1550–70) of Orleans.

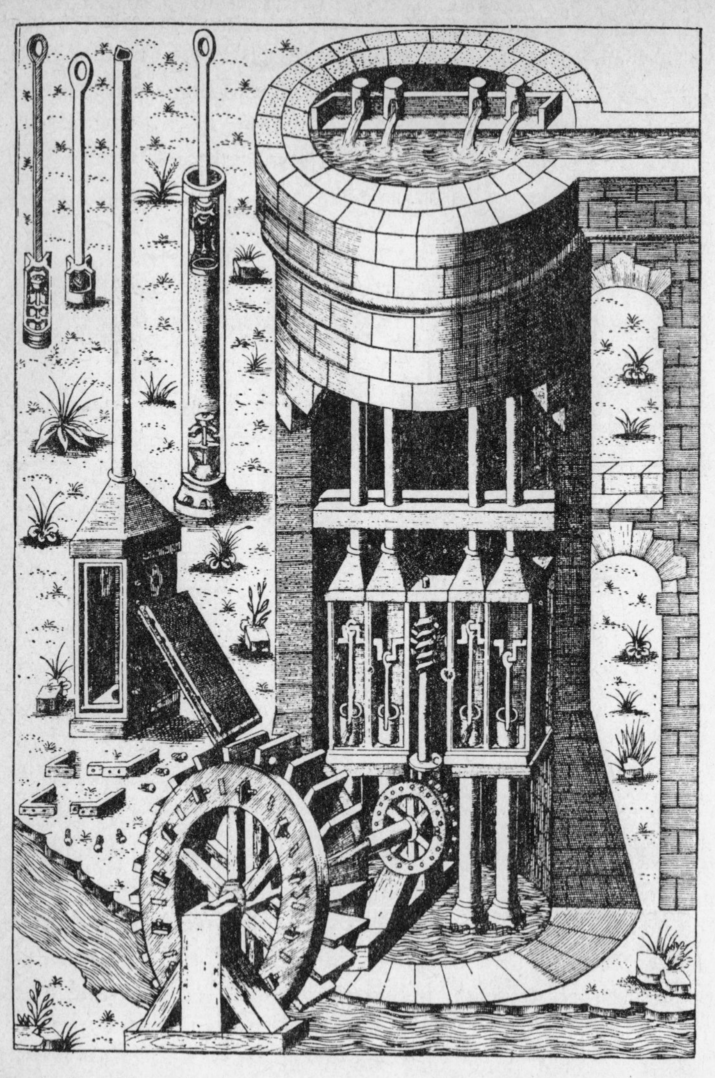

FIG. 89. Ramelli's quadruple action suction-pump (1588). Contrast Fig. 85 and note metallic construction, worm drive, and details of valves in inset. The power is an undershot wooden waterwheel. (Contrast Fig. 83.)

With a technical literature of which these authors were the most distinguished exponents, the seventeenth century opened with a firm basic knowledge of the general properties of minerals and metals. Alchemy, in the old sense, was outdated and the making of scientific instruments had become a recognised profession.

5. ENCYCLOPAEDIC NATURALISTS OF THE SIXTEENTH CENTURY

The sixteenth century brought a combination of events favourable to the accumulation of information on living things. The biological works of Aristotle and of Theophrastus were now fully available in Latin, as was the herbal of Dioscorides (p. 98). Explorers were bringing home new and strange animals and plants which compelled attention to the variety of living things and to the difference between the fauna and flora of the New World and of the East Indies from those known in Europe.

The medicine of the age used mainly vegetable drugs—'galenicals', as they were called, since many were recorded by Galen (p. 98). These differed from the chemicals and minerals—the 'spagyric' remedies—some of which had been introduced by Paracelsus (p. 199), or his followers. Thus physicians came accurately to recognize a much larger variety of herbs, many of exotic origin. Moreover the woodcutters and, later, the copperplate makers were perfecting their techniques. Faithful representations of living things were now appearing in many printed books.

The special development of plant portraiture began in Germany, the home of printing. OTTO BRUNFELS of Mainz (1489–1534) produced the first collection of realistic plant-figures (Strasbourg, 1530). Many of his figures rival those of a good modern flora (Fig. 90) though he still often identifies his Rhineland plants with those of Dioscorides, a native of Asia Minor. The most remarkable early printed herbal (Basel, 1542) was by LEONHARD FUCHS (1501–66). It is intended to guide the collection of medicinal plants. He had an excellent artist and woodcutter whose figures established a tradition traceable to our time.

If plants were favoured by the physician, there was no lack of interest in animal forms. One of the early biologists, as we may now call them, was the traveller PIERRE BELON (1517–64) of Le Mans. His tendency towards comparative studies is too striking to be missed

FIG. 90. Christmas Rose from Brunfels, 1530.

(Fig. 91). His contemporary, GUILLAUME RONDELET (1507–66) of Montpellier, a friend of Rabelais, prepared an illustrated monograph of Mediterranean fishes (1554–6), using Aristotle as a basis but far surpassing the number and variety of species of his master, though verifying some of his most remarkable observations. Stress on Aristotelian biology in the sixteenth and seventeenth centuries had none of the delaying influence of the entrenched Aristotelian cosmology and physics. The explanation is simple: Aristotle was himself a working biologist, but as physicist and astronomer he was a compiler.

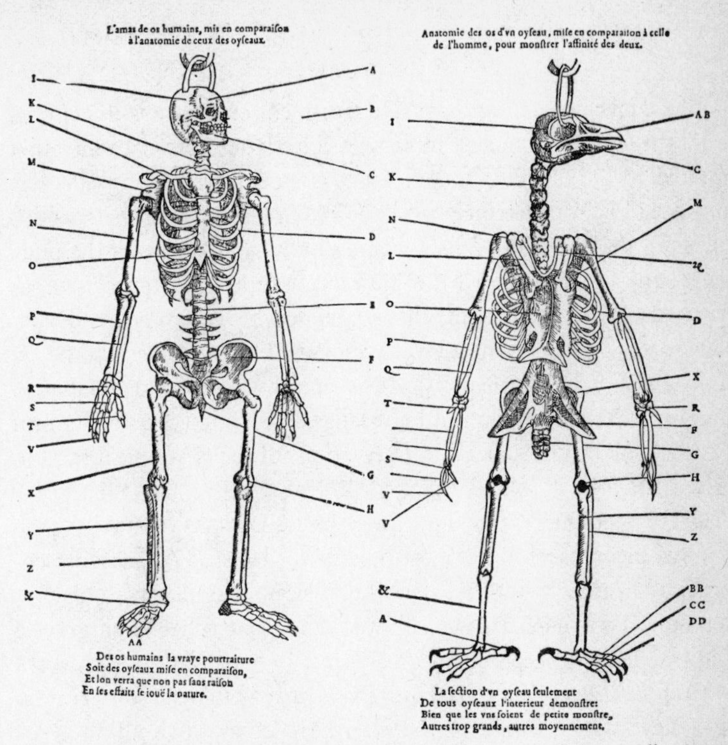

FIG. 91. Pierre Belon compares skeleton of man and bird (1555). Corresponding letters point to bones which he thought homologous. Note treatment of fore-limb and specially of digits.

The ablest of the 'encyclopaedic' naturalists was CONRAD GESNER of Zürich (1516–65). He developed a scheme (1562–8) for a complete description of the animal and vegetable kingdoms in a series of magnificently illustrated volumes but died before their completion. His correspondence with other naturalists fulfilled some of the functions of the scientific journals of later times.

The most encyclopaedic of the encyclopaedists was ULYSSI ALDROVANDO (1522–1605) of Bologna, first director of its botanic garden.

He accomplished his task, similar to that on which Gesner had embarked, in twelve enormous folio tomes (1598–1651), of which several were published from his notes after his death—they covered the whole animal, vegetable, and mineral kingdoms.

6. ANATOMY AND PHYSIOLOGY IN THE SIXTEENTH CENTURY

The encyclopaedic or extensive treatment of nature is in contrast with the monographic or intensive. The latter was naturally first applied to the object of man's most intense interest, himself. The artists of the Renaissance with their new instrument, perspective (p. 178), delighted in realistic representation. In studying the nude they naturally considered the bones and superficial muscles. Leonardo da Vinci (1452–1518) alone extended his researches to deeper structures. His anatomical work remained unpublished, but may well have stimulated the physicians, some of whom, from about 1500 sought to extend anatomical knowledge. That subject gained a firm basis by the printing of Latin translations of the anatomical texts of Galen between 1530 and 1542.

Of the anatomists of the time the Fleming ANDREAS VESALIUS (1514–64) was incomparably the most influential. He studied anatomy between 1533 and 1536 at Paris and there became familiar with Galen's anatomical writings. In 1537 he was called as teacher of surgery to Padua. There he prepared his great treatise *On the Fabric* (= 'working') *of the Human Body* (Basel, 1543). It raised biological observation to a new standard of detailed exactness from which all the life-sciences immediately benefited (Fig. 92). Some of the figures would adorn a textbook of our time while his method of exposition, despite grave stylistic faults, is also distinctly modern. His work contrasts with that of Copernicus whose astronomical treatise was published within a few days of that of Vesalius but is essentially medieval in outlook (pp. 212–15).

The physiology of the age had still hardly advanced beyond that of Galen. Vesalius had the advantage of Galen in being able to clarify his meaning by the use of figures. He records also experiments designed to demonstrate the actions of various organs, notably of the nervous system, but they surpass little those of Galen except perhaps

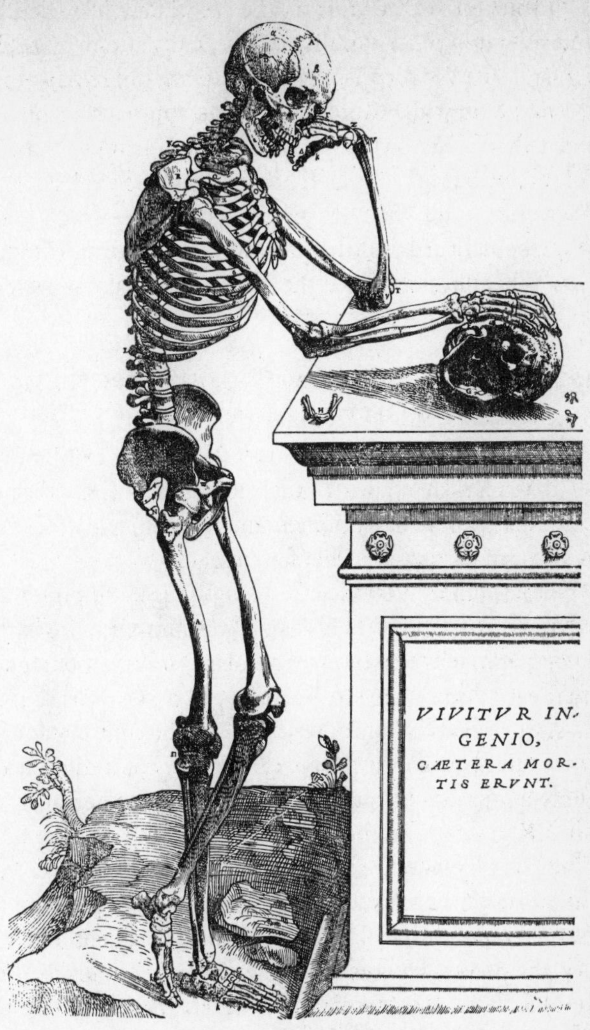

FIG. 92. Human skeleton from Vesalius, 1543.

in exposition of the muscular mechanics of the body. Nevertheless he does express doubts as to the working of Galen's physiology, notably as to the action of the heart.

The word 'physiology' (*physiologia*) was coined for the subject (1542) by the physician JEAN FERNEL of Paris (1493–1558). He shows no more advance than Vesalius on the physiology of Galen but he treats the subject philosophically and attempts the impossible task of bringing Galen's theories into line with Christian thought. Fernel's work was widely studied, and his approach to physiology determined that of his own and several subsequent generations. He was an observer, but not constantly or systematically. His only traceable discoveries—and they are not insignificant—were the central canal of the spinal cord and the push-pull mechanism of the digestive tract. The physiology of the time remained far behind its anatomy.

7. ASTRONOMICAL OBSERVATION AND HYPOTHESIS IN THE SIXTEENTH CENTURY

The astronomy of the earlier sixteenth century exhibits certain activities that mark it off with some definiteness from that of the Middle Ages. The work of Regiomontanus (p. 196) was widely known and was in large part responsible for this.

The Pole, NICOLAS COPERNICUS (1473–1543), despite the vast change that was introduced in his name, was himself more in the line of such comparatively conservative scholars as Regiomontanus than the more revolutionary Leonardo (pp. 197 f.) or Vesalius (pp. 210 f.) He was a student rather than an observer, and he continued to attend university courses until over thirty years of age. He studied at several Italian universities, giving attention to classics, mathematics, astronomy, medicine, law, and theology. It was in Italy that he first discussed the Pythagorean theory with which his name has become associated. Copernicus is said to have had skill in painting which suggests that type of visualizing imagination frequently associated with scientific power. He was not at all active as a practical astronomer. He had, it is true, taken a few observations of eclipses and oppositions of planets, but for the most part his results were obtained in the study.

Copernicus was induced to seek a new theory of the heavenly bodies by finding that mathematicians differed among themselves on this subject. He had considered the various motions of the heavenly bodies according to the old system, and concluded that some essential factor had been missed. He found his hint in the traditions that had survived of the thought of Philolaus the Pythagorean (p. 27) and of Aristarchus (p. 55).

Occasioned by this [he says] I decided to try whether, on the assumption of some motion of the Earth, better explanations of the revolutions of the heavenly spheres might not be found. Thus assuming the motions which I attribute to the Earth . . . I have found that when the motions of the other planets are referred to the circulation of the Earth and are computed for the revolution of each star, not only do the phenomena necessarily follow therefrom, but also that the order and magnitude of the stars and of all their orbits and the heaven itself are so connected that in no part can anything be transposed without confusion to the rest and to the whole universe. (Copernicus, Introduction to *De Revolutionibus.*)

The new or rather renovated scheme of Copernicus retained much of the ancient theory. It still assumed that the universe is spherical and finite, terminating in the sphere of the fixed stars. It still assumed that the movements of the celestial bodies are always circular and always with uniform velocities. It still invoked epicycles. It still demanded the eccentric (p. 91). In fact Milton's description of the Ptolemaic world fits not ill with the attempt to 'save the phenomena' by means of a system of circles and spheres that was made by Copernicus. In *Paradise Lost* the Archangel Raphael tells that

Heaven
Is as the Book of God before thee set
Wherein to read his wondrous works, and learn
His seasons, hours, or days, or months, or years.
This to attain, whether Heaven move or Earth
Imports not. . . .

and then goes on to the 'conjectures' of those who would

model Heaven
And calculate the stars: how they will wield
The mighty frame; how build, unbuild, contrive
To save appearances; how gird the sphere
With Centric and Eccentric scribbled o'er,
Cycle and Epicycle, Orb in Orb.
(*Paradise Lost*, viii. 70–84.)

The simplicity of the Copernican system—sometimes hastily inferred rather from his famous diagram (Fig. 93) than from his book itself—is more apparent than real. Thus while he reduced the number of circles demanded to explain celestial movements, he still invoked no less than thirty-four.

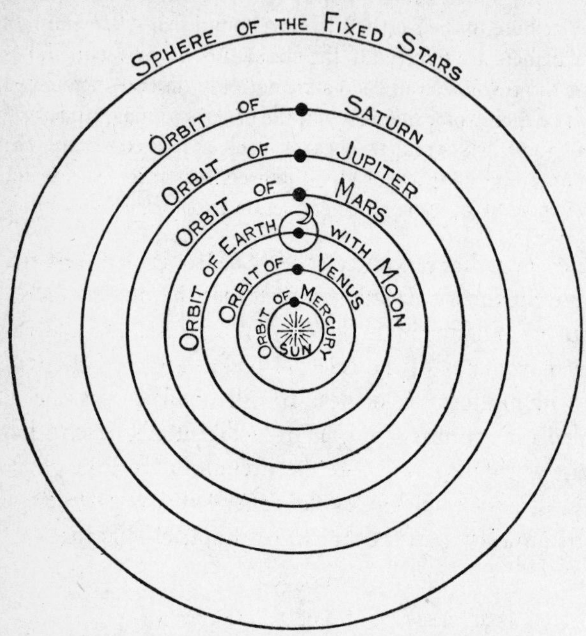

Fig. 93. Copernican world system, 1543.

The immediate influence of the teaching of Copernicus on contemporary thought was much less than has been supposed. Notices of it, for a generation or more, are surprisingly few and not always unfriendly. Religion was the main interest of the day. Religion is, by its nature, conservative, and any scientific advance of the first magnitude disturbs those who profess it. Nevertheless Christian doctrine, further fortified by St. Thomas Aquinas, had adopted the Aristotelian system (p. 168). During the Middle Ages the doctrine of a spherical earth had normally been taught in the schools (pp. 175 f.). A spherical earth is neither more in accord nor less in accord with the biblical account than is a world system of which the sun rather than the earth is the centre. Christian doctrine accommodated itself to the one; it might have accommodated itself, as it did later, to the other. There were, however, certain extraneous circumstances that intervened in determining the reception of the Copernican system.

One group of these had relation with current religious teaching which was greatly disturbed by Giordano Bruno (pp. 218 f.). A second group had to do rather with the contemporary view of the physical universe of which astrology had become part (pp. 166 f.). Now astrology was based on the doctrine that the outer spheres of the universe influenced the inner (Figs. 68, 72). This conception coloured all departments of thought and embedded itself deeply in speech. 'The scheme was conceived under an evil star', 'His fortune is in the ascendant', 'The seventh heaven of delight', 'He has gone to a higher sphere', 'The British sphere of influence', 'Canst thou bind the sweet influences of the Pleiades' (Job xxxviii. 31), 'He has the influenza' are such cases. The ideas on which these phrases were based—and they covered a large part of life—were disturbed by the Copernican view. Remove the earth from her central position among the spheres and the whole astrological system becomes unworkable.

It is too much to expect such disturbance to be accepted calmly. We shall presently consider the reaction against these two provocations. Meanwhile other disturbing astronomical researches were going forward.

The Dane TYCHO BRAHE (1546–1601) was born three years after the death of Copernicus. Unlike Copernicus he was, before everything, a patient and accurate observer. He was provided by his sovereign with an island observatory. It was superbly equipped with instruments perfected with the best techniques of the time. There for ten years and afterwards at Prague, where he was joined by Kepler, he occupied himself in the systematic collection of astronomical observations for the correction of cosmic theories. The records of Brahe were the most extensive and accurate that had been made. His achievements may be summarized:

(*a*) He set forth a planetary system with the earth central to the orbits of moon and sun and central also to the fixed stars. The sun revolves round the earth in twenty-four hours carrying all the planets with it. Of the planets, Mercury and Venus have orbits smaller than that of the sun (Fig. 94). The mathematical result of this system works out as identical with that of Copernicus (1588).

(*b*) In 1572 he observed a new star. Moreover in 1577 in examining a comet he was able to determine its parallax, and thus proved that it too was farther off than the moon and thus outside the sphere of the

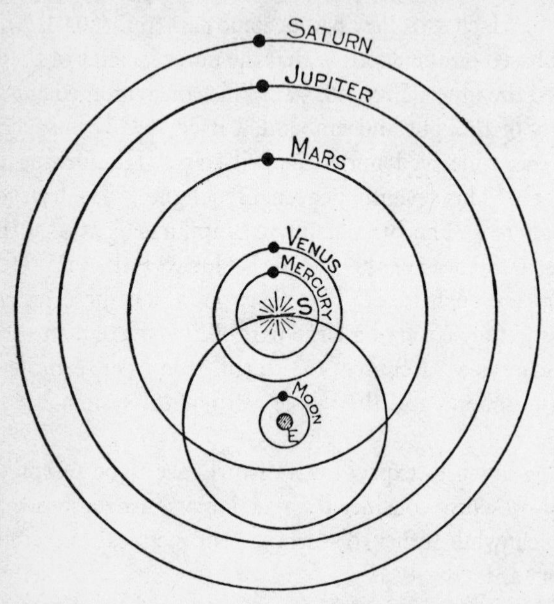

Fig. 94. Tycho's world system. The sphere of the fixed stars is omitted.

'elementary' world (see Fig. 68). This was equivalent to introducing the principle of change into the changeless spheres and therefore contrary to Aristotelian principles.

(*c*) He suggested that the movement of a comet might be 'not exactly circular but somewhat oblong'. This is the first suggestion that a celestial body might move in a path other than circular (1577).

(*d*) He described very accurately perturbations in the moon's motion (1599). These had to await explanation by subsequent generations and new astronomical systems.

(*e*) His numerous observations on the planets enabled Kepler to reveal the true nature of their orbits.

(*f*) He corrected the received value of many astronomical quantities, and in a posthumous work edited by Kepler (Prague, 1602), he gave the position of 719 stars, later increased by Kepler in his *Rudolphine Tables* (1627) to 1005.

Tycho's attempt to represent the structure of the Universe as according to the ideal form of the circle was the last great effort of the Pythagorean spirit save for that of his pupil Kepler. The insurgent century sought for direct evidence as to the nature of the world. The new science concerned itself neither with ideal forms nor with the theory of knowledge nor with the nature of reality nor with the principles of investigation, but with the evidence of the senses.

The mystical poet Henry Vaughan (1622–95) may well have been thinking back to Tycho when, half a century later he wrote:

> I saw Eternity the other night,
> Like a great ring of pure and endless light,
> All calm, as it was bright;
> And round beneath it Time, in hours, days, years,
> Driv'n by the spheres,
> Like a vast shadow mov'd; in which the world
> And all her train were hurl'd.
>
> (Vaughan, *The World*.)

VII

THE INSURGENT CENTURY (1600–1700)

Downfall of Aristotle. New Attempts at Synthesis

The longest Tyranny that ever sway'd
Was that wherein our Ancestors betray'd
Their free-born *Reason* to the *Stagirite*
And made his *Torch* their universal *Night*.
So *Truth*, while onely *One* supplied the State,
Grew scarce, and dear, and yet sophisticate;
Until 'twas bought, like Empirique Wares, or Charms,
Hard words sealed up with Aristotle's Armes.

(John Dryden, 1663.)[1]

I. DOCTRINE OF THE INFINITE UNIVERSE

COPERNICUS worked in Poland, the eastern march of European civilization. It was at the western limit of Europe, in England, where the spirit of the great intellectual revival had not yet obtained full hold, that his message was first translated into philosophic form.

In 1583 there came to London GIORDANO BRUNO (1547–1600), a native of Nola near Naples and a renegade monk. He had already sojourned as a teacher at the universities of Lyons, Toulouse, Montpellier, and Paris. At each his restless and turbulent spirit had combined with an aloofness from the affairs of men to make him unwelcome. Throughout his life he showed a lofty indifference to common sense that cannot fail to command our respect—at a distance. He made a precarious livelihood by lecturing on a barren

[1] Dryden (1631–1700) wrote this in the year of his election to the Royal Society, the foundation of which in the previous year represented for him the downfall of Aristotle in his own country though he was familiar also with the works in the previous generation such as those of Bacon, Gilbert, and Harvey. In the poem: *Longest Tyranny* from Aristotle to Dryden was almost 2,000 years; *Stagirite* is Aristotle, often so called from his birth-place in Macedonia; *Sophisticate* in seventeenth-century English means adulterated; *Sealed up empiric wares* means something near to what we should call patent medicines of undisclosed composition; *Arms* (Latin *arma*) is used in the heraldic or proprietary sense. The general significance is that obscure and corrupt works have been given the authority of a great name.

logical system which he had partly invented. It was intimately linked with an absurd principle of mnemonics which he had partly borrowed. So wayward a genius was predestined for trouble.

During his visit to England Bruno at length developed philosophical coherence. This period of illumination lasted but a few months. In 1584 he published in London, though with the false impress of Venice, three tiny Italian works, *The Ash-Wednesday Supper* (*Cena delle Ceneri*), *On Cause, Principle, and Unity*, and *On the Infinite Universe and its Worlds*. These booklets effectively contain his whole philosophy, which was based in essence on Nicolas of Cusa (p. 178) and in form on Copernicus. Essential parts of his thought are that not only does the earth move round the sun but that the sun itself moves; that there is no such thing as a point absolutely at rest; that the stars are at vast but various distances from the solar system and are themselves centres of comparable systems; that the universe, being itself infinite, can provide no criterion of fixity; that our planetary system is in no sense the centre of the universe; and that 'there is a common soul within the whole to which it gives being and at the same time is individual and yet is in all and every part'.

In such a universe where may Paradise and Purgatory be placed? And is not the 'common soul', which uniformly permeates it, a memory of Neoplatonism? (pp. 133 f.) Bruno's vision of a universe, endless both in time and space, whose soul abides uniformly in every part, differs utterly from the 'created Universe' of medieval Christian philosophy, the Creator of which must, of His nature, be separate from that which He has created. The universe of medieval Christian philosophy was necessarily centred in Man, for into Man alone, among created mundane things, the Divine Spirit had entered. Small wonder that the Church was disturbed by Bruno's thought. His revolution was incomparably greater than any dreamed of by the academic and conservative Copernicus.

The issues involved were not at first generally recognized. Some, stirred by the pagan character of Bruno's thought, fixed on the irrelevant detail of the earth moving round the sun as contrary to scripture. This idea Bruno had certainly taken from Copernicus whose work was not, as yet, prohibited. But Bruno's vision had far deeper implications than a mathematical readjustment of the current world scheme. A finite universe, spherical or not, with or without the earth as its centre, can be conceived as 'created'; an infinite universe cannot be so contemplated. Creation is fundamental to Christianity —at least to the Christianity of that age. In 1600 Bruno was burned at the stake, having passed seven years in prisons of the Inquisition.

Bruno perished without the hope that he had a disciple. And yet much of his view was soon to displace that of medieval Christianity. Before he had been dead for thirty years, the world was, for the man of science, no longer a diagrammatic scheme which required investigation only as regards its details. It had become a world without bounds and therefore of infinite possibilities, and yet a world whose parts were uniformly related by mathematical rules, the physical bases of which were in process of discovery.

It was of course true then, as it is of course true now, that the view of universal law did not and does not occupy the whole mind of all men of science. Men of science reserved, and still reserve, some department of experience in which, consciously or unconsciously, they forbid full play to their vision of universal law. But when they give rein to the scientific mood, then it is bound to displace the mood of faith, nor can the medieval compromise (pp. 168–9) stand against it. Thus the three little tracts of Bruno printed in London in 1584 mark the real change from medieval to modern thought and especially to modern scientific thought. The change was long in coming, longer for some topics than for others, longer in some minds than in others: but the coming of that change was inevitable once these three tracts had got abroad.

It is essential to remember that Bruno's views were not based on experiment or observation. His contribution was a philosophy, not a scientific method or system, and was, in fact, a development of the thought of Nicolas of Cusa (p. 178). The doctrine of relativity in space, in motion, in thought, promulgated by the calm spirit of Nicolas became in the passionate Bruno an ardent and soul-absorbing faith.

It is not easy to trace in detail the progress of the dissemination of the ideas of Bruno. His life was obscure, the propagation of his thought furtive, his influence secret, indirect, unacknowledged. Yet his ideas crop up where they might not be expected. We will consider one such case.

In 1600 there appeared in London a Latin work *On the Magnet and on Magnetic Bodies and concerning that great magnet, the Earth, a New Physiology*, by WILLIAM GILBERT (1546–1603), personal physician to Queen Elizabeth, a man in authority, respectable and respected. His book is the first major original contribution to science published in England. It earned the admiration of Galileo.

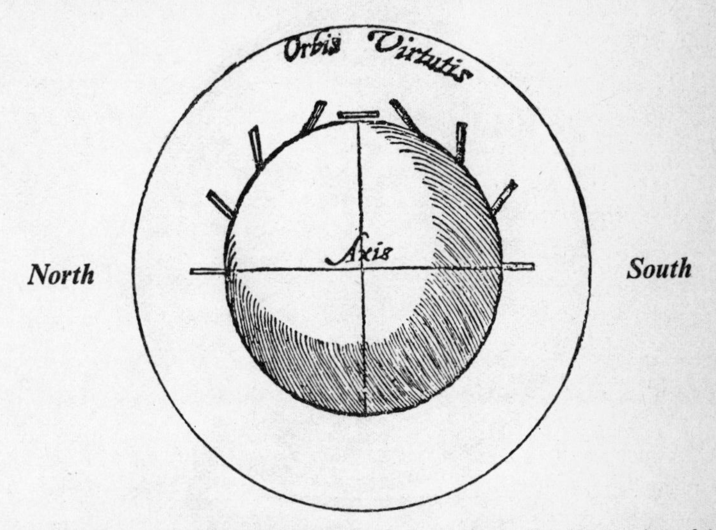

FIG. 95. Diagram of William Gilbert of the earth as magnet, showing his view of the dip of the compass in different latitudes, 1600.

Gilbert's work sets forth his investigations of the properties of the magnet in thoroughly modern form. He has a clear view of the place of observation and experiment, and sharply distinguishes his own experiences from those recorded by others. He was aware of magnetic inclination and declination and held that at the poles a free needle would point vertically, at the equator horizontally, and intermediately between the two. This he explained as due to the earth itself acting as a great magnet, testing this experimentally on a spherical loadstone (Fig. 95). He devised an instrument to measure magnetic dip (Fig. 96), but variability and deviation of the compass he wrongly thought was related to proximity to land masses and to their attractive power. He discerned a magnetic power in all iron and was familiar with methods of intensifying it by touching a needle with the loadstone, by leaving for years an iron rod pointing North–South, or by repeatedly striking an iron bar so held.

The last section of the book is devoted to an exposition of the system of the universe. The universe of Gilbert is that of Bruno,

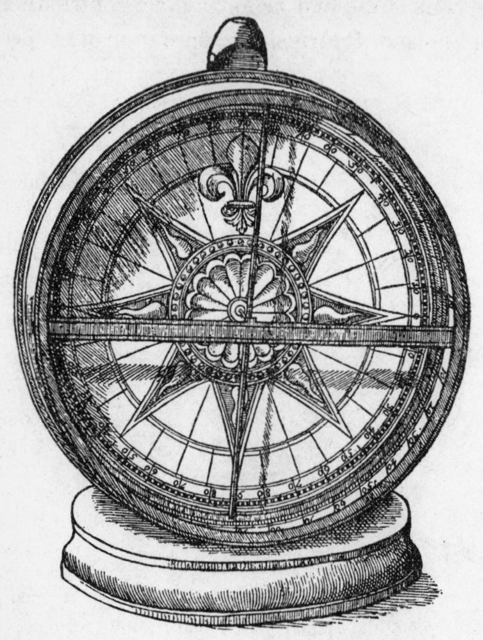

FIG. 96. Gilbert's instrument for measuring dip.

whose name, however, he does not mention. Gilbert must have met Bruno, perhaps in the company of Sir Philip Sidney (1554–86).

Long after Gilbert's death another work by him *On our Sublunary World, a New Philosophy*, was seen through the press (1651). It expounds in detail, quoting Bruno, the idea that the 'fixed stars' are at differing distances from our planetary system and that these stars are the centres of other planetary systems.

The seventeenth century opened lurid with the fires that formed Giordano's shroud in Rome, while the farther reaches of the north were lit by the calmer light of the researches of Tycho and Gilbert, showing Scandinavia and Britain as new stars in the scientific firmament. A glance at the mass of fundamental scientific work of the seventeenth century shows the major departments of science becoming differentiated there. The acceptance of observation and experiment, as the only methods of eliciting the laws of nature, reaches an ever-widening circle. The first scientific generation of the century saw the development of a mathematical technique that became the instrument of the new discoveries.

2. MATHEMATICS BECOMES THE INSTRUMENT OF PHYSICAL INVESTIGATION

The improvement in the means of mathematical expression was a main condition for the development of exact conceptions of a new cosmology and physics. These were an intellectual necessity to replace the tottering Aristotelian scheme. The insurgent century found a qualitative world based on abstract values. It bequeathed a quantitative world based on concrete impressions.

A beginning had been made in the later sixteenth century. Thus the French lawyer FRANÇOIS VIÈTE (1540–1603) was among the first to employ letters to represent numbers. He applied algebra to geometry in such a way as to lay a foundation for analytical trigonometry (1591). At about the same time was introduced the decimal scheme for representing fractions (1586) by the Fleming SIMON STEVIN (1548–1620). This most able man preceded Galileo in proving experimentally the similar rate of fall of bodies of different weight (1586). His name is also associated with the method of resolution of forces, with the distinction of stable and unstable equilibrium, with the law of equilibrium on inclined planes (Fig. 97), and with the 'hydrostatic paradox', that is that downward pressure of a liquid on the base of its containing vessel is independent of its shape and size and depends only on the depth of the liquid and area of the base.

Stevin was able also to calculate the pressure on any given portion of the side of the containing vessel. He laid the essential foundations for the whole science of hydrostatics (1586). He gave much care to analysing into their simplest mathematical elements the traditional mechanical devices in the various trades as, for example, the cranes—widely used in his native Netherlands—which were worked as treadmills (Fig. 98) as in Roman times (Fig. 59).

Stevin's best remembered conclusion, and that to which he

himself attached most importance, was his formal proof of the impossibility of perpetual motion. This arises from the equilibrium of an endless chain suspended around an inclined plane. If the validity of that scheme be denied then the endless chain would rotate continuously—which it manifestly does not (Fig. 97).

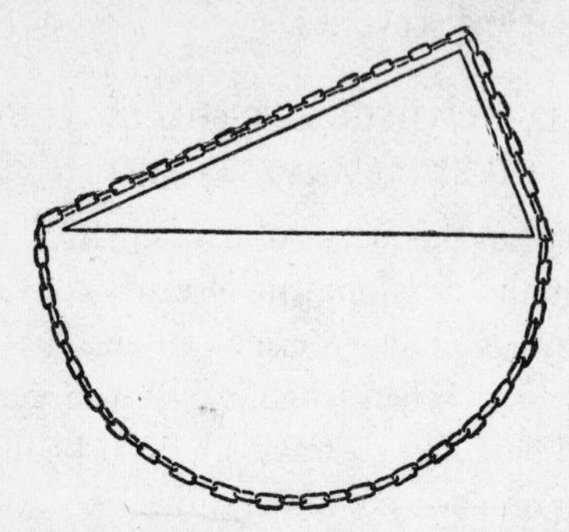

FIG. 97. Stevin's proof of conditions of equilibrium on inclined planes. Around the vertical angle of an upright triangle, of which the opposite side is horizontal, hangs a ring-chain. It will be in equilibrium, for if not it would be in perpetual motion. Remove the suspended loop. Equilibrium remains. Therefore weights on planes inclined to each other are in equilibrium if they are proportional (as are those of the pieces of chain) to the lengths of the planes as cut by the horizontal.

By the use of an improved form of Stevin's decimal notation calculation was much facilitated. Contemporary astronomical activity, however, still carried with it an endless task of computation. No technical advance was more needed than some further alleviation of this deadening burden. Thus the invention of logarithms by Napier was greeted with enthusiasm.

JOHN NAPIER (1550–1617), Laird of Merchiston in Scotland, began his investigations (1573) with an attempt to systematize algebraic knowledge. In his earliest work he says that in considering imaginary roots he discovered a general rule for roots of all degrees. He conceived the principles of logarithms in 1594. The next twenty years were spent in developing the theory and computing the 'canon' or table of logarithms. While thus engaged he invented the modern notation of decimal fractions. His *Description of the Marvellous Canon of Logarithms* appeared in Latin at Edinburgh in 1614. An extension of Napier's long effort to do away with the tediousness of calculation was his *Rabdologia* (1617). It contains the description of 'Napier's bones', devices designed to simplify multiplication and division. These were in use for about a century.

An advance, significant for the whole of subsequent astronomical

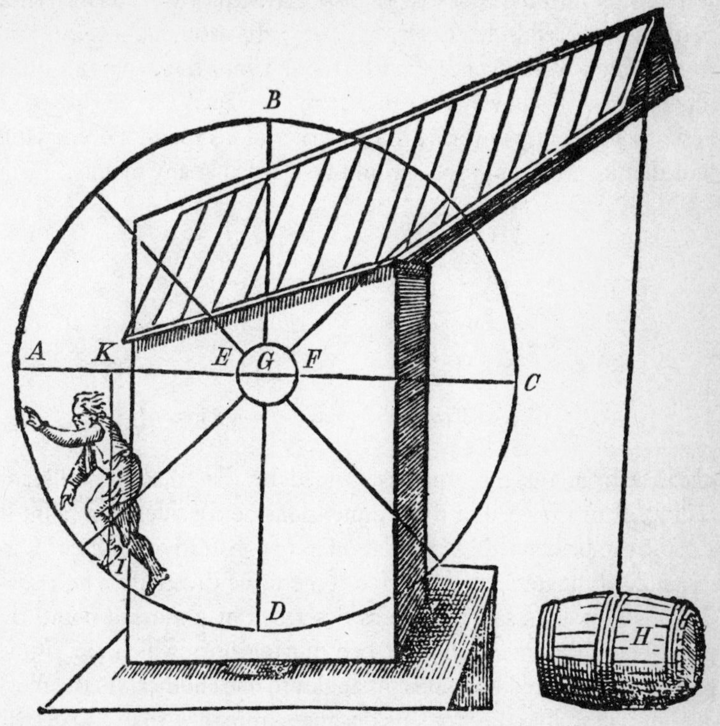

FIG. 98. From Stevin's *Practice of Weighing*, Leyden, 1681: 'Let *ABCD* be the wheel [of a crane] and its axle *EF* with centre *G*. A man (in the wheel) counterbalances the barrel *H*, four times his weight. *IK* is his line of gravity. Then in equilibrium *GK* is to *GF* as 4 to 1. If he proceed towards *A* then *H* will rise for the ratio *GK* to *GF* will increase.'

development, was made by Kepler (pp. 236 f.) in a commentary (1604) on the work of the thirteenth-century mathematician Witelo (p. 170). Kepler regarded conic sections as forming five species passing from the (1) line-pair, through (2) hyperbola, (3) parabola, and (4) ellipse, to (5) circle. In order to indicate the nature of this process Kepler designated as *foci* the fundamental points connected with these curves. The foci of the circle coalesce at the centre. Ellipse and hyperbola have two foci equidistant from the centre. The parabola has two foci, one within it and the other at an infinite distance on the axis (Figs. 21, 30, 99).

Even more fundamental for future mathematical development were ideas introduced by RENÉ DESCARTES (p. 259). His analytical method appeared in his *Geometry* (1637). Its essential novelty is the introduction of the conception of motion into the geometric field. There is a well-known story that lying late abed, as was his wont, and observing a fly hovering in the corner of his room, it occurred to him that its position in space could be defined at any moment by its

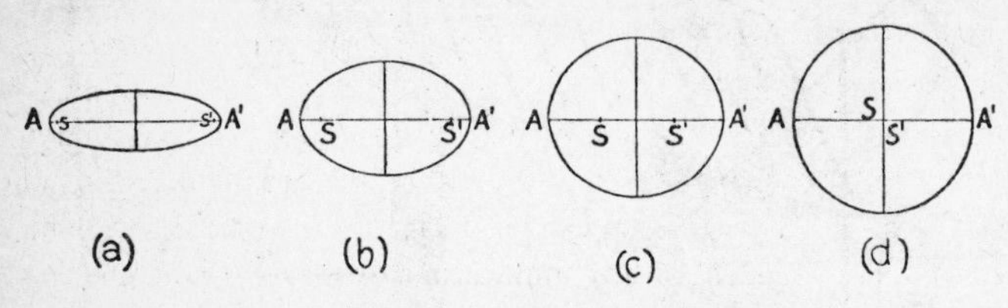

FIG. 99. The circle as special case of ellipse.

distance from the three planes formed by the adjacent walls and ceiling. If two instead of three dimensions be considered, a point in a plane can be defined by its relation to two instead of to three 'Cartesian co-ordinates', as they have come to be called after him.

Thus Descartes saw a curve as described by a moving point, the point being the intersection of two moving lines which are always parallel to two fixed lines at right angles to one another. As the moving point describes its curve, its distances from the two fixed axes will vary in a manner characteristic of that particular curve. An equation between these distances can be formed which would express some property of the curve. The conception has had innumerable developments and has been adopted in every department of science. Its most familiar development is the 'graph'. Important parts of our mathematical notation are due to Descartes.

There is a basic conception in the mathematical attitude of Descartes that is far more significant than any technical addition that he made. His analytical procedure displayed the fundamental correspondences of number and form. Pythagoras and Plato perceived this correspondence (p. 25 f.). The Alexandrians tended to study each in isolation. The later separate development of algebra by the Hindus and of geometry by the Arabs, and the general trend of Western mathematical studies in the Middle Ages and the fifteenth and sixteenth centuries concealed a most essential truth. Descartes, despite professed indifference to historical considerations, called men back to the old paths of Pythagoras and Plato on this most fundamental issue. It is probable that the application of algebraic methods to the geometric field is the greatest single step of the century in the progress of the exact sciences. Descartes himself insists on the unity of the study of mathematics.

> All sciences which have for their end investigations concerning order and measure are related to mathematics, it being of small import whether this measure be sought in numbers, forms, stars, sounds, or any other object; there ought therefore to be a general science, namely mathematics, which should explain all that can be known about order and measure, considered independently of any application to a particular subject. . . . A proof that it far surpasses in utility and importance the sciences which depend on it, is that it embraces at once all the objects to which these are devoted and a great many besides.
>
> (*Rules for Direction of the Mind*, 1628.)

The expression of this Platonic view makes it evident that the reaction against the long reign of Aristotle has begun. (For Aristotle's own criticism on the point, see p. 38.)

About the beginning of the seventeenth century were made the first decided advances since antiquity in synthetic geometry. In this department the leading name is that of Descartes's friend and fellow countryman BLAISE PASCAL (1623–62). He also added much to mathematical theory especially in connexion with probability. Pascal invented one of the first arithmetical machines. He created new geometrical conceptions and would have approved the view of Thomas Hobbes (1588–1679): 'Geometry is the only science that it hath pleased God hitherto to bestow on mortals' (*Leviathan*, 1651). Pascal's important experimental work was done *c.* 1650, published posthumously (1663), and criticized by Boyle in his *Hydrostatical Paradoxes* (1666). In this connexion we note that Galileo strangely failed to link resistance to a vacuum of a column of water ('Nature abhors a vacuum' p. 232) with his knowledge that the air has weight. The synthesis was left to Torricelli (1643, p. 269). But Pascal was the first to demonstrate the barometric effects of altitude on a tower in Paris and in a famous ascent by his brother-in-law of the Puy-de-Dôme of some 4,800 feet.

A versatile mathematician of the age was the learned Oxford Professor JOHN WALLIS (1616–1703). His first great mathematical work, *Arithmetica infinitorum* (1655), contains the germ of the differential calculus. Newton read it early and derived his binomial theorem from it. Wallis wrote the first mathematical work devoted to tides, in which he introduces the assumption 'that, for purposes of calculation, earth and moon can be treated as a single body concentrated at their centre of gravity'. His *Algebra* (1657) contains the idea of the interpretation of imaginary quantities in geometry which was fundamental for the development of analytic methods. Wallis introduced the symbol ∞ for infinity.

A mathematical genius of the highest order was the polymath CHRISTIAN HUYGENS (1629–95). His mathematical skill drew the attention of Descartes, who predicted his future eminence. Before he was twenty he did good work on the quadrature of the circle and conic sections (1651–4). He diffused his mathematical abilities in many departments; optics, astronomy, mechanics, theory of light, and horology, and thus perhaps missed some of the fame that his abilities justified (pp. 245, 286, 302, 341, 367–9).

Mathematical activity early influenced optics, a favourite subject for discussion by mathematicians even during the Middle Ages (p. 170). The leading problem was the nature of the laws of refraction. Ptolemy (p. 89), Alhazen (p. 152), Witelo (p. 170), and their medieval followers were aware that light rays are bent or 'refracted' when passing from a rarer to a denser medium. In a commentary on Witelo, Kepler gives his measurements of the incident and the refracted rays in special cases, but failed to reach a general law (1604). This was successfully elicited (1621, Fig. 100) by the Hollander WILLIBRORD SNELL (1591–1626). Descartes placed the results in a more acceptable form and published them without acknowledging their source (1637).

The nature of the advance can be illustrated by a simple diagram. Rays of light AA′ and BB′ pass at different angles from air into water (Fig. 100). In all such cases they are bent toward the vertical and bent in such a degree that the ratio $\frac{\text{Aa}}{\text{A}'\text{a}'}$ is the same as ratio $\frac{\text{Bb}}{\text{B}'\text{b}'}$. This ratio for water is about 4:3, which is said to be the *refractive index* of water. Each substance has its own characteristic refractive index.

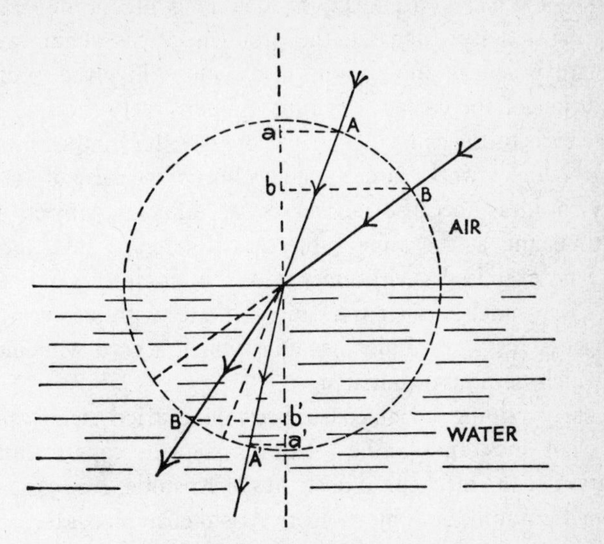

FIG. 100. Diagram illustrating Snell's law.

Different kinds of glass were, for instance, soon found to have different refractive indices.

The laws of refraction of light have been a most important factor in the construction of optical instruments. In that art the effective beginning was made by Galileo (1609, p. 242). The optics of his compound systems of lenses were investigated by Kepler, who first expressed in intelligible mathematical form the action of telescope and microscope (1611). Kepler's work led to further advance by Descartes (1637) who also produced a geometrical theory of the rainbow based on the laws of refraction. The study of refraction occupied Huygens. His knowledge of the subject enabled him to improve lenses and to produce telescopes with much clearer definition than heretofore (1655, pp. 303–4).

If the development of optics was determined by mathematical advance the same is no less true of mechanics. In the latter case, however, the whole body of new teaching was in effect the work of one man, Galileo, to whom we now turn.

3. PHYSICO-MATHEMATICAL SYNTHESIS

GALILEO GALILEI (1564–1642) lived a long life of unparalleled intellectual activity. Many of the products of his genius were of immediate practical application, many more involved profound modification of the current scientific opinions, yet others struck at the very basis of the general beliefs of the day. It is impossible here to treat Galileo's works in chronological order because of his great fertility in ideas, because he worked at different subjects at the same time and at the same subject at different times, and also because he exercised such influence on his disciples and contemporaries that much of their work interdigitates with his. We therefore discuss his work under headings which accord with modern views which he himself initiated.

The early training of Galileo had been along strictly scholastic and Aristotelian lines. In 1585 he began a systematic experimental investigation of the mechanical doctrines of Aristotle. By 1590 he had developed a number of objections to Aristotelian physical teaching. Notably he had accumulated his records of experiments on falling bodies. In a more than doubtfully authentic story they were illustrated from the leaning tower of Pisa. Such experiments were already being made by Stevin (1586, p. 223). They are of the highest importance as unmasking an error of Aristotle, and Stevin's conclusion was certainly reached by Galileo by 1591. Weights of 1 lb. and of 100 lb., dropped from a height, reached the earth together. How then was it possible to maintain with Aristotle that the rate of fall was a function of the weight of the falling object?

The most famous of Galileo's physical investigations resulted in his great astronomical discoveries (pp. 242 f.). It is not, however, in his discoveries, numerous, fundamental, superb though they were, that we sense the full significance of Galileo in the history of thought. It is rather in his initiation of a new attitude toward the objective universe and in his construction of an enduring mathematico-physical scheme that would fit that attitude. More than to any other man, we owe to him the conception of our world in terms of interplay of *calculable* forces and *measurable* bodies. And, moreover, to him more than to any other man we owe the experimental employment of that conception.

> Dynamics [says Lagrange] is a science due entirely to the moderns. Galileo laid its foundations. Before him philosophers considered the forces which act on bodies in a state of equilibrium only. Although they attributed in a vague way the acceleration of falling bodies and the curvilinear movement of projectiles to the constant action of gravity, nobody had yet succeeded in determining the laws of these phenomena. Galileo made the first important steps, and thereby opened a way, new and immense, to the advancement of mechanics as a science.
>
> (*Mécanique analytique*, 1788.)

He set forth these views in his great *Discourses concerning two new Sciences* (1638). The work at one step advanced the subject from the medieval to the modern stage. The two new sciences deal respectively with (*a*) 'Coherence and resistance to fracture' and (*b*) 'Uniform, accelerated, and violent or projectile motions'.

The first part of the work is mainly concerned with the resistance of solids to fracture, and the cause of their coherence. The value of this section lies in the incidental experiments and observations on motion through resisting media. The current belief that machines built on exactly similar designs, but on different scales, are of strength in proportion to their linear dimensions is discussed. It is shown that the larger machine will be less strong and less resistant to violent actions. The machines include animal bodies (Fig. 101 and pp. 251 f.).

After explaining the strength of ropes of fibrous materials, Galileo turns to the cause of coherence of the parts of such things as stones and metals, which do not show a fibrous structure. What prevents a glass or metal rod, suspended from one end, from being broken by a pull at the other? He suggests that it is nature's so-called 'abhorrence of the vacuum' supposedly produced by the sudden separation of two flat surfaces. This idea is extended, and a cause of coherence is found by considering every body as composed of very minute particles, between any two of which is exerted a similar resistance to separation.

This line of reasoning leads to a very important experiment for measuring what is called the force of a vacuum. It will be recalled that the action of pumps was a recurrent problem to the mining and agricultural engineers of the age (Figs. 85 and 89). Galileo remarks that a suction-pump refuses to work over a height greater than 35 feet. This is sometimes told as if Galileo had said, jokingly, that nature's horror of a vacuum does not extend to heights less than 35 feet, but it is plain that the remark was made seriously. He held

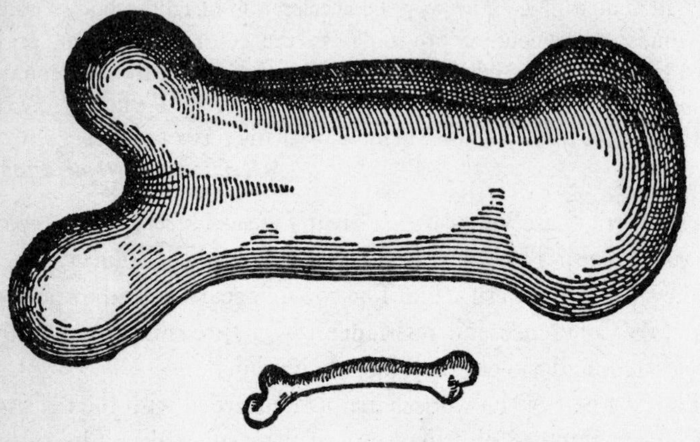

FIG. 101. Galileo's comparison of a normal femur with the form it would need to support an animal each of whose over-all dimensions was thrice the normal.

the conception of suction then current, for he compares the column of water to a rod of metal suspended from its upper end, which may be lengthened till it breaks with its own weight. It is strange that he failed to see how simply this phenomenon could be explained by the weight of the atmosphere, with which he was well acquainted. A fuller explanation had to await Torricelli (p. 269).

Aristotle's ideas on motion and especially that bodies fall with velocities proportional to their weights and inversely proportional to the densities of the media through which they fall, are examined. The result is to substitute, for Aristotle's assumption, that law of the motion of falling bodies which is the foundation of modern dynamics.

The discussion of the strength of beams opens with a consideration of the resistance of solid bodies to fracture. This is very great for a direct pull, but less for a bending force. Thus, a rod of iron will bear a longitudinal pull of, say, 1,000 lb., while 50 lb. of bending force will break it if it be fastened by one end horizontally into a wall.

Galileo assumed, as the basis of his inquiry, that the forces of cohesion with which a beam resists a cross fracture in any section may all be considered as acting at the centre of gravity of the section, and that it breaks away at the lowest point. An elegant result deduced from this theory is that the form of a beam, to be equally strong in every part, should be that of a parabolic prism, the vertex of the parabola being the farthest removed from the point of support. As an approximation to this curve he recommends tracing the line in which a heavy flexible string hangs when supported from two nails.

The curvature of a beam under any system of strains is a subject into which, before the days of Newton, it was not possible to inquire, and even in the simpler problem considered by Galileo he makes assumptions which require justification. His theory of beams is erroneous in so far as it takes no account of the equilibrium which must exist between the forces of tension and those of compression over any cross-section.

The theorems and formulae, deduced geometrically from the phenomena of uniform and accelerated motion, lead to a more detailed statement of the principle of inertia. The definition of uniformly accelerated motion is given as that of a body which so moves that in equal intervals of time it receives equal increments of velocity.

There follows an application of the results. He examines the times of descent down inclined planes, assuming the velocity to be the same for the same height whatever the inclination. This he verified by careful experiments, although he was unable at the time to prove it mathematically.[1] The simplicity of these experiments and of their apparatus is most striking:

A piece of wooden moulding 12 cubits long, half a cubit wide, and three finger-breadths thick, had cut on its edge a channel one finger in breadth, very smooth and polished. A hard, smooth, round bronze ball was rolled along it. Having placed this board in a sloping position, by lifting one end one or two cubits above the other, we rolled the ball along the channel, noting the time of the descent. We repeated this till the deviation between two observations never exceeded one-tenth of a pulse-beat. We now rolled

[1] Viviani relates that, soon after he joined Galileo in 1639, he drew his master's attention to this. The same night, as Galileo lay sleepless in bed, he discovered the mathematical demonstration. It was introduced into the subsequent editions of the *Discourses*.

the ball only one-quarter the length of the channel; and having measured the time of its descent, we found it precisely one-half of the former. Next we tried other distances, comparing the time for the whole length with that for the half, or with that for two-thirds, or three-fourths, or indeed for any fraction. We always found that the spaces traversed were to each other as the squares of the times, and this was true for all inclinations of the plane.

For measuring time, we employed a vessel of water in an elevated position. To its bottom was soldered a pipe giving a thin jet, which we collected in a small glass during each descent. The water collected was weighed, after each descent, on a very accurate balance. The differences and ratios of these weights gave us the differences and ratios of the times, and this with such accuracy that although the operation was repeated many times, there was no appreciable discrepancy in the results [abbreviated].

FIG. 102. Galileo's 'pulsimeter'.

As well as this method of measuring small intervals of time, Galileo used a pendulum adapted as a 'pulsimeter' (Fig. 102). It was merely a weight suspended on a thread which oscillated as a pendulum against a scale. The thread could be so adjusted that the oscillations corresponded to the beats of the pulse, i.e. approximately seventy-two each minute. The length of the thread could be read, on the scale, to a tenth of a pulse beat, that is to one-ninth of a second. Galileo had used this device as far back as 1583. Its clinical application was described (1602) by the physician, Santorio Santorio (1541–1636), Galileo's colleague at Padua. (See also p. 275.)

The mechanical clocks and watches of Galileo's day were insufficiently delicate for his experimental purposes. When he became blind, and shortly before his death, he suggested the application of a pendulum to a clock to his son, Vincenzio Galilei (1600–49) whose sketch of it has survived. Pendulum clocks did not become practical, however, until the publication of Christian Huygens's *Horologium Oscillatorium* in 1673 (p. 303). Watches could hardly be used as delicate measurers until after Galileo. Notable was the adoption by Thomas Tompion (1630–1713), of the balance-spring in 1675 in a form suggested by Robert Hooke (p. 271).

The next section of Galileo's *Discourses concerning two new Sciences* (1638) passes to consideration of the properties of a body whose motion is compounded of two other motions, one uniform, and the other naturally accelerated. Such is the motion of a projectile. The law of the independence of the horizontal and the vertical motions is here laid down. A body projected horizontally would—but for its weight and external impediments—continue to move in a straight line. Again, the effects of gravity acting by itself on the projected body would be entirely downwards. But gravity acting on the projected body can neither increase nor diminish the rate at which it travels horizontally. Therefore, whatever the path or the direction of motion at any moment, the distance travelled horizontally is a measure of the time that has elapsed since motion began. He proves that the path described has the geometrical properties of a parabola (Fig. 103).

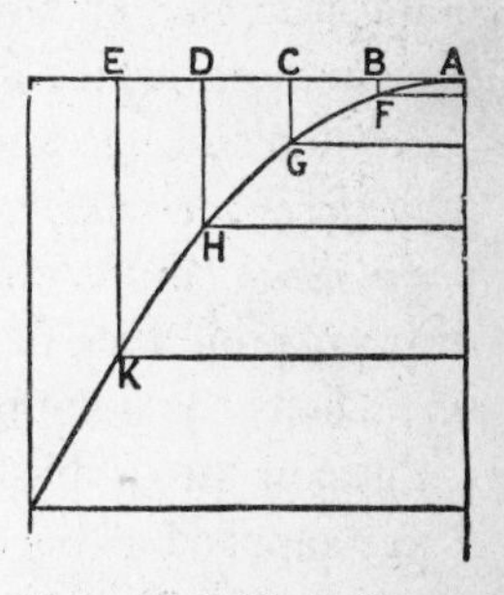

FIG. 103. Galileo's diagram of the course of a body projected horizontally. *AB*, *BC*, *CD* . . . represent equal forward displacements in equal increments of time of a horizontally ejected projectile. The distances of fall *BF*, *CG*, *DH* . . . increase as the square of the time. The actual path *AFGH* is a parabola.

Galileo drew up a table giving the position and dimensions of the parabola described with any given direction of projection. He showed that the range on a horizontal plane is greatest when the angle of elevation is 45°. He was essentially applying the principles of the Differential or Fluxional Calculus. Had pure mathematics attracted him as strongly as its applications, he might have founded the Fluxional Calculus, which is the glory of Newton and of Leibnitz.

No sooner was the manuscript of these dialogues out of his hands in 1636 than Galileo began to occupy himself with new projects which he left unfinished at death. In them he approaches the laws of interdependence of force and motion which appear at the beginning of Newton's *Principia* (1687). But Galileo not only prepared the way for Newton: he supplied him with much of his materials. Thus, Newton's first law—*that a body will continue in a state of rest, or of uniform motion in a straight line, unless compelled to change its state by some force sed upon it*—is a generalization of Galileo's theory of uniform motion. Since all the motions that we see taking place on the surface of the earth soon come to an end, we are led to suppose that continuous movements, such, for instance, as those of the celestial bodies, can only be maintained by a perpetual consumption and a perpetual application of force, and hence it had been inferred that rest is the natural condition of things. We make, then, a great advance when we comprehend that a body is equally indifferent to uniform motion or to rest, and that it perseveres equally in either state until disturbing forces are applied.

Newton's second law—*that every change of motion is proportional to the force that makes the change, and in the direction of that straight line in which the disturbing force is impressed*—is involved in Galileo's theory of projectiles. Before his time it was a commonly received axiom that a body could not be affected by more than one force at a time.

But now the establishment of this principle of the composition of forces supplied a conclusive answer to the most formidable of the arguments against the rotation of the earth. It is employed by Galileo in his *Dialogue on the Two Systems of the World* (1632, p. 247). The distinction between mass and weight was, however, not valued, and consequently, Galileo failed to grasp the fact that acceleration might be made a means of measuring the magnitude of the force producing the motion, that is to say of the mass of the earth.

Of the third of Newton's laws of motion—that *action and reaction are always equal and opposite*—we find traces in many of Galileo's researches, as in his theory of the inclined plane, and in his definition of momentum. It is adumbrated in a little work on mechanics written by him in youth but published after his death. It is developed in his latest ideas on percussion.

4. KEPLER'S RE-FORMATION OF THE HEAVENS

The first to apply mathematics as an empirical instrument in seeking the laws of celestial motion was the German JOHANNES KEPLER (1571–1630). He had strong mystical leanings, and a large proportion of his writings is now unreadable, but a residuum is of the very highest scientific importance. His idea of the universe was, from the first, essentially Platonic and Pythagorean. He was convinced that the arrangement of the world and of its parts must correspond with some abstract conception of the beautiful and the harmonious and, further, must be expressible in numerical and geometric form. It was this belief that sustained him in his almost incredible labours. He spent years chained to the drudgery of computation, without assistance and without the devices, such as logarithms or reckoning machines, that now lighten the computer's task. Nothing but a burning yet steady faith could make such labour endurable.

We gain an insight into the transition state between the old and the new in which Kepler worked when we recall that his professed occupation was largely astrological calculation. Nor was he cynically sceptical as to the claims of astrology, but sought, in the events of his own life, a verification of the theory of the influence of the heavenly bodies.

Kepler adopted the Copernican view from an early date. He turned his mind to the question of the number, size, and relation of the orbits of the planets. He was ever seeking a law binding together the members of the solar system. After trying various simple numerical relations, after attempting to fill the gaps by hypothetical planets and after discarding various other suggestions, he lit, at last, on a device which satisfied him (1596). There are only five possible regular solid figures (i.e. figures with equal sides and equal angles)—'Platonic bodies' as they were called (p. 28)—and there are only five intervals between the six planets that he recognized. As far as the calculations of Kepler extended at that time, the five regular solids could be fitted between the spheres of the planets so that each was inscribed in the same sphere about which the next outer one was circumscribed (Fig. 104. Compare p. 28).

Sphere of Saturn
Cube
Sphere of Jupiter
Tetrahedron
Sphere of Mars
Dodecahedron
Sphere of Earth
Icosahedron
Sphere of Venus
Octahedron
Sphere of Mercury.

For the first time a unitary system had been actually introduced in explanation of the structure of the universe. We may well smile at this instance of human presumption. Kepler soon found that he had wrongly estimated the distances of the planets from their centre.

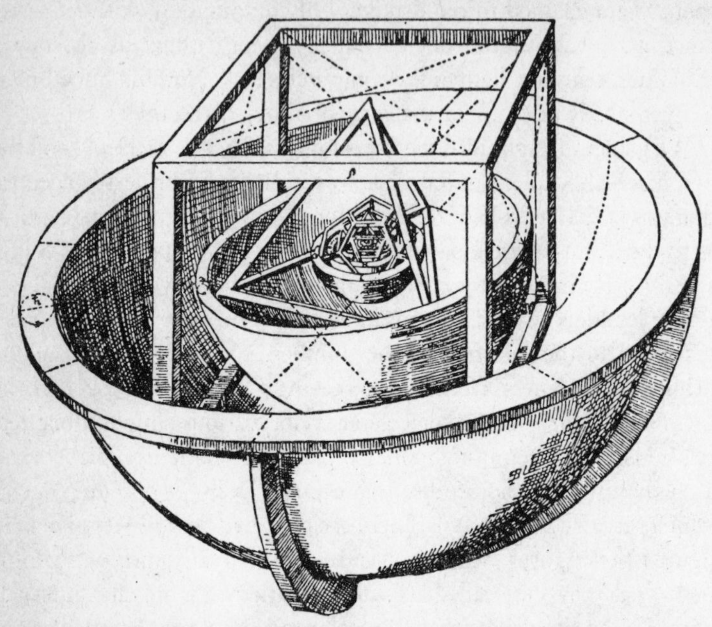

FIG. 104. From Kepler's *Mysterium Cosmographicum* (Tübingen, 1596) illustrating supposed relationships between the five Platonic bodies (p. 28) and the number and distances of the planets. The concentric figures are inscribed within each other thus:

Outermost sphere bearing sign of Saturn
Cube
Sphere bearing sign of Jupiter
4-sided regular pyramid
Sphere bearing sign of Mars
12-sided regular body
Sphere bearing sign of earth
20-sided regular body
Sphere of Venus (hardly traceable)
8-sided regular body
Sphere of Mercury
Innermost central body of the sun.

The unitary nature of this system could not be maintained. But to Kepler, who, like the medieval thinkers, held that the physical universe had a moral plan, these new mathematical relationships came as illustrations of what he conceived to be the divine purpose. The regular solids, he observed, were of two classes: primary (cube, tetrahedron, dodecahedron) and secondary (icosahedron, and octahedron), differing in various ways. What more fitting than that the earth, the residence of man 'created in God's image', be placed between the two kinds of solids? The scheme was confirmatory of the main tenets of his Pythagorean faith (Fig. 15).

That Kepler sought so persistently for a simple mathematical scheme of the material world, and that, having found one, he regarded it as fitting his scheme of the moral world, suggests certain reflections on the workings of the mind itself. Whatever reality may be, we seem to be so made that we aspire towards an interpretation of the universe that shall hold together in a complete and reasonable scheme. The fact that we thus aspire does not in the least prove that such a scheme corresponds to reality. Nevertheless, all great religions attempt to provide such an interpretation. All become skilfully 'rationalized'.

It is because science disturbs part of this already carefully rationalized field that religion resents its intrusion. Most minds recoil from a dualistic universe, and modern rationalized religion usually minimizes even such remnants of dualism as the conception of a spirit of evil. It is easy for us now to regard the opponents of Galileo and Kepler as purblind fools. Base motives certainly prompted some of the opposition; but in essence the opposition expresses the reluctance of the human mind to adopt any teaching which disturbs its unitary conceptions. A reasoned unitary view of the universe, physical and moral, had grown up during the Middle Ages. It would have been a marvel if this had been relinquished without an embittered struggle, for faith is not necessarily accompanied by either wisdom, or learning, or foresight.

Despite the failure of his first attempt, Kepler still pursued his life aim, the foundation of an astronomy in which demonstrable mathematical principles should replace arbitrary hypotheses. He examined the relation of the distances of the planets to their times of revolution round the central sun. It was clear that the time of revolution was not proportional to the distance. For that the outer planets were too slow. But why? There is, he suggested, 'one moving intelligence in the sun that forces all round, but most the nearest—languishing and weakening in the more distant by attenuation of its virtue by remoteness'. How different from the phraseology of modern astronomy which dates from Newton. In such terms as 'moving intelligence', 'languishing of its virtue', &c., Kepler was employing the Aristotelian phraseology that had been adopted by the Middle Ages. The conception was familiar to the medieval philosophers, Christian, Moslem, and Jewish. Aquinas (p. 168), Averroës (p. 155), and Maimonides (p. 155) all had a clear conception of intelligence moving the spheres. They had derived this ultimately from Greek thinkers, and adapted it each to his own theology.

Such expressions as 'Harmony of the Spheres' or 'Angels of the Spheres' were familiar to all in the sixteenth and seventeenth centuries and the Microcosm was the purpose of the Macrocosm. (Figs. 61, 68, 72 and compare with Dryden's verses, p. 289.)

As the sixteenth century turned into the seventeenth, Kepler received a great incentive to work by joining Tycho Brahe (p. 215) as assistant, and on his death in 1601 became his literary legatee. The next nine years saw him largely occupied with the papers of Tycho and with work on optics, developing an approximation to the law of the refraction of light. In 1609 he issued his greatest work, the *New Astronomy with Commentaries on the Motions of Mars*. It is full of important suggestions, notably that the earth attracts a stone just as the stone seeks the earth, and that two bodies near each other will always attract each other if adequately beyond influence of any third body. It also develops a theory of the tides in relation to attraction by the moon. But above all, the work sets forth the cardinal principles of modern astronomy, the so-called first two planetary laws of Kepler:

(*a*) Planets move round the sun not in circles, but in ellipses, the sun being one of the foci.

(*b*) A planet moves not uniformly but in such a way that a line drawn from it to the sun sweeps out equal areas of the ellipse in equal times (Fig. 105).

It was another ten years before Kepler enunciated in the *Epitome Astronomiae* (1618) his third law to the effect that.

(*c*) The squares of the period of revolution round the sun are proportional to the cubes of the distances.

For one who accepted these principles of Kepler the Aristotelian cosmology lay derelict. Its foundations were undermined and their place taken by an intelligible mathematical relationship. The scholastic Aristotelianism was to become as much an embarrassment to official religion as the narratives of miracle became at a later date. It

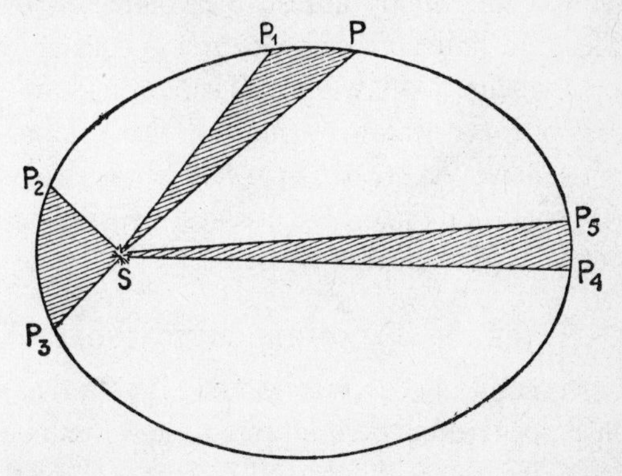

FIG. 105. Planets sweep out equal areas in equal times. PP_1, P_2P_3, P_4P_5 are distances along its orbit around the sun S traversed by a planet in equal times. Areas SPP_1, SP_2P_3 and SP_4P_5 are equal. The path here presented is far more elliptical than those pursued by the planets, which are very much more nearly circular.

was, however, as hard for one section of the Church to rid itself of its scholastic heritage as it was for another, at a later date, to disembarrass itself of the dead weight of miracle.

Certain further reflections on Kepler's work rise to the mind. It is a fundamental error to separate science from learning or, perhaps, it would be best to say from tradition. By the Greeks the study of conic sections had been prosecuted as an intellectual exercise. These figures, ellipse, hyperbola, parabola, existed, so far as they knew, in the mind and in the mind alone except in man's own artifacts. They corresponded to nothing known in the natural world. And then, after two thousand years, Kepler shows that these ancient concepts correspond to something in nature that is also revealed by the use of the senses. Is not the mind then somehow attuned to nature? It has been well said by a great historian of science that 'if the Greeks had not cultivated conic sections, Kepler could not have superseded Ptolemy; if the Greeks had cultivated Dynamics, Kepler might have anticipated Newton' (Whewell, 1794–1866).

Dynamics, as we have seen, was in fact a creation of Kepler's contemporary, Galileo. In character and temper Kepler and Galileo form an extraordinary contrast. The German Protestant, mystic and dreamer rather than observer and experimenter, produced voluminous, numerous, and wholly unreadable volumes. He stands over against Galileo, the Italian Catholic, clear and cold of intellect, unrivalled in experimental skill, witty, and happily endowed with artistic and literary prowess, who wrote no work that was not significant. In sheer genius, however, the two men were not rivals but peers and comrades. On them, in equal measure, rest the foundations of the conception of a mathematical universe.

5. GALILEO'S ASTRONOMICAL DISCOVERIES

Galileo's astronomical activity began in 1604. In that year, in the constellation Serpentarius, there appeared a new luminous body. He demonstrated that it was without parallax, that is to say there was no difference in its apparent position in the heavens, from whatever point it was viewed. Now parallax decreases with increase of distance. In Galileo's time the planets were known to have parallax, but the parallax of the 'fixed stars' was so small, by reason of their vast distance, that it was unmeasurable by the instruments of the day. This new body was thus in the remote region of the fixed stars. Now that outer zone had been regarded by Aristotle and his followers as absolutely changeless (p. 54). New stars, like meteors and comets, had been generally held to be in the lower and less perfect regions near the earth. Tycho's demonstration that the new star of 1572 was beyond the moon (p. 216) was effectively published in 1603. Galileo thus followed Tycho in an attack on the incorruptible and unchangeable heavens and had delivered a blow to the Aristotelian scheme, wellnigh as serious as the experimental proof that the rate of fall is not in direct proportion to weight (p. 232).

In 1609 Galileo made accessible two instruments that had the profoundest influence on the subsequent development of science, the telescope and microscope. His earliest discoveries with the telescope were issued in a little pamphlet of 24 leaves, his *Starry Messenger* ('Sidereus nuntius') in 1610. There are no 24 leaves in all scientific literature that contain more important revelations.

The first half of that famous booklet is occupied by observations on the moon. The surface of the moon, far from being smooth and polished, as it appeared to the naked eye, was now seen to be rough,

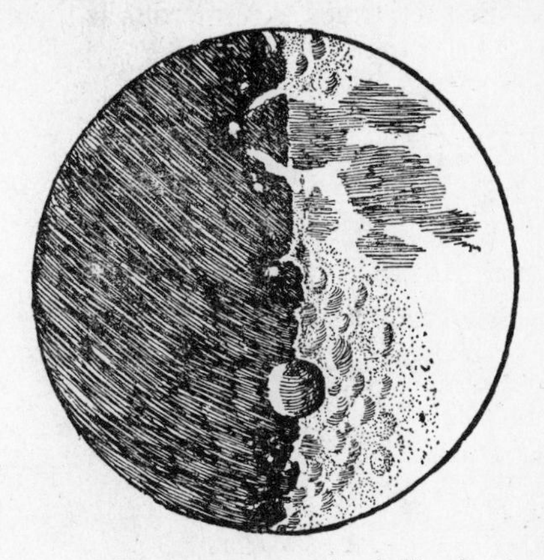

FIG. 106. The moon as seen by Galileo in 1609.

with high mountains and deep depressions. The latter Galileo interpreted as rivers, lakes, and seas. From the shadows of the illuminated mountain tops he could estimate the height of some of them. He found them to rise four or five miles above the general level (Fig. 106).

Galileo's lunar observations have an interesting relationship with English literature. In 1638 he was old and blind and nominally a prisoner of the Inquisition at Fiesole. He was visited there by Milton. The incident has inspired several artists and writers. In 1658, sixteen years after Galileo's death, Milton began his *Paradise Lost*, completing it in 1666. Its cosmology is deliberately Ptolemaic, not Copernician. Nevertheless, *Paradise Lost* does recall the poet's induction into the new astronomy twenty-seven years previously. It describes Satan's shield of which the

> broad circumference
> Hung on his shoulders like the moon, whose orb
> Through optic glass the Tuscan artist views

At evening, from the top of Fesole,
Or in Valdarno, to descry new lands,
Rivers, or mountains, in her spotty globe.
(*Paradise Lost*, i. 286–91.)

Galileo's *Starry Messenger* discusses the revelation by the telescope of an immense number of hitherto unobserved fixed stars. These were seen to be at least ten times as numerous as those that had been

FIG. 107. The Pleiades, from Galileo's *Starry Messenger* 1610: 'I have depicted the six stars known as the Pleiades (I say six because the seventh is seldom visible). Near by are more than forty others, none much more than half a degree away from the six. I have shown thirty-six stars in this diagram, preserving the distinction between old stars and new.'

catalogued. The more conspicuous star clusters were found to contain many stars too faint for recognition by the naked eye. Parts of the Milky Way and some of the nebulous patches in Orion, in the Pleiades and in other constellations were resolved into groups of newly discovered stars of various magnitudes (Fig. 107).

The remainder of the little book is devoted to an account of the satellites of Jupiter which Galileo discovered on one of the first occasions when he used his telescope. The existence of these bodies was of peculiar interest at the time, since the planet was seen to be itself a sort of little model of the solar system, with minor bodies circulating round a great central body. The contemporary discussion as to the 'plurality of worlds' (p. 245), started by Bruno, was given a new turn by this new world on the Copernican model.

There were other observations made by Galileo about this time that were later the subject of much discussion. Important were the observations on the inner planets and notably on Venus. It had been a real objection to the Copernican hypothesis that if the planets resemble the earth in revolving round a central sun, they might be expected to be luminous only when exposed to the sun's rays. In other words, they should exhibit phases like the moon. Such phases were now actually observed in Venus by Galileo.

In the same year the outermost of the known planets, Saturn, was investigated. Peculiar appearances in him were noted, though their interpretation as rings was the work of Christian Huygens (1629–95) at a later date (p. 302).

Soon after this Galileo first observed dark spots on the surface of the sun. These he saw narrowed continuously as they approached the edges of the sun's disk. He rightly regarded the process as foreshortening and as indicating that they were on the surface of the sun's orb which was itself rotating. The date and circumstances of the announcement (1612) were unfortunate, since they involved him in a controversy with a powerful Jesuit rival who not only claimed priority of observation, but also put another interpretation on the spots.

The controversy spread far beyond its original focus. An aspect of the dispute was the question of the habitability of the moon, the planets, and even of the stars, for these, too, some thought to be worlds. His critics believed this a natural corollary of Galileo's development of the 'Copernican' view which he had now openly espoused. The conception of the 'plurality of worlds' gave rise to a very considerable literature. The doctrine, it was believed, was contrary to Aristotelian and Christian teaching. It had been enunciated by the heretical Bruno (pp. 218 f.).

Thus became united against Galileo a variety of interests. The academic Aristotelians had long been fuming against him. Some Jesuits who were active in teaching, as well as many political churchmen, joined them. Pious folk were outraged by the conception of the plurality of worlds. To them were further united many of that intellectually timid class that forms the mass of every population in every age and is by no means rare in university circles. Deeper though less expressed was the great philosophic fear of the infinite universe that Bruno had suggested. The matter came before the Inquisition early in 1616. Galileo was admonished 'to abandon these opinions and to abstain altogether from teaching or defending or even discussing them'. A few days later a decree was issued ordering the work of Copernicus to be 'suspended till corrected'. Copernicus had been dead for 73 years.

6. SOME PHILOSOPHICAL IMPLICATIONS OF GALILEO'S VIEWS

In 1624 Galileo published *Il Saggiatore* ('The Assayer'), a work which contains a conception of great import for the subsequent development of science. This conception, moreover, was destined to colour deeply much of the philosophical thought of later ages. He here distinguishes sharply between those qualities of an object that are susceptible of exact numerical estimation and those which can never be treated in this way.

No sooner [says Galileo] do I form a conception of a material or corporeal substance, than I feel the need of conceiving that it has boundaries and shape; that relative to others it is great or small; that it is in this place or that; that it is moving or still; that it touches or does not touch another body; that it is unique, rare, or common; nor can I, by any effort of imagination, disjoin it from these [*primary*] qualities. On the other hand, I find no need to apprehend it as accompanied by such conditions as whiteness or redness, bitterness or sweetness, sonorousness or silence, well-smelling or ill-smelling. If the senses had not informed us of these [*secondary*] qualities, language and imagination alone could never have arrived at them. Wherefore I hold that tastes, colours, smells, and the like exist only in the being which feels, which being removed, these [*secondary*] qualities themselves do vanish. Having special names for them we would persuade ourselves that these [*secondary qualities*] have a real and veritable existence. But I hold that there exists nothing in external bodies for exciting [*the secondary qualities*] tastes, smells, and sounds, but [*the primary qualities*] size, shape, quantity, and motion. If, therefore, the organs of sense, ears, tongues, and noses were removed, I believe that [*the primary qualities*] shape, quantity, and motion would remain, but there would be no more of [*the secondary qualities*] smells, tastes, and sounds. Thus, apart from the [*percipient*] living creatures, I take these [*secondary qualities*] to be mere words. (Italics and brackets added.)

This distinction between *primary qualities* and *secondary qualities*, as they came afterwards to be called, has been made by men of science ever since. Galileo was the prime mover in that development which is summed up in the phrase *Science is Measurement*. 'He ordered all things by measure, number, and weight.' (*Wisdom of Solomon*, *c.* 100 B.C.)

As to whether men of science have been right or wrong in their view that primary qualities have a reality lacking in secondary qualities, we need not for the moment consider. It is evident that ordinary experience is almost entirely made up of secondary qualities. The fact that men of science have dwelt chiefly on something else, something which ordinary men do not ordinarily consider, has separated them from their fellows. Since Galileo, men of science have formed a sort of priesthood which has been, not infrequently, opposed to another priesthood. Nor has the distinction which Galileo made remained entirely with men of science. Through Thomas Hobbes (1588–1679) and John Locke (1632–1704) in England, and through Marin Mersenne (1588–1648) and René Descartes (1596–1650) in France, it passed into more general thinking.

By 1630, after many years' work, Galileo had at last completed his epoch-making *Dialogue on the Two Chief Systems of the World*, that is the Ptolemaic and the Copernican. Quite apart from the discussion of the relative position of earth and sun in the universe, the *Dialogue* is the consummation of the labours of Galileo in that it seeks to present the doctrine of uniformity in the working of the material universe.

The point of view expressed by the *doctrine of uniformity*, the view that corresponding causes are everywhere producing corresponding effects, is so familiar to us nowadays as to be a part of our manner of thinking. We are brought up to it from our earliest years. The only occasions on which it is ever questioned by educated men of our own time are (*a*) in the discussion of the nature or reality of miracles, and (*b*) in the discussion of the relation of mind and matter. But in the seventeenth century it was not so. The Aristotelian conception of the universe still largely ruled. According to that view the events in the high supralunary spheres—'celestial physics' as we may call them—were of a very different order to our earthly happenings—'terrestrial physics'. A large part of medieval philosophy may indeed be regarded as a debate, prolonged through hundreds of years, of the relation of celestial to terrestrial physics. That there was a difference between the two had hardly yet been questioned, save by Bruno (p. 218). Even Galileo was in no strong position to discuss celestial physics. It is of interest, however, that he throws out a definite suggestion that it can be discussed on the terrestrial basis, thus foreshadowing the doctrine of universal gravitation.

> Since, as by a unanimous conspiracy of all the parts of Earth for the formation of its whole, those parts do congregate with equal inclination and, ever striving, as it were, at union, adapt themselves to the form of a sphere, so may we not also believe that Moon, Sun, and the other members of the Solar System (*corpi mondani*) are likewise of spherical form by a concordant instinct and natural concourse of all their parts? And if any of their parts were violently separated from the whole, might we not reasonably suppose that they would revert spontaneously by natural instinct? May we not therefore conclude that as regards their proper motion, all members of the Solar System (*corpi mondani*) are alike?

Permission to print this dialogue was obtained from the ecclesiastical authorities on the express condition that the subject was to be treated theoretically as a convenient hypothesis and not as representing the facts. It was issued in 1632.

The debate in this work is carried on by three persons, an open advocate of the Copernican doctrine, an obtuse and obstinate follower of Aristotle and Ptolemy, and an impartial participator open to conviction. The conditions of publication are only superficially complied with, and the tone leaves no doubt as to Galileo's real opinions. The work is full of prophecies of the development of cosmic theory.

The Aristotelian in the dialogue is represented as hopelessly stupid, and the Copernican has the best of the dispute. In fact, however, the Copernican passes far beyond Copernicus, notably in his total rejection of the idea of the stars as fixed in a crystal sphere. The stars, as in the works of Bruno, are held to be at inconceivable but differing distances from our earth, and the absence of visible stellar parallax is considered as due to the vastness of this interval.[1]

The *Dialogue* brought matters to a head. Oddly enough, it was not the sweeping generalizations on which Galileo's opponents seized—

[1] The measurement of the parallax of a fixed star was not published until 1838, when it was achieved by Bessel (p. 313), 200 years after Galileo's work.

maybe they did not realize their full significance. It was rather certain details opposed to the current view that were specially suspected. In August 1632 the sale of the book was prohibited and its contents submitted for examination to a special commission. They reported against Galileo. The end is well known.

7. SOME PHYSICAL IMPLICATIONS OF THE GALILEAN REVOLUTION

Galileo, more than any other man, had introduced the change in our manner of thinking that broke with ancient and led on to modern science. Contributions had also been made by Copernicus, by Vesalius, by Harvey, by Tycho, and by Kepler and others. The share of Galileo is, however, so overwhelming that it is not unfair to call it the 'Galilean Revolution'. The change was more than an addition to knowledge. It was more even than an alteration in the conception of the structure of the universe. It was rather a change in mood as to the kind of knowledge that was to be sought. It partook of the nature of a philosophical crisis. Its implications are so fundamental for science that we must attempt to review them. This we can most conveniently do under various headings which, it must be recognized, are incommensurable. They are not divisions of the subject, but themes which suggest themselves in connexion with Galileo's life-work:

(*a*) The Mechanical World
(*b*) Extension of the Senses
(*c*) The Universe as Mathematical and Boundless
(*d*) Religion and Science

(*a*) *The Mechanical World.* The discipline of elementary mechanics, which had already been well treated by Stevin (pp. 223 f.), exists today in substantially the state in which Galileo left it. Its formulation was his real life task. Among his earliest observations were those on the pendulum—made when he was eighteen years of age. In explaining its movements, in the draft of a work on mechanics prepared a few years after these observations, he invoked the action of gravity. Nevertheless Galileo conceived no exact idea of the action of gravity—of which the pendulum is a special case—until many years later. His conclusions on that topic are embodied in his *Discourses concerning two new sciences* (1638) published when he was seventy-four years old. The wide separation in time of these two events illustrates how wholly different is the order and manner of presentation of the thought of a scientific investigator from the order and manner in which he reaches his conclusions.

In this, his final work, the results of his investigation extending over more than half a century are placed in logical or rational order. Thus their historic sequence is concealed. The process of setting forth a scientific discovery involves of necessity the covering up of its true historic sequence. This is one of the reasons that make the history of science difficult to master.

Of all Galileo's contributions to mechanical conceptions perhaps the most fundamental was that the continuous application of a force produces either an increment or decrement of velocity *at every moment*. The conception of acceleration as a constantly changing velocity *accompanying the application of force* was in contradiction to the Aristotelian principle that terrestrial bodies tend *of their own nature* to come to rest at a level which is natural for them. Acceleration, as we understand it, was one of Galileo's fundamental contributions. It involves the conception of the indefinite splitting up of time and thus of the application to time of the doctrine of limits as Archimedes had applied it to space. Through his mathematical teaching concerning moving bodies Galileo leads on to Newton.

Again, the philosophers of the Middle Ages and the mathematicians of the sixteenth century had found great difficulty in conceiving a body as the subject of several simultaneous movements. For them the type of 'perfect' motion was to be seen in the supposedly circular path of the heavenly bodies. Galileo by introducing the idea of acceleration, and especially of acceleration as natural to falling bodies, made familiar the idea of compounded motion. By his analysis of the path of projectiles (p. 235) he introduced that view into curvilinear as well as rectilinear motion. Thus he paved the way for Newton's synthesis of terrestrial and celestial mechanics.

Moreover, Galileo's developments of the science of mechanics were applicable to all visible and tangible objects. His conception of a mechanical universe swiftly reacted even on the biological sciences. In the rebound of sentiment against Aristotle, biologists sought to explain the animal body as a machine. The first important biological works of the seventeenth century—for example those of Santorio (p. 275), of Harvey (pp. 275 f.), or Descartes (pp. 277 f.)—all sought thus to explain the body. Though Galileo in general eschewed the investigation of living things, in this matter he was himself a pioneer.

Arising out of this principle he shows it to be impossible for a swiftly moving land animal to increase its size, retaining its proportion of parts, and at the same time to maintain its agility. Increase in size increases weight as the cube of the length, but areas of cross-section of bones or muscles only as the square of the length. He writes:

> You cannot increase the size of structures indefinitely in art or nature. It would be impossible so to build the structures of men or animals as to hold together and function, for increase can be effected only by using a stronger material or by enlarging the bones and thus changing their shape to a monstrosity. In illustration I sketch a bone [a femur] whose length has been increased three times but the thickness multiplied until, for a correspondingly large animal, it would perform the same function as the small bone for a small animal. You can see how out of proportion it is. To maintain in a giant the limb-proportion of an ordinary man you must find a stronger material or he will fall and be crushed under his own weight. If the body-size be diminished, however, the strength is not proportionally lessened; indeed the smaller the body the greater the relative strength. A small dog could carry two or three dogs of his own size on his back but I doubt if a horse could carry even one of his own size. [Much abbreviated from *Discourses concerning two new sciences*, 1638.] (Fig. 101.)

But Galileo also saw that if an animal be immersed in water, then the weight is counterpoised to the extent of an equivalent volume of water. Under such circumstances the character of the physical barrier to increase in size is altered. Consider the grace of a swimming seal.

The importance of this principle was for long not adequately appreciated. Each species has, of its physiological nature, a limit of growth, which is enormously higher for certain water animals, as whales among vertebrates and cuttle fish among invertebrates, than for any land animals. The change in the proportions of the parts during growth had, in fact, already been made a subject of special study by the artist Albrecht Dürer a century earlier. But Dürer had not been able to subject the process to basic mathematical analysis as did Galileo (p. 251 and Fig. 101).

In sum Galileo produced a conception of a world in which search might reasonably be made for mechanical principles alike in the movements of the heavens and the changes on the earth, in the circulation of a planet's satellites as in the structure of a minute insect. It is an increasingly mechanic world with which men of science have henceforth to deal. Astrology had laid sacrilegious hands on the heavens. The new determinism was to be a much more intimate thing which concerned the stars no less than men, and men no more than mice. This was evident enough to the lofty genius of Spinoza, but these complications of the mechanical conception of the world were almost wholly missed by Galileo's leading opponents. They saw in him, as they had seen in the astrologers, merely another disturber of traditional religion. Had the real nature of the Galilean revolution been realized, it would have fared even worse than it did with its author and his followers.

We may here say a word concerning Galileo's opponents. They have been the objects of contumely because of his humiliating treatment by those in high authority in the Church. We need not stop to defend these inane pomposities, nor need we pause to denounce the foolishness of other of his opponents. But not all those who were opposed to Galileo were fools or rogues. A great body of not unreasonable opinion hesitated to accept his physical philosophy. It is fair to remember that a complete system of philosophy, weaving into one vast scheme the moral and physical, the terrestrial and celestial worlds, had been built up during the Middle Ages. This satisfied the need of the day. The fact that Galileo had made a breach in that scheme was no clear reason to abandon the whole. Would the fact that recognized scientific laws were shown to be inapplicable to some particular group of phenomena be a reason nowadays for abandoning the scientific method of exploring nature? Remember that Galileo had to offer his audience no complete system even of physical philosophy—that was reserved for successors of Newton. Even if the contemporary critic were a specialist in physics—and such were few in those days—a reasonable attitude would surely have been one of friendly and non-committal suspension of judgement not so much as to Galileo's findings, but as to their implications.

It is true that the older astronomical position had been shaken also by Kepler's demonstration that the movements of the planets are more easily understood if we suppose them to follow elliptical and not circular courses. But Galilean physics and Keplerian astronomy had hardly yet been linked. That again was reserved for Newton's generation. Moreover many of the best exemplars of Aristotle's science were taken from the world of life. Aristotle's biological system was still valid though it was the Aristotelian physical system that Galileo was attacking. Further, as things then stood, abandonment of the Aristotelian scheme of the universe meant abandonment of much religious teaching. We are entitled to expect that judges should be both just and merciful. The judges of Galileo were doubtless neither. But the facts of human nature offer no warrant for the hope that most men will have the insight and understanding of a great master's immediate following. *Suspension* of judgement as to the validity of Galileo's arguments was thus a necessary consequence of the imperfection of human nature. To note this is not to justify the ignorance, the duplicity, or the cruelty of some of his opponents.

Apart from professional theologians on the one hand and Spinozists on the other, most reasonable men in the later seventeenth century were content with a compromise. 'The heavens are the heavens of the Lord; But the earth hath he given to the children of men.' This was the atmosphere in which arose and flourished the great scientific movement of the age. There is something in its favour still.

(*b*) *Extension of the Senses.* Galileo is best remembered for his wonderful astronomical observations. But at the back of these lay his effective invention of the telescope and his successive improvements in the construction of that instrument. And at the back of that lies yet another movement, the introduction of the skilled mechanic into the service of science. In this movement, too, Galileo may be said to be an important figure.

Apart from the striking changes, artistic, literary, intellectual, during the fifteenth and sixteenth centuries, there were other changes, less dramatic but affecting even more closely and deeply the lives of men. One of these was the refinement of craftsmanship incident on the greater accessibility of good steel for tools. This was conditioned by the general replacement of the old bloomeryhearth by the blast-furnace, the chief medieval technical advance. The houses, the furniture, the apparatus of life of, say, 1600, show great improvement on 1450 (pp. 200 f.). An exhibition of that improvement were seagoing ships which had made transoceanic exploration possible. One reason for the forwardness of Germany and of Germans in the art of printing was the excellence of German metalwork. Regiomontanus left Hungary for Nuremberg (1470) because he could there obtain good workers for his astronomical instruments (p. 196). But until the seventeenth century highly skilled craftsmen were seldom invoked by the man of science.[1] No small part of Galileo's success as an experimenter was due to his employment of trained mechanics. That class became essential to the advancement of science in the centuries which followed. Compound optical apparatus had been constructed by others before Galileo. The results obtained were negative till the great discoverer perfected the method of manufacture. They were felt not least in England where a rising instrument-making profession was related to the movement which led to the Royal Society and to overseas trade.

With such instruments in his hand Galileo was in a position to observe with an accuracy and a detail that had been previously quite unknown. He is the effective inventor of the telescope and the father of modern observational astronomy. There is, however, another aspect of his optical discoveries that is less often recalled. He is the inventor also of the compound microscope, and, indeed, revelations of that instrument are mentioned in one of the earliest independent accounts of his work. The minute world revealed by this instrument was almost as wonderful as the new discoveries in the starry sphere. The heavens had always been recognized as vast almost beyond the power of thought. But the incomparable complexity of life and of matter close at hand was a wholly new conception. That beings, minute beyond the powers of our vision, could have structures as complete and complex as ourselves was a truly startling thought. If there was world beyond world in the heavens there was world beyond world within us.

[1] A prominent exception was Tycho Brahe. Other exceptions are the German artists and woodcutters employed by the anatomists and botanists.

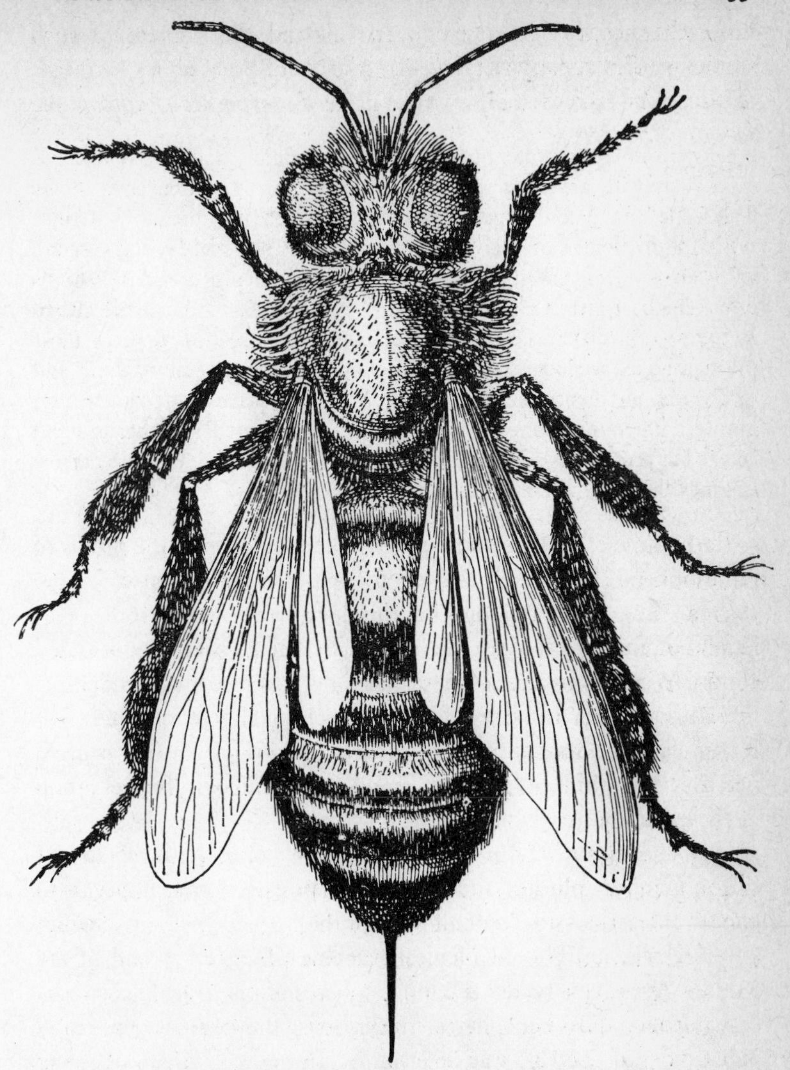

FIG. 108. Stelluti's figure of a bee of original size (1625).

It is interesting to see how these matters looked to the first generation of professed microscopists. One of Galileo's microscopes had been used by a colleague as early as 1625 to illustrate the minute structure of insects (Fig. 108). For the next forty years the microscope was treated as a mere curiosity but in the sixties it became a serious instrument of research. In England the pioneer of such studies was HENRY POWER (1623–8), disciple of Sir Thomas Browne, of Norwich (1605–82), who writes in his *Experimental Philosophy*:

> Dioptrical Glasses are but a Modern Invention neither do Records furnish us with anything that does antedate our late discoveries of the Telescope and Microscope. The want of which incomparable Artifice made the Ancients not onely erre in their fond Coelestial Hypothesis and Crystalline wheelwork of the Heavens but also in their nearer observations of the smallest sort of Creatures which have been perfunctorily described as the disregarded pieces and huslement of the Creation. . . . In these pretty Engines are lodged all the perfections of the largest animals: . . . and that which augments the miracle, all these in so narrow a room neither interfere nor impede one another in their operations. Ruder heads stand amazed at prodigious and Colossean pieces of Nature, but in these narrow Engines there is more curious Mathematicks. (1663.)

In the time of Galileo atomic views had long been in the minds of philosophers. There was as yet no experimental evidence for the existence and nature of atoms but this doctrine seemed to fit the revelations of the microscope. Were the tiny beings that they revealed atoms? Were atoms alive? These questions gave rise to a considerable literature which, since it led nowhere, has been almost forgotten. Yet it stimulated current speculation, and the curiosity which it aroused had a very definite influence in directing the biological observation of the generations which followed.

(*c*) *The Universe as Mathematical and Boundless.* With the advent of the Galilean physics and the Keplerian astronomy, it began to appear at least possible that all parts of the universe were mechanically interrelated. The astrological teaching of antiquity and of the Middle Ages had treated the inner spheres of the world scheme as dependent on the outer spheres. In this sense the extreme expression of astrological doctrine was determinist. But now Galileo, following Bruno and Gilbert, thought of the world as boundless. In such a universe no part could be said to be inner, none outer, none centre, none circumference. In such a universe the mechanics of one part are presumably the mechanics of another, though evidence for this had to await Newton. Of such a boundless universe no beginning in time can be intelligibly predicated. This difficulty is with us still.

The implications of this view represent a series of enormous changes some of which we have already suggested. Especially it affected the conception of the task of the man of science.

The physical world, in the thought of Galileo, was a separate and mathematical conception, a piece of machinery, the action of any part of which was calculable. It was thus quite separate from the moral world with which it had been united in the medieval scheme. The knowledge of the world as a whole—philosophy—was thus divided into two categories, *natural philosophy* and *moral philosophy*, a distinction which is still recalled in the naming of the departments of the university where Newton taught. In the main we may say that the division has held from Galileo's day to our own.

A further implication of the conception of a boundless physical universe and the separation of natural from moral philosophy is the movement known in modern times as 'scientific specialization'. Science, natural philosophy, proceeds on the information given by the senses. The line of its attack is thus limited and we cannot hope that anything but limited objectives can be reached. Science does not profess to solve ultimate problems. On the other hand it does seek to solve its limited problems with a known degree of accuracy and a known margin of error. The desire for exact expression and for the translation of observation into terms of measurement has penetrated every department of science from the time of Galileo onwards. Even the biological sciences have been affected. The physico-mathematical form in which the biological works of Santorio (1600), Harvey (1628), and Descartes (1664) are cast may be contrasted with the beautiful but not mathematically controlled works of the 'German fathers of botany' (1530–42, pp. 207 f.) or of Vesalius (1543, pp. 210 f.). Since the work of Galileo there has always been a group of biologists that has sought to represent biology as a department of physics.

(*d*) *Religion and Science*. Medieval philosophy had presented a view of the world as a whole. Looking back on it, from our modern point of view, we can see two breaches of continuity. One is between the celestial and the terrestrial, the other between the living and the not-living. These two gaps were, however, well concealed from all but the most acute, until displayed in the seventeenth century by the work of such men as Galileo and Harvey. But thought could not rest content with the multiple system thus revealed. There is an insatiable demand for explanation of the world on a unitary basis. Law must reign, and if not divine law then physical law. This call for an explanation of all things in terms simpler than themselves was first met in modern times by the philosophy of Descartes (pp. 259 f.).

The conception of a mechanical and mathematical universe affected other philosophers whose world schemes have endured better than that of Descartes. The model suggested by the new science of mechanics involves the belief that any event in one part of the world must, of necessity, have its consequences in another part. Each event gives rise to its own chain, circle, sphere of events. Events are never without consequences which go on like waves from the dropping of a pebble in still water, producing ever widening if less apparent circles which are reflected and reflected again from the margins of the pool. This view of the world was essential to the thought of Spinoza (1632–77). It does not allow for the annihilation of either matter or energy. The belief in the conservation of both was implicit in all of Galileo's work though not expressed until he had been dead for two hundred years. It is implicit also in the thought of many who would be as astonished as Galileo at the characters of the company they keep:

> All things by immortal power,
> Near or far,
> Hiddenly
> To each other linkéd are,
> That thou canst not stir a flower
> Without troubling a star.
>
> (Francis Thompson (1859–1907), *Oriental Ode*.)

The whole question leads on to the philosophical problem of 'causality' where we cannot follow it. But science, true to its principle of limited attacks and limited objectives, has its own working rules of causality. It follows Galileo in agreeing to discuss only certain particular types of sequence and treating them as related, the relation being regarded as cause and effect. Thus the physicist will deal only with physical, the chemist only with chemical sequences, the biologist only with biological sequences. In the course of this process new relationships may be discerned or become more apparent, as for instance in the physical state of the heavenly bodies or of the relative constitution of parents and offspring. Thus will arise new sciences—astrophysics and genetics—which will limit their scope to the relations in their particular fields. All departments will agree, however, that only those sequences shall be considered that can be measured or at least estimated. From Galileo's day onward we see science as measurement, though there remain still important scientific departments in which measurement takes a secondary place.

But since science must limit its objectives, the world based on science, as Galileo the artist well knew, is not a complete world. The appearance of our world depends on how we look at our world—that is, on our 'mood'. We may be in a scientific, an artistic, an emotional, a social mood. The resultant of all the ways that we have of looking at our world—the resultant of our moods—is, in effect, our religion. Galileo founded a new conception of the world—he almost founded a mood in which to regard it. In doing so he certainly affected the religion of all men who are able to accept or partake of his mood. But to say that that mood was all of Galileo, to say that the universe as he looked at it was wholly mathematical and physical, is not only going beyond his teaching but also going beyond all that we can learn of the nature of the man. Reasons are doubtless at hand for the rejection of any established religious formula, but it would be perverting the historical record to ascribe the desire to do so to Galileo or to men of science in general.

8. PROPHETS OF SCIENCE

RENÉ DESCARTES (1596–1650), the 'first modern philosopher' and most dominant thinker of the seventeenth century, made striking contributions both to scientific theory and practice (p. 226).

(*a*) He set forth views as to how science should be prosecuted.

(*b*) He was the first in modern times to propound a unitary theory of the universe that became widely current.

(*c*) He made important contributions to mathematical, physical, and physiological science.

These three activities of Descartes are not as essentially connected as he would have wished. In 1633 he was about to publish his cosmic view in a work which he termed *The World*, when he heard of the condemnation of Galileo. He promptly withdrew the book. In the event his first publication was the *Discourse on Method* (1637).

(*a*) *Descartes on Scientific Method*

From an early date Descartes felt great dissatisfaction with the results of the usual studies of his time. It seemed to him that there was no clear distinction between facts, theories, and tradition. Want of clarity was abhorrent to him. He attempted to divest himself of every preconceived notion and then to build up his knowledge. With this end in view he tells us in his *Discourse* that he made certain resolutions:

(i) Never to accept anything for true which he did not clearly know to be such, avoiding precipitancy and prejudice, and comprising nothing more in his judgement than was absolutely clear and distinct in his mind.

(ii) To divide each of the difficulties under examination into as many parts as possible.

(iii) To proceed in his thoughts always from the simplest and easiest to the more complex, assigning in thought a certain order even to those objects which in their own nature do not stand in a relation of antecedence and sequence—i.e. to seek relation everywhere.

(iv) To make enumerations so complete and reviews so general that he might be assured that nothing was omitted.

He believed that such truth as is ascertainable is so only by the application of these principles. These, he thought, are the true principles of science, and only by their application can science advance. They apply, he held, as much in the sphere of religion as in mathematical or physical matters. In essence, therefore, revealed religion, in the ordinary sense, is superfluous. For him the fundamental test of truth is the clearness with which we apprehend it. *I think*, *therefore I am*, is the most clearly apprehended of all truths, and, therefore, personality cannot be an illusion. Similarly, to him, the conception of the soul as separate from the body was clear and even obvious; therefore, he maintained, it must be true. Moreover, he considered that the mind could not create something greater than itself. Therefore, the conception of infinite perfection, transcending humanity, must have been put into our minds by infinite perfection itself; that is, by God.

It is noteworthy that in reaching his scientific results he did not employ the method that he advocates. It is doubtful if anyone actively and successfully prosecuting scientific discovery has ever or could ever proceed on the lines that he lays down. It may, indeed, be doubted whether scientific discovery ever follows any prearranged system. 'The spirit bloweth where it listeth' and discovery is a thing of the spirit. There is no one method of discovery but as many methods as there are discoverers. There is no human faculty or power that has not at times been pressed into the service of scientific discovery. There *is* a method of scientific *demonstration*, but that is a very different thing from a method of discovery. The setting forth of the one must almost necessarily conceal the nature of the other. We, therefore, consider Descartes separately as a scientific discoverer and as a prophet and critic of science.

Of the achievement of Galileo, Descartes formed no high estimate. Galileo was eliciting mechanical laws. Descartes belittled this effort since it included no analysis of the basic conceptions with which Galileo was dealing, force, motion, matter, space, time, number, extension, and the like. The obvious retort is that had Galileo done these things, philosophy might have been richer, but science would certainly have been poorer in being deprived of the most successful experimenter and the most acute exponent of natural law that has yet arisen.

(*b*) *Descartes's Cosmology*

We may now turn to the conception of the material universe as formed by Descartes. Here, too, we may honour him as a pioneer, while we regret that he is less critical of himself than of others. The form of the world, according to him, is inevitable, in the sense that, had God created more worlds, 'provided only that He had established certain laws of nature and had lent them His concurrence to act as is their wont, the physical features of these worlds would inevitably form as they have done on ours'. Descartes accepts the probability of creation of matter as a momentary act, but holds that this act of creation was the same as that by which creation is now sustained.

Descartes regards the universe as infinite and devoid of any empty space. The primary quality of matter is extension, but there are also *derived* (not *secondary*) qualities of divisibility and mobility, which are created by God. We may connect the assertion of Descartes that divisibility and mobility are derived qualities with the formulation of the law that matter, in so far as it is unaffected by extraneous forces, remains in motion or at rest.

He regarded matter as uniform—i.e. made of the same basic stuff—though divided and figured in endless variety. Matter is closely packed, without any vacuum. Therefore, the movement of any part of matter produces movement of all matter. It thus follows that throughout the universe there are movements or vortices of material particles that vary in size and velocity. If one considers any limited part of the universe, the particles in it, as they whirl around their vortices, get their corners rubbed off. These being rubbed finer and finer become a minutely divided dust which tends to centripetal action. This fine dust is 'first matter' and forms the sun and the stars. Ultimately these spherical globules acquire a contrary or centrifugal action. They then form 'second matter', which constitutes the atmosphere enveloping first matter. The centrifugal tendency of the second matter produces rays of light which come in waves from the sun or the stars to our eyes. In the process of vortex formation particles are liable to get detained on their way to the centre. These settle round the edge of the sun or star, like froth or foam. This 'third matter' can be recognized as the sun-spots (p. 245) and certain other celestial phenomena. Major vortices are responsible for planetary movements, minor vortices for terrestrial phenomena. The action of gravity is identified with centripetal action of a vortex.

The theory of vortices failed to explain a multitude of known phenomena, including Kepler's laws of planetary motion (pp. 240–1). It became, however, very fashionable. It was elaborated and a whole system of physics and cosmology erected on it. It survived in France until near the middle of the eighteenth century though it had less influence in other countries, and it was made untenable by the work of Newton. It exemplifies one of the many great blind-alleys of science.

(*c*) *Descartes on the Nature of Man*

For the completeness of his system it was necessary for Descartes to include the phenomena presented by living things. Here, too, his work was of a pioneer character though he invented a number of structures and functions that had no existence outside his mind. The analogies that he draws, however, are sometimes both striking and valuable:

> I remained satisfied that God first formed the body of man wholly like to ours, as well in outward shape as in inward conformation, and of the same matter; that at first He placed in it no rational soul, nor any other principle, beyond kindling in the heart a flameless fire similar, as I think, to the heat generated in damp hay, or to that which causes fermentation in must. (*L'homme* published 1664, written before 1637).

Descartes was here trying to integrate combustion, metabolism, respiration, and fermentation.

> For, when I examined the kind of functions which might, as consequences of this supposition, exist in this body, I found precisely all those which may exist in us independently of all power of thinking, and consequently *without being in any measure owing to the soul*; in other words, to that part of us which is distinct from the body, and from that of which it has been said above that the nature distinctly consists in thinking—functions in which the animals void of reason may be said wholly to resemble us; but among which I could not discover any of those that, as dependent on thought alone, belong to us as man, while, on the other hand, I did afterwards discover these as soon as I supposed God to have created a rational soul, and to have annexed it to this body. (Italics inserted.)

He thus considered that man once existed without a rational soul and that animals are still automata. He knew, for instance, William Harvey's account of the circulation of the blood, and he based upon it a most elaborate and carefully worked-out theory of the action of the animal body. Man, however, at least in his present state, Descartes considered to differ from animals, in the possession of a soul. This he believed to be especially associated with a particular part of the body, the pineal gland, a structure within the brain which, in his erroneous opinion, was not found in animals. In the pineal gland two clear and distinct ideas produce an absolute mystery (Fig. 112).

The Cartesian philosophy was the first complete and coherent system of modern times. It rapidly found adherents, spread in every country, and was popular for several generations. In Descartes's native land it won its way even among churchmen. Gradually, however, the numerous physical errors which it involved were exposed. Towards the end of the century the theory of vortices became quite untenable. It was then shown to be inconsistent with astronomical observation, and to harmonize neither with the cosmical system of Newton nor with the revived atomic theory. As an explanation of cosmic phenomena it could no longer be held. Important scientific works that professed to be based on the Cartesian system appeared, however, as late as the middle of the eighteenth century.

Further, the advance of physiological knowledge exposed basic errors of Descartes in the interpretation of the workings of the animal body. Descartes, however, had laid the foundations of modern philosophy, and from his time on there has been a continuous chain of thinkers who have claimed to interpret the world by the unaided powers of their own minds.

(*d*) *Francis Bacon*

Less adapted than Descartes by his powers, his temper, and his outlook to make a great philosophical synthesis was FRANCIS BACON Lord Verulam (1561–1626). The Englishman was, moreover, less efficient in the handling of scientific material and incomparably below Descartes in mathematical capacity. Despite the fact that Bacon was the older of the two, his influence made itself felt somewhat later than that of Descartes. Bacon's experimental ineffectiveness dims his light for us. But the generation after him thought only of his prophetic power, though his own might have endorsed Harvey's shrewd estimate of his patient. 'He writes philosophy [i.e. science] like a Lord Chancellor.' While no one ever worked on the scientific principles laid down by Bacon, yet he inspired many for the founders of the Royal Society (*c.* 1662) who held themselves to be his disciples. They held in special respect Bacon's *Of the Proficiencie and Advancement of Learning* (1605) and the *Instauratio magna* or *Novum Organum* (1620).

Let us consider Bacon's attitude towards the investigation of

Nature as set forth in these works. What was this new scientific process which he practised worse than he preached? Bacon was for conducting his investigations by collecting all the facts. This done, he thought, the facts might be passed through a sort of automatic logical mill. The results would then emerge. But this method cannot be applied in practice, since facts, phenomena, are infinite in number. Therefore, we must somehow choose from among them, though Bacon thought otherwise. How then shall we choose our facts?

Experience shows that only they choose profitably who have a knowledge of how their predecessors have succeeded or failed in their choosing. In other words, the process of choosing facts is an *act of judgement* on the part of the *learned chooser*, the man of science. So it is also with the process of choosing words on the part of the word-chooser whom we call a poet. The choice of the man of science, as of the poet, is controlled by knowledge of his art—of 'his subject' as we are wont to call it at the universities or in the laboratories. The man of science, like the poet, exercises his judgement to select those things which bear a certain relation to each other. And yet no skill in reasoning, however deft, no knowledge of the nature of scientific method, however profound, no acquaintance with his science, however complete, will make a scientific discoverer. Nor, for that matter, will any learning in the lore of metre or in the nature and history of poetry make a poet. Men of science, like poets, can be shaped, but they cannot be made. They need that incommunicable power of judgement.

The scientific man in the prosecution of his art of discovery has to practise three distinguishable mental processes. These may be considered as firstly, the choosing of his facts; secondly, the formation of an hypothesis that links them together; and thirdly, the testing of the truth or falsehood of the hypothesis. When this hypothesis answers numerous and/or stringent tests, he has made what is usually called a 'scientific discovery'. It is doubtless true that the three processes of choosing facts, drawing an hypothesis or conclusion, and testing the conclusion, are often confused in his own thinking by the man of science. Often, too, his demonstration of his discovery, that is the testing of his hypothesis, helps him, more or less unconsciously, to new acts of judgement, these to a new selection of facts, and so on in endless complexity. But essentially the three processes are distinct, and one might be largely developed while the others were in a state of relative arrest.

In this matter scientific articles, and especially scientific text-books, habitually give a false impression. These scientific works are composed to demonstrate the truth of certain views. In doing so they must needs obscure the process by which the investigator reached those views. That process consists, in effect, of a series of improvised judgements, or 'working hypotheses', interspersed with imperfect and merely provisional demonstrations. Many hypotheses and many demonstrations have had to be discarded when submitted to a further process of testing. Thus a scientific article or book which tells nothing of these side issues, blind alleys, and false starts tends, in some sort, to conceal the tracks of the investigator. For this reason, among others, science can never be learned from books, but only by contact with phenomena.

The distinction between the *process* of discovery and the *demonstration* of discovery was constantly missed during the Middle Ages. On this point, in which our thought is separated from that of the men of those times, Bacon remained in darkness. He succeeded, indeed, in emphasizing the importance of the operation of collection of facts. He failed to perceive how deeply the act of judgement must be involved in the effective collection of facts.

As an insurance against bias in the collection and error in the consideration of facts, Bacon warned men against his four famous *Idols*, four false notions, or erroneous ways of looking at Nature. There were the *Idols of the Tribe*, fallacies inherent in humankind in general, and notably man's proneness to suppose in Nature greater order than is actually there. There were the *Idols of the Cave*, errors inherent in our individual constitution, our private and particular prejudices, as we may term them. There were the *Idols of the Market-place*, errors arising from received systems of thought. There were the *Idols of the Theatre*, errors arising from the influence of mere words over our minds (*Novum Organum*, 1620).

But did not Bacon himself fail to discern a fifth set of idols? These we may term the *Idols of the Academy*. Their worship involves the fallacy of supposing that a blind though learned rule can take the place of judgement. It was this that prevented Bacon from entering into the promised land, of which but a Pisgah view was granted him.

Yet despite Bacon's failure in the practical application of his method, the world owes to him some conceptions of high importance for the development of science.

(*a*) He set forth the widening intellectual breach which separated his day from the Middle Ages. He perceived the vices of the scholastic method. In the clarity and vigour with which he denounced these vices, he stands above those of his contemporaries who were striving toward a new form of intellectual activity.

(*b*) He perceived, better than any of his day, the extreme difficulty of ascertaining the facts of nature. He forecast the critical discussion that characterizes modern science. He missed, however, the important point that the delicate process of observation is so closely interlocked with discussion that both must almost necessarily be performed by the same worker.

(*c*) English writers of the later seventeenth century concur in ascribing to the impetus of Bacon's writings the foundation of the Royal Society. Thomas Sprat (1635–1713), Bishop of Rochester, the first historian of the Society, assures us of this (1677), as do Oldenburg and Wilkins, its first secretaries. The opinion is fully confirmed by Robert Boyle (1627–91), the most effective of its founders, and by John Locke (1632–1704), the greatest of English philosophers.

(*d*) It is, perhaps, in the department of psychological speculation that the influence of Bacon has been most direct. The basic principle of the philosophy of John Locke is that all our ideas are ultimately the product of sensation (*Essay concerning Human Understanding*, 1690). This conception is implicit in Bacon's great work, his *Novum Organum* (1620). Through the 'practical' tendency of his philosophy and especially through Locke, Bacon was the father of certain characteristically English schools of thought in psychology and ethics. These have affected deeply the subsequent course of scientific development.

Whatever his scientific failures, we may thus accord to Bacon his own claim that 'he rang the bell which called the wits together'. The 'wits' were the Fellows of the Royal Society.

9. CHARACTER AND CONDUCT OF MATTER

Our word *matter* is derived from the Latin *materia*, which in its turn is connected with *mater*, 'mother'. Originally *materia* was a general term for the stuff of which things are composed and especially things employed in buildings. So in the medieval nomenclature and in that of the alchemists *materia prima* was the stuff of which all things were built, the 'primal matter' that lay at the back of all four elements. Both alchemists and the medieval philosophers were prepared to believe that matter of one type, by a mere rearrangement of its four elements, could be transformed into matter of another type. Nor were they convinced that, in some circumstances, matter might not appear 'out of the air' or out of nothing. They did not in general regard air as possessing weight, and some of them would have claimed that it had 'negative weight' since like fire it tended to rise. Nevertheless, it is not exactly true to say that the medieval writers had no idea at all of what we call the 'conservation of matter'. Had that been so, no trade that used weights would have been possible. With no idea of constancy in weight such stories as that of Hiero's crown (p. 70) would have been meaningless, the alchemists and assayers would have had no use for delicate scales, and Stevin and Galileo could not have begun their work. We would rather say that in the Middle Ages the idea of conservation of matter was indefinite, inexact, unexpressed, and implicit, whereas now it became definite, exact, formulated, and explicit. Three centuries of application of experimental methods made this difference.

There was one particular aspect of matter that had special bearings on the early development of modern ideas on the subject. The question as to the nature of the air that we breathe and whether or not it has weight had been debated since antiquity. One of the most popular of the pagan systems of physical philosophy—to which Galen adhered—held that the 'pneuma' of the world soul is inhaled during the act of breathing, which on that account is necessary for life. On the cessation of breathing the individual soul joined again the world soul (p. 99). Such a view was contrary to the medieval Christian attitude. Medieval Christian thought paid little attention to the question whether the air has weight. In the fourteenth century Peter of Abano, on theoretical grounds, held that it had. In the fifteenth century Nicolas of Cusa (p. 178), despite his stress on the use of the balance ignored the possibility that the growing plant might draw anything material from the air. The subject was given a new aspect by van Helmont.

The Belgian, JAN BAPTIST VAN HELMONT (1577–1644), was a pious mystic who devoted his life to the investigation of chemical processes, basing himself primarily on the views of Paracelsus (pp. 199 f.). He published little. Soon after he died his son, who occupied himself with similar pursuits, collected his father's writings and issued them as *The Fount of Medicine* (*Ortus medicinae*, 1648). These writings are in extremely obscure language. Moreover, the alchemical school, to which van Helmont belonged, was justly despised by clear thinkers, such as Galileo and Descartes, who were attacking Aristotelianism and contributing to the upbuilding of the new physical philosophy. Thus van Helmont exerted little influence on scientific writings until his works were translated into the vernaculars and interpreted in the sixties of the seventeenth century.

Van Helmont concluded from a repetition of the experiments of Nicolas of Cusa (p. 178) that plants draw their whole substance from water. (He did not, of course, know the part played by the atmosphere, especially by carbon dioxide, in the growth of plants.) Nevertheless, he showed that vapours, though similar in appearance, may be very different in character and conduct. In other words, there are many kinds of 'gas'. The idea is so familiar to us that it is hard to realize it as an innovation. Yet the very word *gas* was invented by van Helmont. Etymologically it is *chaos* phonetically transmuted in his native Flemish speech.

Galileo himself was aware that the atmosphere has weight. Nevertheless, he too failed to invoke it to explain the failure of a suction pump to lift water higher than 35 feet. The explanation was adduced by Galileo's pupil and secretary, EVANGELISTA TORRICELLI (1608–47). He reasoned that as mercury is about 14 times as heavy as water the atmosphere should support $\frac{35}{14}$, i.e. about $2\frac{1}{2}$ feet of mercury. He selected a glass tube of $\frac{1}{4}$-inch calibre and 4 feet long and closed at one end. This he filled with mercury, applied his finger to the open end and inverted it in a basin of mercury. The mercury sank at once to 2½ feet above the basin, leaving 1½ feet apparently empty (1643). This was the *Torricellian vacuum*, as it came to be called.

Torricelli had in fact constructed a barometer (Greek 'weight measurer'). He observed that at times his barometer stood higher than at other times. He inferred that when the barometer stood high the air was heavier; when low, lighter. Descartes predicted that at greater altitudes the mercury column would stand lower since there was less atmosphere to support it. Pascal (1646) confirmed this (1623–62). The matter was further investigated by Boyle, Huygens, Halley, Leibniz, and others. The barometer has since been greatly improved, but in essence it is still that suggested by Torricelli.

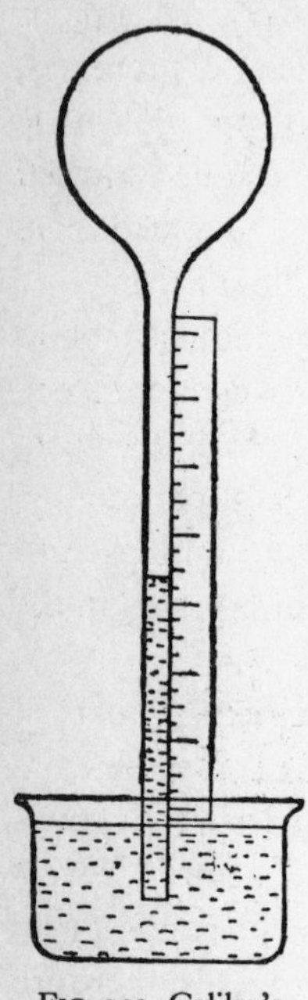

FIG. 109. Galileo's thermometer.

The thermometer has had a somewhat different history. An air thermometer was invented by Galileo. It consisted of a glass bulb containing air connected to a glass tube dipping into a liquid (Fig. 109). It was very sensitive to temperature changes, but was very inexact as it was also subject to barometric changes. About 1612 Galileo invented the modern type of sealed tube with glass bulb filled with liquid. Technical difficulties in construction, however, prevented an accurate instrument from being made until the eighteenth century (pp. 348 ff.).

Very great advances in our knowledge of physical and chemical states are due to the air-pump. This instrument was invented in 1656 by OTTO VON GUERICKE (1602–86), burgomaster of Magdeburg in Prussia. With it he gave a convincing demonstration of the force of atmospheric pressure by the 'Magdeburg hemispheres' which, though easily separable under normal conditions, could not be separated by two teams of eight horses each when he had drawn out the air with his air-pump. Guericke also invented the first electrical machine. It consisted of a globe of sulphur which was made to rotate. Pressure of the hands upon the rotating globe charged it electrically. He also showed that bodies charged with the same kind of electricity repel each other.

The air-pump of Guericke was considerably improved (1658–9) by ROBERT HOOKE (1635–1703) working at Oxford for his employer ROBERT BOYLE (1627–91). Hooke was one of the most skilful and in-

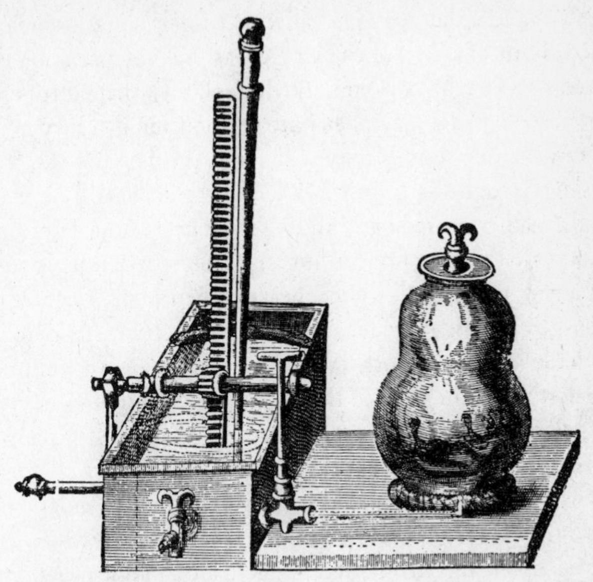

FIG. 110. One of Boyle's air-pumps. A cat in the receiver shows signs of asphixia when the air is exhausted by the pump.

genious of physical experimenters, Boyle one of the ablest and most suggestive of scientific investigators. A large part of the foundations of the modern sciences of chemistry and physics in their various departments was laid down by these two men.

By means of the air-pump Boyle and Hooke examined the elasticity, compressibility, and weight of the air (1660). The necessity of air for respiration and combustion was later demonstrated by the same instrument (1662, Fig. 110). Finally, Boyle showed that a part only of the air was used in the process of respiration or combustion. The matter was well expressed by Hooke in his great work *Micrographia* (1665):

The dissolution of sulphureous bodies is made by a substance inherent and mixt with the Air, that is like, if not the very same, with that which is fixt in Salt-peter. . . . That shining body which we call flame is nothing else but a mixture of Air and volatile parts of combustible sulphureous bodies which are acting upon each other whilst they ascend.

This substance 'inherent and mixed with the air' we call oxygen. In this sense Hooke and Boyle may be regarded as its discoverers.

Boyle's name is familiarly recalled in 'Boyle's law' which states that the volume of a gas varies inversely as the pressure upon it, provided temperature be constant. Boyle took a U-shaped tube with a shorter closed and a longer open limb. By pouring mercury into it he cut off air in the short limb and, by shaking, the mercury was brought to the same level in both limbs. The air in the short limb was now under atmospheric pressure. Adding mercury to the long limb he could increase the pressure continuously, thereby reducing the bulk of contained air. Thus when the barometric pressure stood at 30 inches, by adding mercury in the long limb till it stood 30 inches above the level in the short limb, the pressure on the imprisoned air was doubled. The bulk of that air was then found to be reduced to one half. Under three times the atmospheric pressure it was reduced to a third, and so on. Moreover, he could reverse the process.

Boyle's more purely chemical investigations and speculations were of high importance. His most famous work, the *Sceptical Chymist* (1661), opens the modern period of chemistry, and marks the end of the doctrine of the four elements of the Aristotelians.

To prevent mistakes [he says] I must advertize to you, that I now mean by Elements . . . certain Primitive and simple . . . bodies; which not being made of any other bodies, or of one another, are the Ingredients of which all those call'd perfectly mixt Bodies are immediately compounded and into which they are ultimately resolved.

This, in effect, is the modern definition of an element. There can be little doubt that he derived his view of chemical elements in part from the modest German teacher, JOACHIM JUNG (1587–1657) of Hamburg. Jung had enunciated similar views as early as 1634 and published them in 1642. Boyle had received a draft of Jung's physical philosophy in a letter received by him in 1654.

Among other important contributions of Boyle must be included the suggestion of chemical 'indicators' for testing the acidity or alkalinity of liquids, and his isolation of elemental phosphorus. He was extremely active in the scientific life of the later seventeenth century. Almost every aspect of contemporary science is discussed in the course of his numerous and diffuse works.

There is one doctrine popularized by Boyle to which we must pay especial attention. In his *Origin of Forms and Qualities* (1666) he definitely 'espoused an atomical philosophy, corrected and purged from the wild fancies and extravagancies of the first inventors of it'. He assumes the existence of a universal matter, common to all bodies, extended, divisible, and penetrable. This matter consists of innumerable particles, each solid, imperceptible and of its own determinate shape. 'These particles are the true *prima naturalia*.' There are also multitudes of corpuscles built up from several such particles and substantially indivisible or at least very rarely split up into their *prima naturalia*. Such secondary 'clusters' have each their own particular shape. 'Clusters' and 'prima naturalia' may adhere together. They thus form characteristic and similar groups which are not without analogy to molecules and atoms in the more modern use of these terms. Nevertheless, the resemblance of Boyle's atomism to either modern or ancient atomism is far from close.

Boyle had certainly derived his atomic views from the French philosopher, PIERRE GASSENDI (1592–1655), 'the reviver of Epicureanism'. Gassendi adapted that system of thought to the exigencies of the philosophy of his time. Boyle's nomenclature is taken direct from Gassendi who devoted at least twenty years to his great work on atomic philosophy (1649).

Some form of corpuscular philosophy was widely accepted by Boyle's contemporaries, especially in England, where it was espoused by the philosopher, JOHN LOCKE (1632–1704). The corpuscular philosophy, however, though much discussed, was not developed on the experimental side for more than a century. Chemical observations were collected in plenty and chemical science became overwhelmed by a vast number of disconnected facts and records, inadequately linked by generalizations.

An idea of the estimate which seventeenth-century thought placed upon corpuscular (or atomic) hypothesis can be gathered from John Locke's *Essay concerning Human Understanding* (1690). Whenever he deals with the ultimate physical cause of secondary qualities (p. 246) and of powers of material substances, it is to 'the corpuscularian hypothesis' that he appeals. 'These insensible corpuscles', 'the active parts of matter and the great instruments of nature', are for him the source of all secondary qualities. He maintains that if the figure, size, texture, and motion of the minute constituent parts of any two bodies could be known, then the mutual operations of those bodies could be foretold. Thus 'the dissolving of silver in *aqua-fortis* and gold in *aqua-regia* and not vice versa, would then, perhaps, be no more difficult to know than it is to a smith to understand why the turning of one key will open a lock and not the turning of another'.

10. MECHANICAL CONCEPTS IN BIOLOGY

(i) *First Application of Physics to Physiology*

Biological science, it is often said, always lags behind physical science and is always in a more elementary stage. The statement is hardly borne out by history. It depends for any truth that it may possess upon a particular conception of the nature of science. In antiquity, in the hands of Aristotle, biological science was far ahead of physical. Again the earliest modern scientific works of a monographic character, the great books of Belon, Rondelet, Vesalius, Gesner, are exclusively biological. The treatise of Copernicus is medieval by comparison, and contains very few original observations. To justify the doctrine of the relative backwardness of biological science it is necessary to postulate that the aim of biology is to represent biological phenomena in physical terms. Thus expressed the statement becomes a self-evident proposition for, if the postulate be granted, biology can never advance beyond its physical interpretation. A large school of biological thinkers does not accept this postulate. Even more numerous are those who, taking no philosophical stand in the matter, hold that, until we know far more of the chemistry and physics of the so-called 'life processes', living organisms can be profitably treated as things apart from the inorganic world. Nevertheless, it is true that the most significant biological advances of the insurgent century were attempts to express biological findings in physical terms.

The first to apply the new physical philosophy to biological matters was SANTORIO SANTORIO (1561–1636), a professor of medicine at Padua, in his little tract *De medicina statica* (1614). Inspired by the methods of Galileo who had been his colleague at Padua, he sought to compare the weights of the human body at different times and in different circumstances. He found that the body loses weight by mere exposure, a process which he assigned to 'insensible perspiration'. His experiments laid the foundation of the modern study of

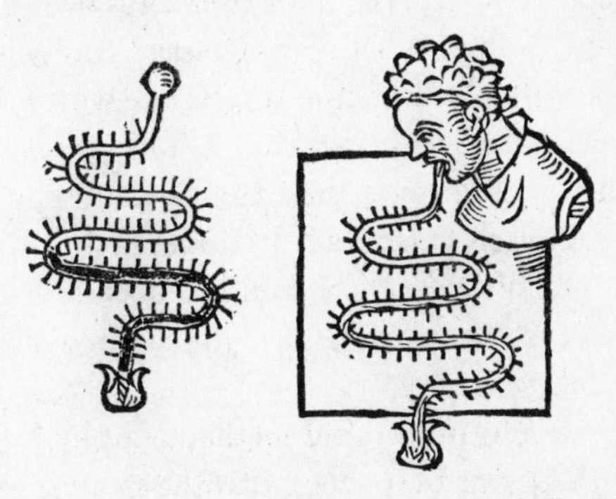

FIG. 111. The principle of Galileo's thermometer applied clinically. The man who seems to be swallowing a centipede has the bulb of a graduated thermometer in his mouth. From Santorio 1626.

'metabolism'. Santorio also adapted Galileo's thermometer to clinical purposes (Fig. 111). It marks the medieval character of much of the thought of the day that his account of this (1626) is concealed in a commentary on a work of Avicenna (p. 147).

The Englishman, WILLIAM HARVEY (1578–1657), is also to be regarded as a disciple of Galileo though he himself was, perhaps, little aware of it. Harvey studied at Padua (1600–2) while Galileo was active there. By 1615 he had attained to a conception of the circulation of the blood. He published his demonstration in 1628. The story of that discovery is very accessible. We would emphasize that the essential part of its demonstration is the result not of mere observation but of the application of Galileo's principle of measurement. Having shown that blood can leave the ventricle of the heart only in one direction, he turns to measure the capacity of the ventricle. He finds it to be two ounces. The heart beats 72 times a minute so that in the hour it throws into the system 2×72×60 ounces = 8,640 ounces = 540 pounds, that is to say about three times the body weight! Where can all this blood come from? Where can it all go to? The answer to that is that the blood is a stage army which goes off only to come on again. It is the same blood that is always returning.

The knowledge that the blood circulates has formed the foundation on which has since been built a mass of physical interpretation of the activities of living things. This aggregate forms the science of *physiology*. The blood is a carrier, ever going its rounds over the same route to return whence it came. What does it carry? And why? How and where does it take up its loads? How, where, and why does it part with them? And what does it bring back? The answering of such questions as these has formed the main task of physiology since Harvey's time. As each generation has obtained a more complete answer for one organ or another, so it has been possible to form a clearer picture of some part of the animal body as a working mechanical model.

Yet despite triumphs of physical methods in physiology, we cannot suppose, with Descartes (p. 260), that the clearest image—which is certainly at first sight the most satisfying—is of necessity also the truest, for the animal body can be shown on various grounds to be no mechanical model. A machine is made up of the sum of its parts. An animal body, as Aristotle perceived, is no more the sum of its parts than is a work of art. The Aristotelian world-system was falling. The Aristotelian biology held. Perhaps it still holds.

Nevertheless, the physical discoveries of Galileo and the demonstrations of Santorio (p. 275) and of Harvey (p. 275) gave a great impetus to the attempt to explain vital workings on mechanical grounds. A number of seventeenth-century investigators devoted themselves to this task. The most impressive exponent of physiological theory along these lines was Descartes himself. His account of the subject written by 1637 appeared posthumously in 1664.[1] It is important as the first modern book devoted to the subject of physiology.

Descartes had not himself any extensive practical knowledge of

[1] Descartes wrote his book *L'Homme* in French and saw to its illustration. The earliest edition, that of 1662, is a translation into Latin with very different figures. The edition of 1664 is in the author's French with the original figures. It is thus the more authentic.

physiology. On theoretical grounds he set forth a very complicated apparatus which he believed to be a model of animal structure. Subsequent investigation failed to confirm many of his findings. For a time, however, his ingenious scheme attracted many. A strong point in his physiological teaching was the stress laid on the nervous system, and on its power of co-ordinating the different bodily activities. Thus expressed, his view may sound modern, but it is grotesquely wrong in detail.

An important part of Descartes's theory is the position accorded to man. He regarded man as unique in his possession of a soul. Now in the view of Descartes the special prerogative of the soul is to originate action. Animals, he thought, are machines, automata. Therefore, given that we know enough of the works of the machine, we can tell how it will act under any given circumstances. But the human soul he regarded as obeying no such laws, nor any laws but its own. Its nature he believed to be a complete mystery for ever sealed to us. Descartes conceived that the soul governs the body through the action of the nervous system, though how it does so he again left as a mystery. The two insoluble mysteries come, he believed, into relationship to each other in a structure or organ in the brain, known to modern physiology as the 'pineal body' (Fig. 112). This organ he wrongly believed absent in animals other than man. All their actions, even those which seem to express pain or fear, are automatic. It is modern 'behaviourism' with man excluded. Anatomically he was refuted by Steno (p. 279) in a famous lecture at Paris in 1665.

The word 'mystery' is not popular among modern men of science. It is, therefore, right to point out that the processes by which a sensory impression passes into sensation, by which sensation educes thought, and by which thoughts are followed by acts, have been in no way elucidated by physiological science. In these matters we are in no better case than Descartes. If we have abandoned his terminology we are no nearer a solution of his leading problems. The basic defects of Descartes's system were errors in matters of fact. It was on account of these that he ceased to have a physiological following in the generation after the publication of his essay *On Man*.

Descartes is often credited with the idea of reflex action which simply means the automatic response of the body to certain external stimuli without consciousness being (necessarily) involved. This conception, however, was implicit in the works of Vesalius (p. 210), Galen (pp. 98 f.), Herophilus (pp. 67 f.), and perhaps of yet earlier writers. Nevertheless, Descartes certainly emphasized and illustrated reflex action. Moreover, like most of his contemporaries and pre-

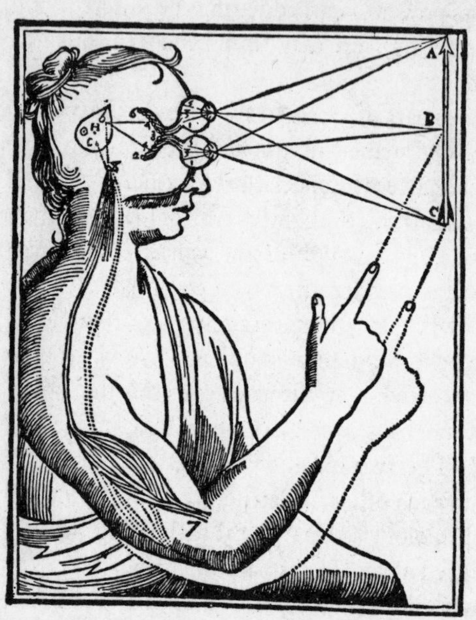

FIG. 112. Descartes's conception of the relation of a sensory impression and a motor impulse. The image of the object *ABC* passes to the eye and is formed on the retina. Owing to the optical properties of the eye, it is there inverted. The image is reinverted within the skull at the crossing of the optic nerves. Thence it passes to the pineal body *H* at point *b*. The position and character of the image formed on the retina determines the nature and distribution of its effect on the pineal body. According to the nature and distribution of that effect is the result on the motor nerves and, through them by the passage of nervous fluid, on the muscles. The movement in the nerve is initiated at the point *c*. The relation between *b* and *c* is an insoluble mystery in which, according to Descartes, is enwrapped the very nature of the soul.

decessors he believed the nerves to be tubular and that they conveyed a 'nervous fluid', distension by which inflated and thus shortened the muscles. He imagined that the movements of this fluid were controlled by an elaborate system of valves (Fig. 113). The existence of such a fluid was disproved by his younger contemporary, Swammerdam (p. 283).

(ii) *The Physiological Sects*

One of the ablest critics of the physiological system of Descartes was the Dane, NIELS STENO (1648–86), whose scientific work was done mostly in Italy and France. Steno, like Descartes, was a mechanist, but unlike Descartes applied himself to the exploration of bodily structure. He found a pineal gland like that of man in other animals, and he could not persuade himself that it had the connexions,

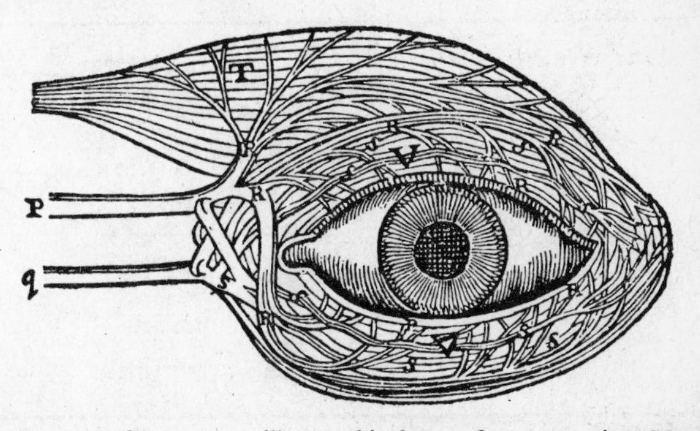

FIG. 113. Diagram of Descartes to illustrate his theory of nervous action. *PR* and *qs* are nerves which supply the muscles of the eye *T* and *VV*. Descartes held that nerves are hollow and provided with valves, which can be seen at the point at which the *PR* and *qs* first branch. These valves were controlled by little fibrils (which can be seen in the main stems of *PR* and *qs* and in certain of their branches). Movements of the fibrils therefore control the movement of the fluid within the hollow spaces of the nerves. Additional complication is lent to the scheme by intercommunications of *PR* and *qs* at certain points. The view of Descartes, and all such theories of nervous fluid, were destroyed by the experiment of Swammerdam which, however, long remained unpublished.

material or spiritual, described by Descartes. His criticism of Descartes in detail was very damaging (p. 277).

More constructive was the achievement of GIOVANNI ALFONSO BORELLI (1608–79), an eminent Italian mathematician, astronomer, and polymath, a friend of Galileo and Malpighi. Borelli's work *On Motion of Animals* (1680) is the classic of what is variously called the 'iatrophysical' or 'iatromathematical' school. It stands as the greatest early triumph in the application of the science of mechanics to the working of the living organism. Stirred by the success of Stevin and Galileo in giving mathematical expression to mechanical events, Borelli sought to do the like with the animal body. In this undertaking he was very successful. That department of physiology which treats of muscular movement on mechanical principles was effectively founded and largely developed by him. Here his mathematical and physical training was specially useful. He endeavoured, with some success, to extend mechanical principles to such activities as the flight of birds and the swimming of fish. His mechanical analyses of the movements of the heart, or of the intestines, were less successful, and he naturally failed altogether in his attempt to introduce mechani-

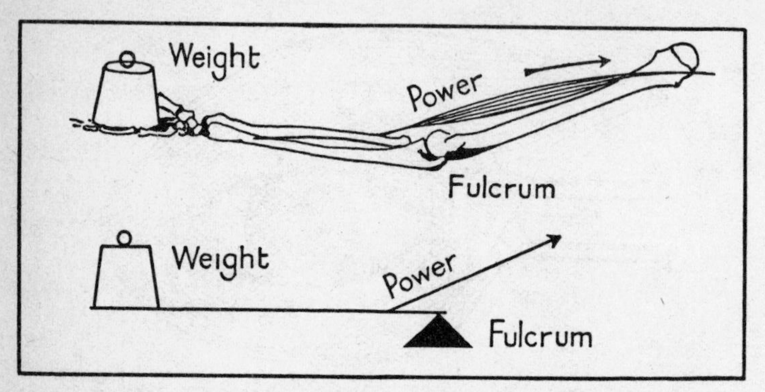

FIG. 114. Modified from Borelli to illustrate muscular action as mechanism.

cal ideas in explanation of what we now know to be chemical processes, such as digestion.

Just as Descartes and Borelli sought to explain all animal activity on a mechanical basis, so others resorted to chemical interpretation. Forerunners of this point of view were Paracelsus (p. 199) and van Helmont (p. 269). A more coherent attempt was made by FRANCIS DE LA BOE (Sylvius, 1614–72), professor of medicine at Leyden. That university had become by the second half of the seventeenth century the most progressive scientific centre. It was the seat of the first university laboratory, which was built at his instance.

Sylvius devoted much attention to the study of salts, which he recognized as the result of the union of acids and bases. Thus he attained to the idea of chemical affinity—an important advance. With a good knowledge of anatomy and accepting the main mechanistic advances, such as the doctrine of the circulation of the blood and the mechanics of muscular motion, Sylvius sought to give a chemical interpretation to other vital activities, expressing them in terms of 'acid and alkali' and of 'fermentation'. In this attempt he made no clear distinction between changes induced by 'unorganized' ferments, as gastric juice or rennet, and changes induced by micro-organisms, as alcoholic fermentation or leavening by yeast. Nevertheless, he and his school added considerably to our knowledge of physiological processes, notably by their examination of the body fluids, especially the digestive fluids such as the saliva and the secretions of stomach and pancreas.

The views of yet another group of biological theorists were best expressed by another expert chemist, GEORG ERNST STAHL (1660–1734). He is remembered in connexion with *phlogiston* (pp. 338–9) and also stands as protagonist of his age of that view of the nature of the organism which goes under the term *vitalism* (p. 48). Though expressed in obscure and mystical language, Stahl's vitalism is in effect a return to the Aristotelian position and a denial of the views of Descartes, Borelli, and Sylvius. To Descartes the animal body was a machine, to Sylvius a laboratory. But for Stahl the phenomena characteristic of the living body are governed neither by physical nor chemical laws, but by laws of a wholly different kind. These are the laws of the *sensitive soul*. This sensitive soul in its ultimate analysis is not very dissimilar from the *psyche* of Aristotle. Stahl held that the immediate instruments, the laboratory tools, of this sensitive soul were chemical processes, and his physiology thus develops along lines of which Aristotle could know nothing. This does not, however, alter the fact of his hypothesis being essentially of Aristotelian origin.

Most of the physiological discussion of the seventeenth century turned on the vital process of animals and especially those of man. The plant physiology of the age was of a more elementary character. Van Helmont had shown that plants draw something of nutritive value from water (p. 269). This was contrary to the Aristotelian teaching that plants draw their food, ready elaborated, from the earth. The generation following van Helmont sought to erect a positive scheme of plant physiology without, however, very much success. MARCELLO MALPIGHI (1628–94), the great Bolognese microscopist (p. 283), held wrongly that the sap is brought to the leaves by the fibrous parts of the wood. The leaves, he thought, form from the sap the material required for growth. This, he knew, is distributed from the leaves to the various parts of the plant. He conceived a wholly imaginary 'circulation of sap' comparable to the circulation of the blood in animals. The respiration of plants, he falsely believed, is carried on by the 'spiral vessels' (Fig. 115) which bear a superficial resemblance to the breathing tubes or *tracheae* of insects with which he was familiar.

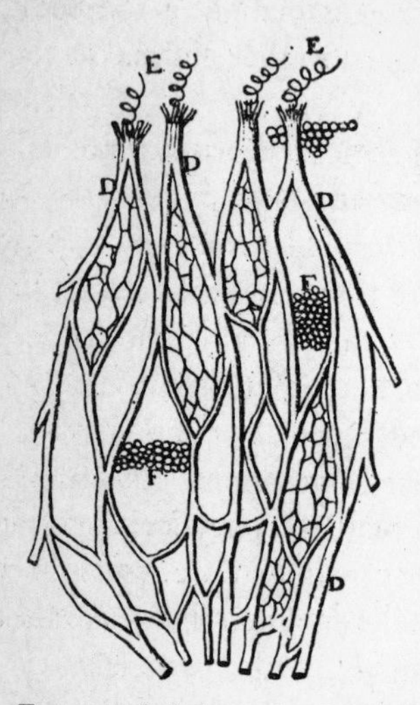

FIG. 115. Spiral vessels of a fig tree from Malpighi. To be compared to the tracheal vessels of insects.

The earliest experimental work on the physiology of plants was that of the French ecclesiastic, EDMÉ MARIOTTE (died 1684). This able physicist observed the high pressure with which sap rises. This he compared to blood pressure. To explain the existence of sap pressure he inferred that there must be something in plants which permits the entrance but prevents the exit of liquids. He held that it is sap pressure which expands the organs of plants and so contributes to their growth (1676).

Mariotte was definitely opposed to the Aristotelian conception of a vegetative soul (p. 49). He considered that this conception fails to explain the fact that every species of plant, and even the parts of a plant, exactly reproduce their own properties in their offspring, as with 'cuttings'. He was, so far as plants are concerned, a complete 'mechanist', and, therefore, anti-Aristotelian. All the 'vital' processes of plants were for him the result of the interplay of physical forces. He believed, as a corollary to this view, that organisms can be spontaneously generated (p. 285).

(iii) *The Classical Microscopists and Spontaneous Generation*

The interpretation of vital activity in chemical and physical terms has had a continuous history to our own time. It is far other with the very striking microscopical researches with which the second half of the seventeenth century is crowded. Five investigators of the front rank, MARCELLO MALPIGHI (1628–94) at Bologna, ROBERT HOOKE (1635–1703) and NEHEMIAH GREW (1641–1712) in London, JAN SWAMMERDAM (1637–80) at Amsterdam, and ANTONY VAN LEEUWENHOEK (1632–1723) at Delft, all busied themselves with microscopic investigations on the structure and behaviour of living things. Their results impressed their contemporaries as deeply as they have some modern historians. But none of these microscopists inspired a school, and we have to turn to the nineteenth century for their true continuators. On this account the 'classical microscopists' must be accorded a less prominent place in a general history of science than the great interest of their biological observations might suggest. We briefly consider the general ideas that they initiated.

(*a*) The infinite complexity of living things in the microscopic world was nearly as philosophically disturbing as the unexpected complexity and ordered majesty of the astronomical world which Galileo and Kepler had unveiled to a previous generation. Notably the vast variety of minute life gave at once new point and added new difficulty to the conception of 'Creation'. This was specially the effect of Hooke's pioneer *Micrographia* (1665, Fig. 117).

(*b*) In a few notable respects the microscopic analysis of the tissues of animals aided the conception of the living body as a mechanism. Thus Harvey had shown that the blood in its circulation passed from arteries to veins. The channels of passage were unknown to him. They were revealed as 'capillary vessels' by Malpighi (Fig. 116) and Leeuwenhoek (Fig. 118). These observers discovered the corpuscles of the blood, the secretory functions of 'glands', and the fibrillary character of muscles, thus helping to complete details of the 'animal machine'.

(*c*) The nature of sexual generation had long been in dispute. The discovery (1679) in the male element of 'animalcules'—'spermatozoa', as we now call them—aroused new speculations. The sperm then was organized. But how? The eye of faith, lit within by its own light, looking through an imperfect lens lit without by a flickering flame (Fig. 117), saw many a 'homunculus' in many a spermatozoon and even the piercing eye of a Malpighi or a Leeuwenhoek saw that which was not (Fig. 119). The faith of others claimed that the homunculus should be carried by the female element, by the germ rather than by the sperm. That, too, was seen by the eye of faith. The more conservative Harvey insisted that the complex embryo in the simple substance of the egg was a 'new appearance', excited by that imponderable ghost, the 'generative force'.

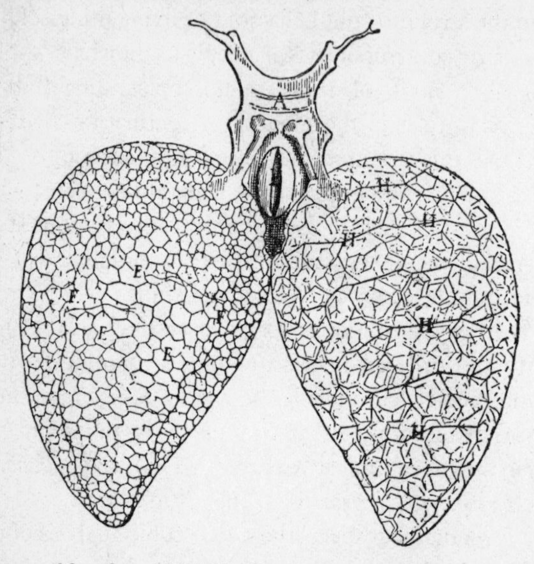

FIG. 116. Lungs of frog from Malpighi 1661. *EEE* outer surface of the lung where network of capillary vessels can be seen arising from the arteries *FF*. On the other side the lung has been cut open and is seen to be lobulated, the lobules being separated by a series of veins *HHH* into which the network of capillaries empty.

(*d*) Microscopic analysis revealed some similarity between the structures of plants and animals. False analogies were drawn and carried at times to fantastic lengths. The 'loves of the plants', on which poets had dwelt, were not wholly fables. It was slowly realized that flowers contain sexual elements, and a parallel was perceived between their reproductive processes and those of animals.

(*e*) Lastly, there is an aspect of minute life that came to the fore in the later seventeenth century that requires some special discussion. It is the theme of *spontaneous generation* of living things, that is, the origin of living things from non-living matter.

Neither ancient nor medieval nor renaissance scientific writers doubted that spontaneous generation took place on occasion. Corpses were said to 'breed' worms, dirt to 'breed' vermin, sour wine to 'breed' vinegar eels, and so forth. Spontaneous generation is often fathered on Aristotle and is in his writings, but it was not so much a doctrine as a universal assumption. When the reality of spontaneous

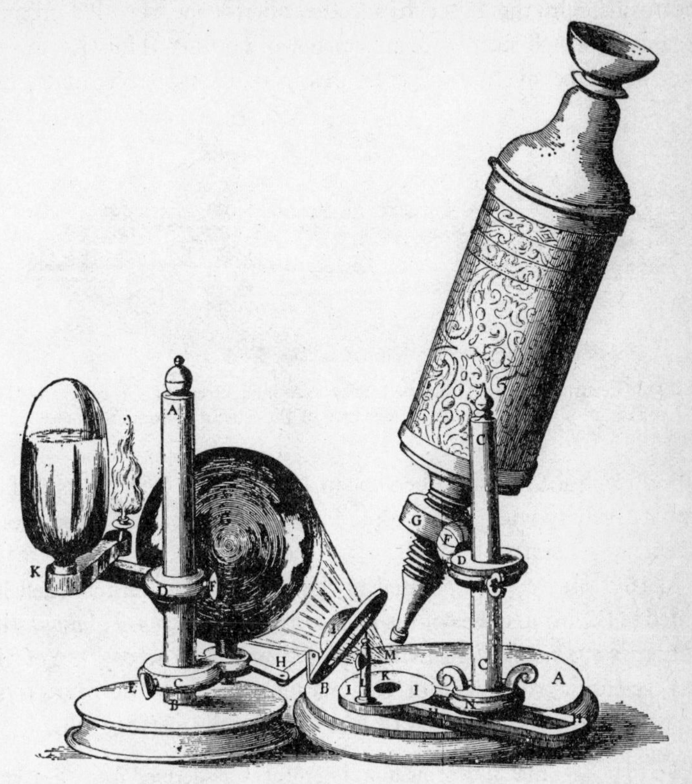

FIG. 117. Hooke's compound microscope from his *Micrographia* 1665.

generation was first questioned, the authority of Aristotle—or rather the contemporary misunderstanding of him—was a very real obstacle to scientific advance. It is also true that Aristotle gave spontaneous generation a place in his biological scheme. But his error was shared by every naturalist until the seventeenth century, and it is hard to see how any other view could then be taken.

With the advent of effective microscopes in the second half of the seventeenth century, new tendencies set in. On the one hand, exploration of minute life showed many cases of alleged spontaneous generation to have been falsely interpreted. Hooke watched the formation of spores in moulds (1665), Huygens (*c.* 1680) and others (1702) saw the division and multiplication of certain microscopic organisms. On the other hand, the microscope revealed minute organisms which seemed to appear out of nothing. Thus Leeuwenhoek saw excessively small creatures in infusions of hay and other

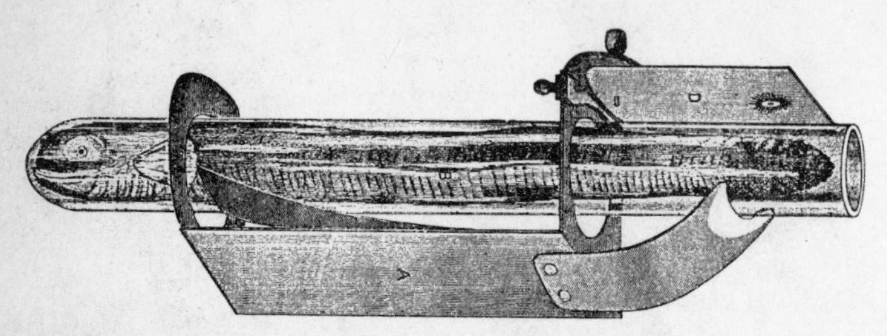

FIG. 118. Leeuwenhoek's microscope set for examining circulation in eel's tail. The screw on plate *D* controls the distance of the lens in it from the object.

substance. Such infusions become in a few days or even hours turbid with actively moving microscopic forms. These seemed to be spontaneously generated.

In 1651 Harvey had pointed to the heart of this matter when he added to the frontispiece of his book *On the Generation of Animals* the words *Ex ovo omnia* ('All living things come from an egg'). Yet the first scientific treatment of the matter was made by FRANCESCO REDI (1621–97), a physician of Florence. He tells us (1668) that he

> began to believe that all worms found in meat were derived from flies, and not from putrefaction. I was confirmed by observing that, before the meat became wormy, there hovered over it flies of that very kind that later bred in it. Belief unconfirmed by experiment is vain. Therefore, I put a (dead) snake, some fish, and a slice of veal in four large, wide-mouthed flasks. These I closed and sealed. Then I filled the same number of flasks in the same way leaving them open. Flies were seen constantly entering and leaving the open flasks. The meat and the fish in them became wormy. In the closed flasks were no worms, though the contents were now putrid. Outside, on the cover of the closed flasks, a few maggots eagerly sought entry. Thus the flesh of dead animals cannot engender worms unless the eggs of the living be deposited therein.

Since air had been excluded from the closed flasks I made a new experiment to exclude all doubt. I put meat and fish in a vase covered with gauze. For further protection against flies, I placed it in a gauze-covered frame. I never saw any worms in the meat, though there were many on the frame, and flies, ever and anon, lit on the outer gauze and deposited their worms there. [Abbreviated.]

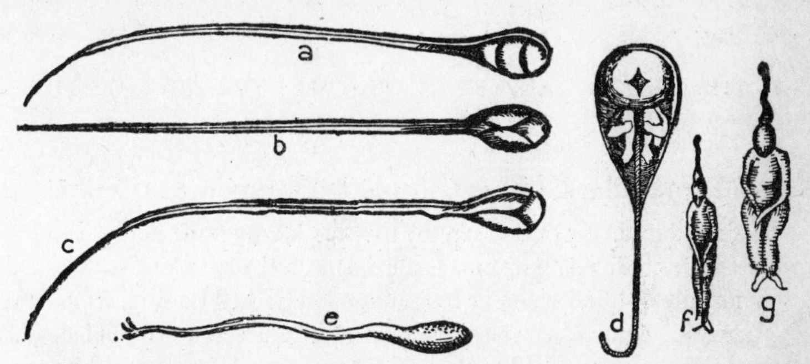

FIG. 119. Spermatozoa as seen in the seventeenth century: *a*, *b*, *c*, by Leeuwenhoek (1679), *d*, by Hartsoeker (1694), *e*, *f*, *g*, by Plantades (1699).

Redi continued to believe that gall insects were spontaneously generated. This subject was taken up by another eminent Italian physician, ANTONIO VALLISNIERI (1661–1730). Correcting Malpighi, he showed that the larvae in galls originate in eggs deposited in the plants (1700). Vallisnieri compared the process of gall formation, as well as infection of plants by aphides, to the transmission of disease. Other investigators showed that fleas and lice are bred only by parents like themselves.

Thus the matter stood in the early eighteenth century with the general balance of opinion against spontaneous generation. The possibility had been disproved—so far as a universal negative can be disproved—for larger organisms. The question was still open for the minute creatures encountered in infusions.

In summary we may say that for Biology the Insurgent Century closed with a strong mechanistic bias. The microscopic world, however, remained an enigma, a land of wonders where all laws seemed at times to be broken. *De minimis non curat lex*, 'The law does not concern itself with the most minute things' was often quoted, but *lex* of the lawyer was quite different from *lex naturae*.

VIII

THE MECHANICAL WORLD

Enthronement of Determinism (1700–1850)

I. THE NEWTONIAN KEY TO THE MATHEMATICS OF THE HEAVENS

ST. AUGUSTINE, about A.D. 427, that is 1300 years before Newton.

This glorious doctor, as he went by the sea-side studying on the Trinity, found a little child which had made a little pit in the sand, and in his hand a spoon. And with the spoon he took water and poured it into the pit. And St. Augustine demanded what he did. And he answered: 'I will lade out all the sea into this pit.' 'What?' said St. Augustine, 'How may it be done, sith the sea is so great, and thy pit and spoon so little?' 'Yea,' said he, 'I shall lightlier draw all the water of the sea and bring it into this pit than thou shalt bring the mystery of the Trinity into thy understanding, for it is greater to the comparison of thy wit than is this great sea unto this little pit.' And therewith the child vanished. [Abbreviated from *The Golden Legend*, as englished by William Caxton in 1483.]

ISAAC NEWTON, A.D. 1727, shortly before his death.

I do not know what I may appear to the world, but to myself I seem to have been only like a boy playing on the sea shore, and diverting myself in now and then finding a smoother pebble or a prettier shell than ordinary, while the great ocean of truth lay all undiscovered before me. [From the *Anecdotes* of Joseph Spence (1699–1768).]

Nothing emerges more clearly from a survey of the history of science than the lasting and essential sameness of the human spirit. The same aspiration for a coherent and comprehensive plan of his universe has characterized the mind of man from his very dawn and has survived a thousand defeats. It is therefore by no means strange that two men widely separated in time, genius, and mood should take refuge in the same image to express their thought of infinity.

St. Augustine (354–430) marks the effective beginning of a great epoch—a space of thirteen centuries—of which the effective end is marked by the arrival of ISAAC NEWTON (1642–1727). In his *Confessions* Augustine says that the sole fundamental truth lacking to the 'Platonists'—by which he means his Neoplatonic teachers (p. 34)—was the doctrine of the Incarnation. It was Augustine who determined that Christian thought should be cast in a Neoplatonic mould, the impress of which it has borne to our own day. It was his specifically Christian contribution to award to man a unique dignity that was denied by certain pagan philosophers.

The Augustinian Neoplatonist is still working in John Dryden (1631–1700). He is still straining his ears to hear the 'music of the spheres' in the very year in which Newton's greatest work appeared:

> From harmony, from heavenly harmony,
> This universal frame began.
> From harmony to harmony
> Through all the compass of the notes it ran.
> The diapason closing full in Man.
> (*A Song for St. Cecilia's Day*, 1687.)

In the Neoplatonic Christian world there was a hierarchy of existences from purely spiritual to purely physical, the whole linked together in God's heavenly harmony. The centuries rolled on, and still that music of the spheres lulled man's mind to sleep while his spirit waked. At last 'Aristotle'—a strangely changed Aristotle—was recovered by the Latins from his Arabian custodians (pp. 164 f.), and Scholasticism was born. Thus the ancient cosmic scheme was enlarged by a Neoplatonic Aristotelianism and the 'Dark Ages' of Faith gave place to the 'Middle Ages' of Reason. Yet the spell of Plato, through his mouthpiece Augustine, still remained unbroken. The spiritual realm of the medieval Christian stretched to the infinite, aspiring to the timeless God. But the Christian's material world, the world of Augustine, of the Neoplatonists, of the Stoics, and of 'Aristotle' remained limited by those flaming ramparts beyond which even thought could hardly penetrate.

The change came with the sixteenth century. Copernicus put Earth from her ancient seat (pp. 212 f.) in a new form of an old convention. But it was Bruno who proclaimed a universe of world beyond world, without centre or circumference, in which all place and all motion were relative. For him the stars were no longer fixed and the frontiers of the universe were an idle dream. Next Kepler reduced the movements of the planetary bodies to intelligible mathematical rules. Galileo developed the system of earthly mechanics with which, he hinted, the heavenly bodies must somehow show accord. The conduct of matter was explored by Boyle and the rising experimental school in a new and exact spirit, without the older presuppositions. While Harvey, Descartes, Borelli, expounded the living body as a mechanical system, Malpighi, Hooke, Grew, Leeuwenhoek, Swammerdam revealed, with their microscopes, vast and unsuspected regions and forms of life and the endlessly complex structure of even the minutest living things whose very existence had not been conceived.

In the first half of the seventeenth century the basic philosophic idea that delayed the full modern scientific awakening was the limitation of the universe. This view carried with it the medievally accepted corollary that the field for scientific exploration was also, and of necessity, limited. Bruno was disregarded. Kepler held timidly back, feebly defending the medieval scheme. Galileo, for once, did not make himself clear on which side of the fence he stood. Though John Wilkins (1614–72, afterwards secretary of the Royal Society and bishop) anonymously in 1638, and William Gilbert (p. 222, if indeed it was he) posthumously in 1651, expressed a view of the stars as scattered through boundless space, nevertheless this point remained unsolved in men's minds until the second half of the century (with Newton and Huygens).

In the third quarter of the seventeenth century learned societies in France, England, and Italy became centres for the exchange of scientific ideas. Perhaps the greatest achievement of these societies was the development and perfection of the manner of presenting inquiries. Thus the form of scientific communications became standardized and the demand for rigorous demonstration insistent. To quote authority was useless. *Nullius in verba* ('On the words of no man') stands on the crest of the Royal Society, whose publications began in 1664.[1] The demand for observation, for tangible evidence

[1] The motto is presumably modified from a line of Horace *Epistles* (14) *Nullius addictus jurare in verba magistri*, 'I am committed to the words of no master'.

and for experience that can be repeated at will, had created science as we know it.

A fruitful source of misunderstanding of the aims and methods of the new science has been the unfortunate necessity that its technique of presentation must conceal the investigator himself. With the advent of the 'scientific journal' it becomes increasingly difficult to reach behind the text to the mind of the author. The new method of scientific publication does not allow us to see the trial attempts and tentative views of the men who write these books and papers. The point comes out admirably in the career of Newton himself.

The demonstrations of Galileo and Kepler, while they banished the earth-centred universe, did not at once destroy the conception of a sun-centred universe. No one had proved that the fixed stars were at various distances from our planetary system, and that view was not generally expressed. Nevertheless, such an opinion was held in scientific circles. The varying brightness of some stars, the occasional appearance of new stars and many other phenomena, suggested that the stars were of the same order as our sun, or earth, and the planets of our system. In 1686, the year before the publication of Newton's *Principia*, appeared the very famous work *On the Plurality of Worlds* by the French writer Le Bovier de Fontenelle (1657–1757). There were many who were thinking the same thought. 'I am of like opinion with all the great philosophers of our age', wrote Huygens, 'that the sun is of the same nature as the fixed stars. And may not every one of the stars or suns have as great a retinue of planets with moons to wait upon them as has our own sun?' (1698). The earth, then, being but a moving particle in space, space itself must be infinite, as Bruno had claimed. The Cosmos, not Man, must be the prime existence. In that new-found Cosmos the philosophers vied with one another in tracing laws, but the music of the spheres grew more distant and, at times, even discordant.

The change was at first one of degree rather than of kind. Law had been traced in the heavens from of old. The rules of planetary and stellar motion had been gradually developed from the astronomical theories of antiquity. Even in the Middle Ages a few new mathematical relationships of the heavenly bodies had been discerned. In the sixteenth century, astronomy, under Tycho (pp. 215 f.), put her house in order for the Great Instauration (p. 264) of the coming age. And then Galileo startled the world with his proof of change in the uttermost heavens (1604) in the very region held by the Aristotelian and Platonic schemes to be utterly changeless (p. 242).

By 1618 Kepler had enunciated his 'three laws of planetary motion', bringing these movements into an intelligible relation with each other (pp. 240–1). Then Galileo (pp. 235–6) determined the rule of action of gravitation and came near to the 'three laws of motion' which we associate with Newton. Others, Hooke and Wallis among them, were feeling their way in the same direction. But it was Newton who first affirmed these laws and succeeded in linking them with Kepler's laws of planetary movement. Before Newton, no man had shown, or clearly and demonstrably perceived, how the complex movements of the planetary bodies were analogous to the natural succession of earthly phenomena. Reason no less than Faith would have been against such a view. Newton's unique achievement was to prove that this analogy amounted to identity. It was Newton who moved men to see that the force that causes a stone to fall is that which keeps the planets in their paths. It was Newton who first enunciated a law the writ of which ran no less in the heavens than on the earth. With Newton the universe acquired an independent rationality, quite unrelated to the spiritual order or to anything outside itself. The Cosmology of Aristotle, of Augustine, of the theologians was doomed.

Newton knew that if a stone be let drop, its weight—which is another name for earth's attraction—will cause it to fall a certain measurable distance in the first second of its fall. He came early to suspect that the force which kept the moon in her orbit was none other than this terrestrial attraction. The period of the moon's revolution round the earth, and the dimensions of her orbit, were alike susceptible of estimation, so that her velocity could be calculated. Now the moon, like any body pursuing a curved course, is moving at any particular moment in a direction tangential to her orbit. But the moon, as we know, does not continue to move along the tangent, but is constrained to follow her elliptic path round the earth. At the end of the second, she, like the stone, has 'fallen' a certain distance toward the earth (Fig. 120). The earth has drawn her to herself.

Now, from Kepler's laws, Newton had reason to suspect that the attractive power of the earth on any body decreases as the square of the distance from the centre of the earth. If the conjecture were correct, he had the equation:

$$\frac{\text{Distance fallen by moon}}{\text{Distance fallen by stone}} = \frac{(\text{Distance of stone from earth's centre})^2}{(\text{Distance of moon from earth's centre})^2}.$$

When Newton first approached this problem (1666) he found that the moon's 'fall' was but seven-eighths of what he expected. But he

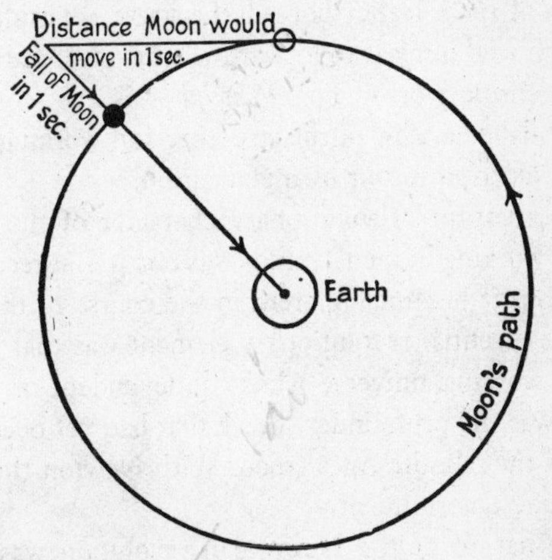

FIG. 120. Illustrating the orbit of the moon as compounded of tangential and centripetal movements.

had seized on the conception of universal gravitation, that is, that every particle of matter attracts every other, and he suspected that the attraction varied directly as the product of the attracting masses, and inversely as the square of the distance between them. It was still years before he was armed with the knowledge and means to show that the 'fall of the moon' had the value required by his theory. By that time (1671) he had developed the wonderful mathematical method of dealing with curves which has since, with another nomenclature, become familiar under the name of 'Calculus' (p. 309).

The action of gravity on the earth and in the heavens was now seen to be the same, at least for a particular case. Newton's grand hypothesis was launched, though not yet worked out in detail. We owe it to the astronomer EDMOND HALLEY (1656–1742)—whose name is recalled periodically by his comet (pp. 305 f.)—that Newton undertook to attack the whole problem of gravitation. He had years of labour before he could show that the attraction of a spherical body on an external point was as if the spherical body were concentrated at its centre (1685). He had no expectation of so beautiful a result till it emerged from his mathematical investigations. With this theorem in his hands, all the mechanism of the universe lay spread before him. The vision was set forth in the *Philosophiae Naturalis Principia Mathematica* of 1687. Halley bore all the stress, set aside his own researches, sacrificed himself to forward what is regarded as the greatest of all scientific works. The *Principia*—as the work is usually called—established a view of the structure and workings of the universe which survived to our own generation.

The full extent and revolutionary character of the change that Newton was working in men's minds was not at first recognized even by himself, but it became apparent in the course of the eighteenth century. The essential revolutionary element was that Newton had conceived a working universe wholly independent of the spiritual order. This was the profoundest break that had yet been made with all for which the Middle Ages stood. With Newton there set in an age of scientific determinism.

But if the nature of the Newtonian revolution was not at first apparent, the scientific importance of the *Principia*, as of Newton's other contributions, was recognized immediately on publication. Newton wrote for mathematicians, and his full significance was beyond the comprehension of any others. He needed interpreters. Of these the ablest and most effective was VOLTAIRE (1694–1778), who spent the years 1726–9 in England. To him we owe the well-known story of Newton and the falling apple. Voltaire was aided in the preparation of his version of the Newtonian philosophy by his mistress, Émilie de Breteuil, Marquise du Chastelet (1706–49), who was a competent mathematician and herself translated the *Principia* into French (published posthumously 1759). Voltaire's delightful and lucid exposition (1737–8) marks the real victory of the Newtonian philosophy and the final submergence of Aristotelianism.

The changes in method and outlook introduced by Newton were

so great that their general conformity as members of an series is sometimes lost to view. The issue is further obscured by use or misuse of certain well-worn phrases. Newton's phrase 'I invent no hypotheses' is often quoted. The prestige of his name led to the assertion that 'whereas his predecessors *described* the motions of the heavenly bodies, Newton was the first to *explain* them'. Scrutiny of these statements throws light on the nature of scientific process.

Newton's famous phrase *Hypotheses non fingo* occurs at the end of the *Principia.* 'I have not yet been able to deduce from the phenomena the reason of these properties of gravitation and *I invent no hypotheses*. For whatever cannot be deduced from the phenomena should be called an *hypothesis*.'

Now Newton is here giving to the word hypothesis its exact original meaning. In the works of Plato[1] as well as in yet earlier works bearing the name of Hippocrates the word 'hypothesis' is used for a postulated scheme or plan which must be accepted if discussion is to take place. It is literally a 'foundation' (Greek *hypo thesis*, 'a thing placed under'). We have such hypotheses constantly before us in law. Some are mere legal fictions, as that 'the King can do no wrong'; others are convenient presentations of a remote possibility, as 'the lease that runs for 999 years'; others refer to procedure, as that 'a man is innocent [i.e. *treated* as innocent] until proved guilty'. All these are hypotheses in the Platonic, Hippocratic, and Newtonian sense. None are deduced from the phenomena. None are verifiable. All are parts of a working scheme into which certain phenomena can be conveniently and tidily fitted. In this use of the word Newton was certainly right when he said 'I invent no hypotheses'.

But if *hypothesis* be taken to mean what we nowadays understand by a scientific hypothesis, that is a generalization drawn from a series of observations which, it may reasonably be hoped, will be confirmed by yet further observations, then we must say that Newton was constantly both inventing and employing hypotheses. His application to the movements of the moon of the doctrine of gravity as he knew it on earth (Fig. 120) was an obvious example. Once he had such an 'hypothesis' that would fit the moon, he could and did apply it to

[1] e.g. *Phaedo*, 101 D, E.

other members of the planetary system. Its verification from the planets strengthened his conviction of the value of his first inference. The whole of his scientific activity was remarkable for invention of hypotheses. The successful invention of hypotheses is indeed the mark of his scientific eminence.

As regards the distinction between description and explanation, the position is somewhat the same. Newton knew that a property which we call *gravity* is associated with all matter of which we have direct experience. Having reached an exact conception of this property, he proceeds to examine the motions of the planetary bodies and finds that they may be re-expressed in terms of gravity. To do this is to give a description, not an explanation. It may reasonably be claimed that 'description is the true aim of science'. Let us apply the claim to some of Newton's predecessors.

Ptolemy represented the apparent movements of the heavenly bodies in terms of eccentrics and epicycles. This was his method of description. If asked 'Why were the eccentrics and epicycles thus disposed?' he could have given no answer or the merely evasive one 'The Divine Order made it so.' He described; he did not explain.

Copernicus displaced the geocentric scheme. He expounded the appearances more simply and fully by ascribing them to the motion of the earth round a sun that was at rest. If asked 'Why does the earth move so?' he could have given no answer or, again evasively, 'It is part of God's Harmony.' He described; he did not explain.

Kepler represented the appearances more simply and fully by a system of ellipses. If asked 'Why should this form have been chosen?' he could have given no answer though he might have invoked Pythagorean 'Necessity'. He described; he did not explain.

Newton's completer scheme was based on the mutual attraction of bodies. If asked 'Why do they mutually attract each other?' he could have given no answer.[1] If, therefore, his account of the planetary system may be called an explanation, then such an explanation is indistinguishable from a description. The distinction between description and explanation cannot perhaps be ultimately maintained. It is

[1] Newton did attempt to give an answer. He sought to 'explain' gravitation in terms of ether. Even had his attempt been successful, which it was not, it would have been of the nature of a re-description.

the function of science to describe in terms that are as simple as possible. Ultimately the description must be in terms that defy further analysis, if such terms there be.

There is a significant change in nomenclature that expresses epigrammatically the change that came into men's minds with the acceptance of a mechanical world. For thirteen centuries, between St. Augustine and Newton, the Christian philosophic synthesis had reigned supreme; undisputedly at first, a little uneasily at last. But during the succeeding centuries, 1700–1900, the results of investigation of Nature appeared to fit less and less neatly with the accepted philosophic scheme. Changes in the meanings of words are sometimes straws that tell how the winds of thought are blowing. It is no accident that, precisely during these two centuries, certain kinds of 'philosophical inquiries'—as Newton and his contemporaries always described their labours—came gradually to be known as 'scientific researches'. Science, the knowledge of nature, was separated from philosophy, the search for the key to the universe. The change represents a fragmentation of interests that has lasted beyond the period that we are considering. For this reason, among others, it is peculiarly difficult to present the history of modern science as a coherent whole. From now on, our narrative, to become intelligible, needs a minuter subdivision. Science does not describe the world as a whole, but only a bit of it at a time, each science choosing its own bit. This departmentalism becomes increasingly self-conscious.

2. MORPHOLOGY OF THE UNIVERSE

Investigations on the general structure of the cosmos associated with Newton's conceptions fall naturally under three heads:

(i) *Observational astronomy*, that is, the direct investigation of the heavenly bodies by means of the telescope.

(ii) *Dynamical astronomy*, that is, the reduction to mathematical form of the movements of the heavenly bodies and the prediction, on a gravitational basis, of the movements of those bodies based on the mathematical expressions thus reached.

(iii) *Astrophysics*, that is, the investigation of the physical and chemical constitution and state of the heavenly bodies.

(i) *Observational Astronomy*

Astronomical advance, from the seventeenth to the nineteenth century, depended largely on improvement of the telescope. Galileo's instrument of 1610, with concave eyepiece and convex objective, gave an upright image of 30 magnification but a very narrow field. Kepler suggested a convex eyepiece (1611). This yielded a wider

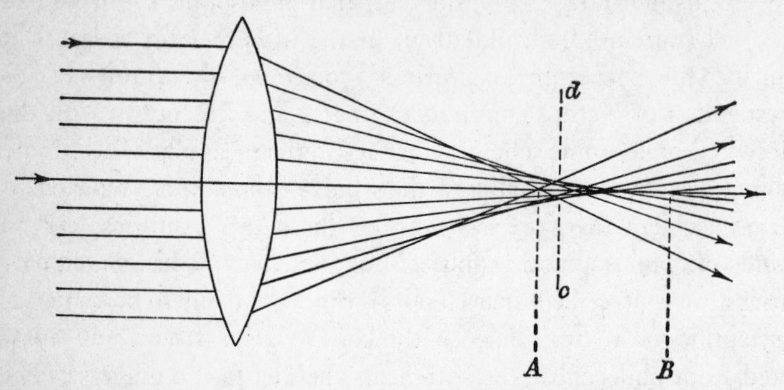

FIG. 121. To illustrate spherical aberration. *A* is focus for rays incident on marginal zone. *B*, focus of rays incident on axial zone, *cd* position of circle of least confusion.

field but inverted the image. For terrestrial purposes inversion could be avoided by using a concave lens as eyepiece.

The images given by all early telescopes suffered from the defects now known as spherical and chromatic aberration. The cause of spherical aberration was understood and could have been avoided by using parabolic lenses but efforts to make these failed. A lens with spherical surfaces focuses the more peripheral incident rays on its axis nearer to itself than it focuses the more central rays (Fig. 121), and nowhere can the image be brought into exact focus.

Chromatic aberration was not understood until Newton explained it (1663). The coloured ring which disturbed the picture yielded by early telescopes is due to the different refrangibility of rays of different colour, blue rays at one end of the spectrum being more refracted than red at the other end (Fig. 122). The eye is most sensitive to yellow rays in which the solar spectrum, as observed on the earth, is specially rich (Fig. 123). Thus the plane on the axis which corresponds to the focus of these, gives the best picture.

Newton's views on light involved a break with older theories. Aristotle held that white light was basic, while colours were mixtures of light and 'darkness'. The philosophy of Descartes, which prevailed until near the end of the century, denied this and regarded light as instantaneously propagated in the cosmic *continuum*, colours being due to different rates of rotation of the corpuscles involved. Colours

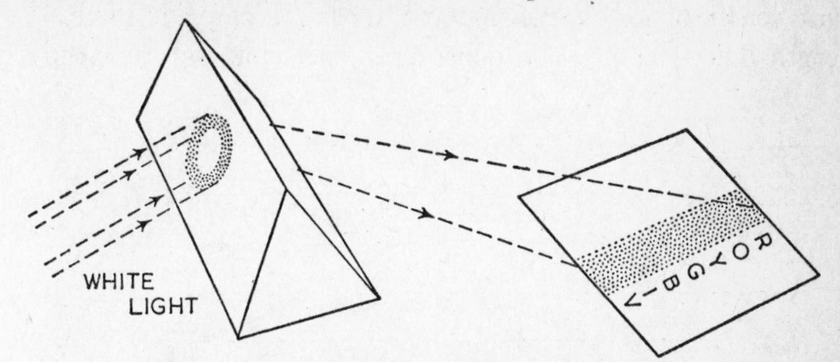

FIG. 122. To illustrate Newton's prism observation.

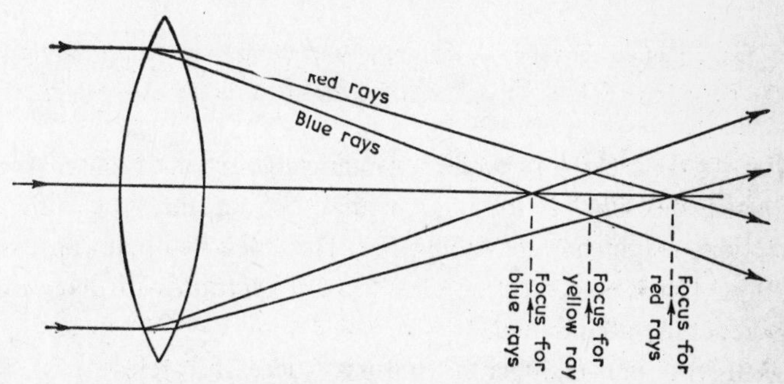

FIG. 123. To illustrate chromatic aberration, showing focus for blue rays, focus for red rays, and focus for yellow rays.

were thus, in Cartesian theory, measurable primary qualities (p. 246). Newton, like Descartes, but unlike Galileo, sought a physical basis for colour that was beyond mere sensation. He showed that light to be white must be a combination of coloured lights in specific proportions. He recognized that lights of different colours are differently refracted, as demonstrated with a prism (Fig. 122). The curved surfaces of a biconvex lens may be considered as a series of prisms with apical angle diminishing as one approaches the axis from the edge. With such a lens the plane on the axis that gives the best result is in the region of the yellow-green rays (Fig. 123).

Newton got rid of chromatic aberration by abandoning a lens as objective and substituting a concave mirror (Fig. 124). Spherical aberration remained, for he could not make parabolic mirrors. By introducing a second and flat reflector and an eyepiece (Fig. 125) he produced a successful telescope. Other telescopic devices of his time with lenses of long focus involved apparatus of many metres in length. They had to be abandoned as impracticable and are scientific

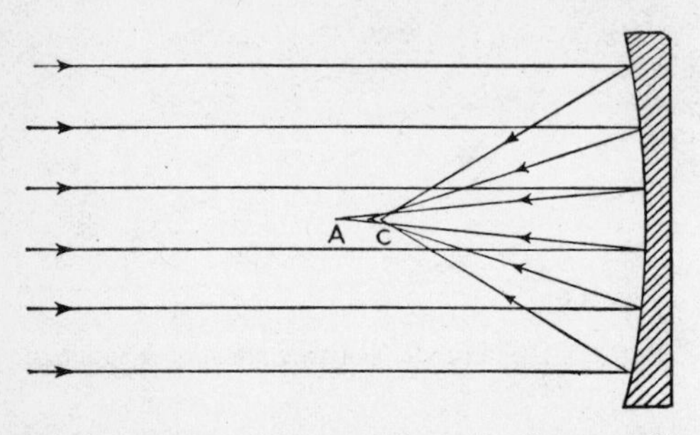

FIG. 124. Spherical aberration of concave mirror. Parallel rays falling marginally are reflected to *C*, those near axis to *A*.

curiosities. In seeking to evade chromatic aberration of lenses, Newton tried differing contiguous media, hoping that the different refractions might correct each other. He failed to obtain any such solution. Hope of correcting chromatic aberration with lenses was thus generally abandoned.

Until the mid-eighteenth century, successful telescopes were therefore on Newton's model. The possibility of correcting both spherical and chromatic aberration was suggested (1747) by the Swiss mathematician, Euler (p. 310), who based his views on the supposed achromatism of the eye, itself a series of media of different refractive powers. A Swedish mathematician pointed out to the London instrument maker, JOHN DOLLOND (1706–61), that Newton's negative results were based on prisms of small apical angle and that the matter needed further inquiry. Dollond, having mastered the mathematical theory, succeeded, after years of experiment, in making an 'achromatic' system (1758).[1] He computed, ground, and polished object-lenses of

[1] An achromatic system had been made in 1733 by another worker who did not publish his results.

crown glass with water between. Ultimately, he replaced the water by a lens of flint glass. He thereby reduced chromatic distortion by using lenses of the two kinds, flint and crown, so chosen that the chromatic aberration of the one was equal to that of the other. Since that of the one, being convex, was in the opposite direction to that of the concave, their aberrations neutralized each other. Dollond's

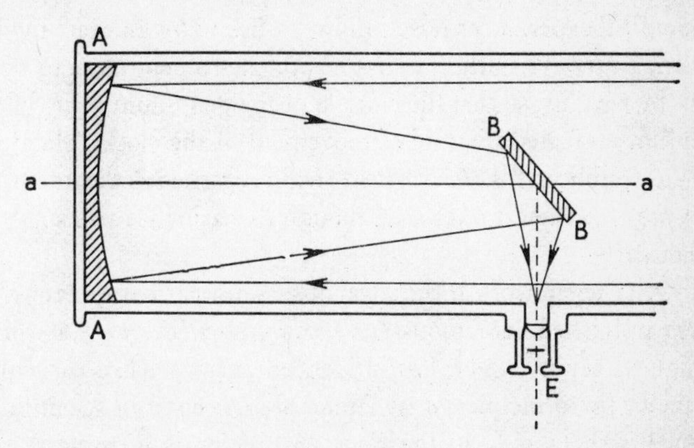

FIG. 125. Diagram of Newton's telescope. At one end of an open tube is a concave mirror *AA* with axis *aa*. Parallel rays that fall on it are reflected to the plane mirror *BB* and thence at right angles to the axis. They form an image in the focus of the plano-convex eyepiece *E* set in a hole in the tube.

principles came to be followed in the construction of telescopes and, on the whole and save for special reasons, refracting telescopes rather than reflectors, held the field till the end of the nineteenth century.

At the command of Louis XIV the great scientific architect CLAUDE PERRAULT (1613–88) built the first State observatory of modern times. It was intended to provide facilities there for men of science, whatever their country. Soon the Frenchman Jean Picard, the Hollander Christian Huygens, the Dane Olaus Roemer, and the Italian G. D. Cassini were all at work there. (See Frontispiece.)

JEAN PICARD (1620–82) was an exact and careful observer, remembered for his measurements of the dimensions of the earth (1671, p. 317). These formed the basis of Newton's calculations. He recognized the astronomical value of the pendulum clock of Huygens, and was the first to introduce the systematic use of telescopic 'sights'.

CHRISTIAN HUYGENS (1629–95, pp. 228, 309, &c.), before coming to the new observatory, had already completed much important scientific work. Thus, he had improved the telescope, and had proved that the changes in the appearance of Saturn—its 'horns' as Galileo called them—were due to a ring inclined at 28 degrees to the ecliptic (1653–6). The *micrometer*, a telescopic device for measuring small angular distances, was effectively introduced by him (1658).[1] His astronomical experiences raised in him a desire for an exact mode of measuring time. With this in view he attached a pendulum to a clock driven by weights, so that the clock kept the pendulum going but the pendulum regulated the rate of movement of the clock. The device was made public in his *Horologium* (1658), regarded as the foundation of the modern clock-maker's art though its basic idea was suggested by Galileo.

Huygens began work at the royal observatory at Paris in 1671, and in 1673 published his famous *Horologium Oscillatorium*,[2] a work of the highest genius which has influenced every science through its mastery of the principles of dynamics. It is second in scientific importance perhaps only to the *Principia*, which is in some respects based on it. It is primarily a mathematical analysis of the principles of the pendulum clock. It devotes attention to the composition of forces in circular motion. It also expounds Huygens's view that a pendulum to be truly isochronous must swing in a cycloidal arc,[3] which is slightly narrower than the corresponding arc of a circle. He introduced 'cheeks' to secure cycloidal movement (Fig. 126). A memorable sentence in the work is the formulation of what has since become known as Newton's 'first law of motion' Huygens writes: 'If gravity did not exist nor the atmosphere obstruct the motions of bodies, a body would maintain forever, with equable velocity in a straight line, the motion once impressed upon it.' The work presents the modern view of the nature of inertia with great clearness.[4]

[1] The micrometer had been invented about 1640 by the Englishman William Gascoigne (1612–44). Huygens's device was improved about 1666 by the Frenchman Adrien Auzout (d. 1691).

[2] Not to be confused with the *Horologium* of 1658.

[3] A cycloid is a curve traced by a point on a circle as the circle rolls along a straigh line. The end of a string wrapped round a cycloid describes a second cycloid, the *evolu* and this was the curve used by Huygens in designing the 'cheeks' to his pendulur

[4] The ideas of mass and of inertia were implied by Huygens in his statement o

Huygens measured the acceleration due to gravity by experiments with a seconds pendulum, that is to say, a pendulum the oscillations of which occupy exactly one second. It is possible to calculate this

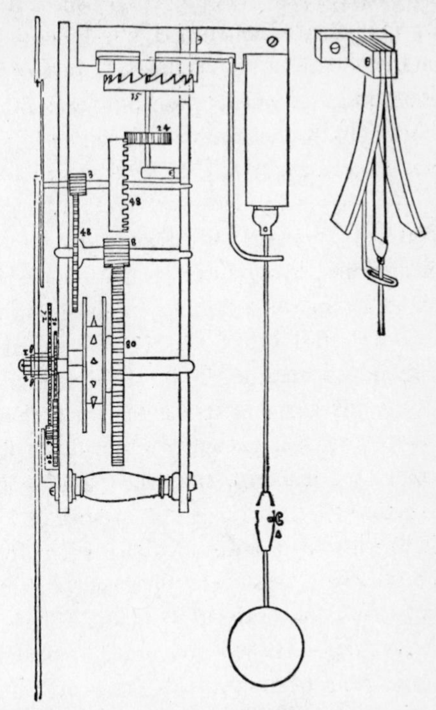

FIG. 126. Works of Huygens's clock from his *Horologium Oscillatorium* 1673. The maintaining power is a weight which has been removed. The cycloidal cheeks of the pendulum are seen in inset as at right angles to the main figure

acceleration at any spot of the earth's surface from the accurate measurement at that spot of the distance between the point of suspension and the centre of gravity of a seconds pendulum. Huygens's own result was 32·16 feet per second.

In 1681 Huygens returned to Holland and devoted himself once more to optical investigations and devices. He introduced a principle of optical construction which obviated much of the difficulty of

laws governing the collision of elastic bodies as presented to the Royal Society in 1669. In this matter he had been preceded to some extent (1668) by Wallis and Christopher Wren (1632–1723).

spherical aberration by employing lenses of enormous focal distance for his very long 'aerial telescopes'. The 'Huygenian eyepiece' invented by him is still in use.

OLAUS ROEMER (1644–1710) was the first to show that light has a definite velocity (1675). His conclusion was based on his observation that the intervals between the eclipses of Jupiter's moons were less when Jupiter and Earth were approaching each other than when they were receding. His discovery was of the highest importance, but it was rejected by the conservative Cassini, the astronomical dictator of the age.

G. D. CASSINI (1625–1712) began life as an engineer in the papal service. He established an astronomical reputation by his writing on comets (1652) and by his observations of the rotation periods of Jupiter, Mars, and Venus (1665–7). He was called to Paris by Louis XIV in 1669 and became the most influential figure in the observatory. Under his auspices it was shown that the earth was flattened towards the poles, a discovery that had important astronomical implications. Under him, too, the parallax of Mars was measured. This led to an estimate of the distance of Mars from the sun (1673). His estimate of the distance of the sun from the earth, though the best to its date, was some 7 per cent. in error.

Cassini was a man of conventional piety and—remarkable at that date—an anti-Copernican. He was succeeded at the Paris observatory by three generations of descendants. The Cassini régime at the observatory lasted for a century and a quarter (1671–1794) and their lives extended over more than two centuries (1625–1845). Their conservative bias gradually weakened as the dynasty came to an end, but it was very injurious to French science.

In England, interests were increasingly maritime, and a scheme for finding longitude at sea was propounded in 1675. JOHN FLAMSTEED (1646–1719), already recognized as a promising astronomer, showed this to be impracticable without a more accurate knowledge of the positions of the fixed stars than was then available. Charles II, hearing of this, declared that 'he must have them anew observed, examined and corrected for his seamen'. An observatory was erected for Flamsteed at Greenwich. His industry there was enormous, and between 1676 and 1689 he determined the positions of many thousand fixed stars. His best observations were made with a mural arc, which he erected in 1689. This marked a great instrumental advance, and made possible far more accurate determinations than had before been attempted.

Flamsteed was succeeded at Greenwich (1720) by EDMOND HALLEY (1656–1742). This remarkable man had detected discrepancies between the observed and the theoretical paths of Jupiter and Saturn before he was twenty. Perceiving that observations in the southern hemisphere were needed for the adjustment of these differences, he embarked for St. Helena (1676), where he observed for eighteen months. During this period he improved the seconds pendulum (p. 303) and determined the positions of 341 stars of which no accurate record then existed. At the same time he made many other contributions to science and, notably, a series of meteorological observations. These led to his publication of the first map of the winds of the globe (1686) and an attempt at their explanation (p. 321). He also made the first complete observation of a transit of Mercury.

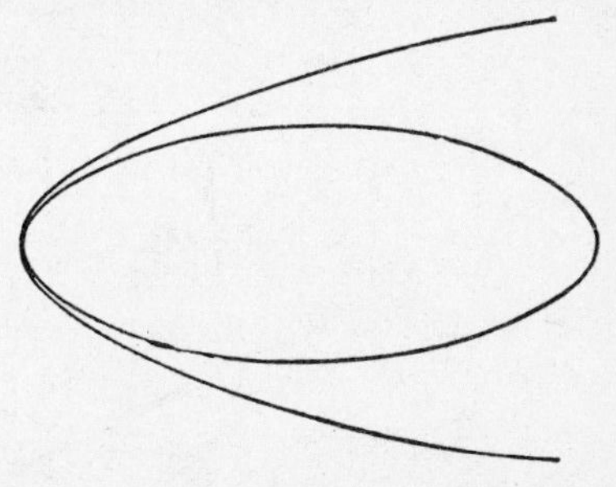

FIG. 127. Parabola and elongated ellipse, showing how they become indistinguishable from each other as they approach their common focus.

In 1680 Halley began the study of the orbits of comets. In 1682 a comet appeared, the course of which was watched by several observers. Newton had suggested that comets might move in very elongated ellipses, indistinguishable from parabolas—as such ellipses must be—when near the sun (Fig. 127). Halley calculated the form, position, and measurements of the path of the comet of 1682, and noted their likeness to those of similar comets of 1531 and 1607. He inferred that his comet was a return of these. Other returns were traced. In 1705 he expressed the view that his comet returns every seventy-five and a half years, following an immensely long elliptical orbit extending far beyond the orbits of the planets (cp. Fig. 175). Halley's comet is now known to have reappeared at about that interval from 12 B.C. to A.D. 1910—twenty-six appearances in all. A famous appearance of this comet was that of 1066, which under-

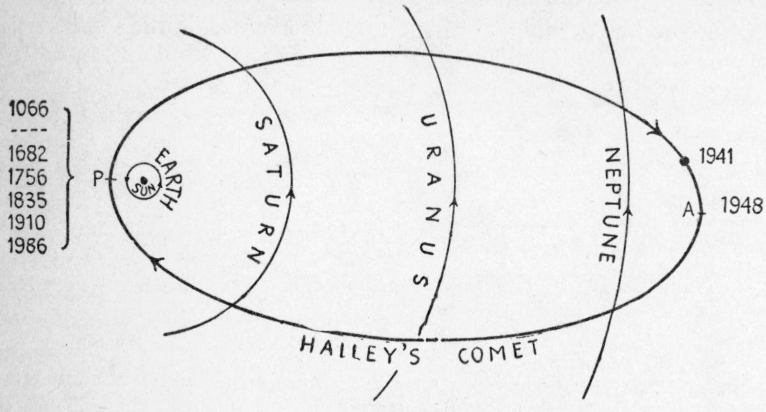

FIG. 128. Path of Halley's Comet. The position at various dates, with reference to the Perihelion, P, and Aphelion, A, is indicated.

mined Harold's morale, being interpreted as indicating his defeat by William the Conqueror. It is represented in the Bayeux tapestry.

Halley was succeeded at Greenwich by JAMES BRADLEY (1693–1762), who contributed to observational astronomy two important conceptions, *aberration of light* (1729) and *nutation of the earth's axis* (1748).

The aberration of light is most simply explained by the very illustration which suggested the idea to Bradley himself. Imagine travelling in a boat in a wind and with a flag at the mast-head. If the course be changed, the flag alters its apparent direction. Replace, in imagination, the wind by light coming from a star, and the boat by the earth moving round the sun and ever changing its direction. The result must be a cyclic change in the apparent position of a star. This Bradley was the first to observe and to explain.

Nutation (Latin 'nodding') of the earth's axis is an undulatory movement grafted on to its circular movement which yields the precession of the equinoxes known since Hipparchus (134 B.C., p. 83). Thus the movement of the axis is not in a circle, as it would be if the precessional movement were uncomplicated, but in a figure of crenated outline (Fig. 129). Since Bradley's time many astronomers have studied the conduct of the earth's axis. It has transpired that nutation is only one of a whole series of complications of its motion.

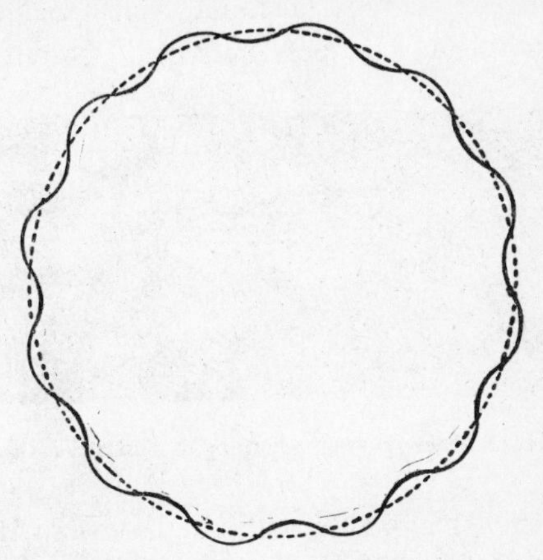

FIG. 129. Precession and Nutation. The axis of the earth moves, in the course of centuries, in such a way that a point on it, the North Pole for example, describes a circle (dotted line). This produces the phenomenon known as 'precession of the equinoxes'. Added to this motion, as Bradley showed, was another, that of 'nutation', producing waves in the circle, in fact a 'gently undulating ring'. In the figure the undulations are relatively enormously exaggerated.

The most impressive figure among eighteenth-century observational astronomers was WILLIAM HERSCHEL (1738–1822). Born in Hanover—which was then a possession of the British Crown—he came to England as a musician (1757), turned to astronomy, and acquired great technical skill in making instruments. He conducted four complete reviews of the heavens, with telescopes of increasingly greater power. The second review revealed Uranus (1781), the first new planet to be discovered in historic time. Further improvements in his instruments led to his discovery of the satellites of Uranus (1787) and of Saturn (1789).

Herschel's industry and accuracy as an observer were unrivalled and his skill as an instrument maker was of the highest order. His most striking investigations were directed to the distribution of the stars. He concluded that the entire sidereal system is of lens shape, the edge being formed by the Milky Way.[1] The diameter of the lens is about five times its thickness. Our sun is not far from the centre of this lens (Fig. 130).

Closely linked with Herschel's conception of the form of the universe was his immense series of observations on nebulae, of which he discovered many hundreds. He found, as had Galileo before him, that some of the nebulous appearances could be resolved into star

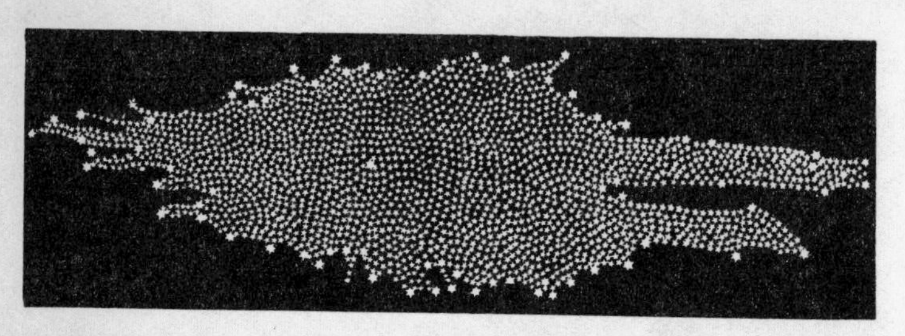

FIG. 130. Section of the universe according to Herschel's 'Lens-theory'.

clusters by instruments of sufficiently high power. At first he considered that all nebulae were of this nature and that they represented 'island universes' outside our own. Later, however, he concluded that some nebulae, at least, were composed of 'a shining fluid, of a nature totally unknown to us' (1791). He finally came to the conclusion that such shining fluid might gradually condense, the points of condensation forming stars and the whole forming a star cluster which might pass into a single star or star group (1814).

Linked also with his conception of the general form of the sidereal system was his view as to the movement within it of the solar system. It had been known since the time of Halley that certain stars move relatively to each other. Basing his opinion on the nature of their apparent movement, Herschel concluded that the entire solar system is itself progressing towards a point in the constellation Hercules (1805).

Herschel always emphasized the fact that stars are not merely scattered at random. In considering their distribution he noted that

[1] A similar conclusion had been reached in 1750 by Thomas Wright (1711–86) and in 1755 by the philosopher Immanuel Kant (1724–1804).

many were in closely contiguous pairs, 'double stars'. On an average the less bright would be the more distant. Owing to the orbital displacement of the earth, such pairs can be viewed, at intervals of six months, from two points 180 million miles apart. The perspective relations thus involved make it theoretically possible to estimate the relative distances of the two members of a pair. Herschel pursued this idea with extraordinary tenacity over a period of many years, mapping out the places and aspects of numerous double stars. At last (1805) he was able to show that some of these stars circulate round each other. In their manner of doing this they follow the mathematical formulae of the laws of gravitation. Those laws, enunciated by Galileo for bodies on our earth and shown by Newton to rule the solar system, were thus demonstrated to rule among the distant stars.

(ii) *Dynamical Astronomy*

In the eighteenth century, in the absence of any knowledge of the exact distances and movements of the stars, mathematical analysis could be applied only to the solar system. The distances from each other of the members of this system as well as their proportional sizes became fairly well known. The demonstrations by Newton for certain of them had left a presumption that all attracted each other according to the law of gravitation. The problem was to fit the exact consequence of that law to the movements which were revealed by progressively more exact observation.

This was the main task of the mathematicians of the age. Among them a foremost place must be accorded to the German philosopher and statesman G. W. LEIBNIZ (1646–1716), a man of very varied talents. His mathematical and scientific activity began after a visit to Huygens in Paris (1672) and to Boyle and others in London (1673). During three years' subsequent residence in Paris he devoted himself to mathematical study under Huygens. From this there resulted the conception of the 'differential calculus' on which the work of subsequent mathematicians was based.

The first formal publication of the method (1684) was preceded and followed by many years of controversy in the learned world on the question as to whether the priority rested with Newton (p. 293) or Leibniz. In fact, however, the nomenclature adopted by subsequent investigators was that of Leibniz.

LEONHARD EULER of Basel (1707–83), who early became blind, showed that certain irregularities in the earth's movement between the time of Ptolemy (pp. 89 f.) and his own were best explained by supposing that our planet is moving in a path which is a 'varying ellipse'

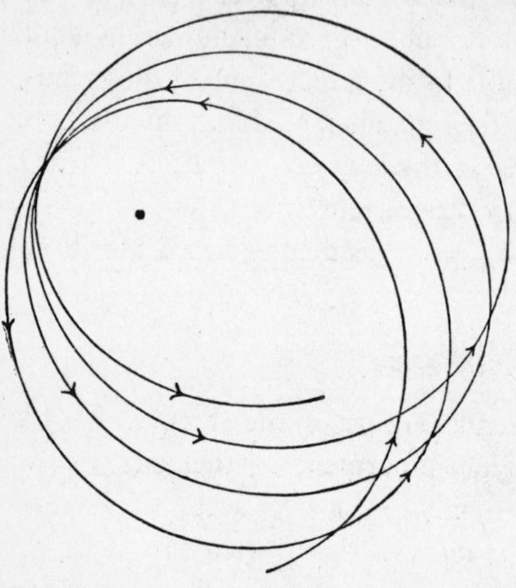

FIG. 131. Illustrating the path of a point moving in a varying ellipse around a focus.

and not a fixed one (1756, Fig. 131). This variation had pursued such a course that the axis of the earth's orbit had altered about five degrees since the time of Ptolemy.

J. L. LAGRANGE (1736–1813), of Turin and Paris, one of the greatest mathematicians of all time, made an important contribution concerning certain irregularities in the moon's motion. It had been known since Galileo that while the moon always turns the same face to us, yet there are parts near her edge that are alternately visible and invisible to us. Lagrange showed that this was best explained on the assumption that neither earth nor moon is truly spherical. Neither could therefore be treated as though the force of gravity acted at its centre (1764), as Newton had thought (p. 294).

Lagrange distinguished two types of disturbance of members of the solar system: (*a*) *periodic*, which complete a cycle of changes in a single revolution or a few revolutions of the disturbing body, and (*b*) *secular*, in which a continuous disturbance acts always in the same direction and presents no evidence of a cyclic factor. The disturbance of one member of the solar system by another depends both on the relative position of the two bodies and also on their orbital sizes, shapes, planes of movements, &c., that is the quantities known mathematically as the *elements* of the orbit. The relative position of the planets is constantly changing. Thus they produce changing disturbances one upon the other, the effects going through periodic cycles. But apart from these, there are disturbing forces based on the orbital elements themselves which give rise to changes in the orbital elements of other bodies. These secular changes in the orbital elements are in general very small, but they accumulate continually.

In the discussion of the periodic and secular movements of the members of the solar system there was a constant interdigitation of the work of Lagrange and that of P. S. LAPLACE (1749–1827). That remarkable man spent his life at Paris, pouring out a stream of books on astronomical and mathematical subjects. He did not permit his activities to be greatly interrupted either by the Revolution or by later successive governmental changes. His first major contribution was to show that an observed, very slow increase in the moon's rate of motion round the earth is explicable as due to a corresponding slow decrease of the eccentricity of the earth's orbit. This change in its turn is being produced by the gravitational action of the planets (1787). The order of change is such that the length of the month decreases by about $\frac{1}{30}$ second per century.

As long ago as 1650 irregularities in the motion of Jupiter and Saturn had been suspected. Halley had noted them (1676). They were thought to be of a secular nature. Laplace, working on suggestions of Lagrange, showed that the inequalities corresponded to a period of about 900 years. This was the starting-point of a series of most remarkable investigations by Lagrange and Laplace on secular inequalities (1773–84). The final result was the following general law:

Take for each planet the product

$$\text{mass} \times \sqrt[2]{(\text{axis of orbit})} \times (\text{eccentricity})^2.$$

Add together these products for all the planets.

The resulting sum is then invariable, except for periodic inequalities.[1]

This law establishes the existence of a constant stock or fund of eccentricity for the solar system. The total of this fund cannot be altered. If the eccentricity of one planet be increased, that of another must be diminished. (In fact nearly the whole fund is absorbed by Jupiter and Saturn.) The law forms a sort of guarantee of the stability of the solar system.

The work of the eighteenth-century astronomers was summed up by Laplace in his great *Celestial Mechanics* (1799–1825). Its object he declares to be 'to solve the great mechanical problems of the solar system and to bring theory to coincide so closely with observation that empirical equations should no longer be needed'. It is the most comprehensive attempt of its kind ever made. With its completion the Newtonian problem seemed solved. The movements of the known members of the solar system were deducible from the law of gravitation. The discrepancies were so small, compared to those which had already been removed, that the impression was created that they too would be removed by more careful observation or by some correction of calculation.

Laplace's name is indissociably linked with his 'nebular hypothesis' which appeared in his popular but nevertheless scientifically valuable *Essay on the System of the World* (1796). He pointed out that the motions of all the members of the solar system—some thirty to forty motions—were in the same direction.[2] All the motions were in planes but slightly inclined to each other, and the orbits of none were very far from circular. Attention was at the time being drawn to the nebulae by Herschel (p. 308). Laplace suggested that the whole solar system had condensed out of a vast rotating atmospheric mass, a huge gaseous nebula, that filled the bounds of the present solar system. The conception struck the imagination of the age and has remained an integral part of general thought concerning the cosmos.

[1] 'Eccentricity' is the technical term for the ratio, in an ellipse, of the distance between the foci to the whole length of the major axis. For ellipses approaching a circle it is very small and it approximates to unity as the ellipse lengthens (see Fig. 30).

[2] The motion of the satellites of Uranus is, in fact, in the opposite direction, but this had not emerged very clearly at the time Laplace was writing.

The death of Laplace took place just a century after that of Newton. The two events provide convenient landmarks in the history of science (1727 and 1827).

Two most remarkable observations, the direct result of theoretical considerations, were made in the first half of the nineteenth century. The first of these was made on the basis of the numerical sequence known as 'Bode's law' (J. E. BODE, 1747–1826) which had been set forth as early as 1772. If to each member of the sequence 0, 3, 6, 12, 24, 48, 96 (each figure after the first being double the previous) we add 4, producing 4, 7, 10, 16, 28, 52, 100, we obtain approximately the proportionate distances from the sun of Mercury, Venus, earth, Mars, Jupiter, Saturn with blank for the number 28. Unsuccessful search was long made for this missing planet. In 1801 GIUSEPPE PIAZZE (1746–1826) of Palermo found a very small planet, which he named *Ceres*, about a quarter the size of the moon, at the required distance. This directed the general attention of astronomers to the possibility of finding more such small bodies. Since that time over a thousand of these 'minor planets' or asteroids have been found, most of them in very similar orbits to that of Ceres and nearly all circling between the orbits of Mars and Jupiter. It is suggested that they represent an exploded larger planet of which meteors may also have been parts.

The second and more famous of these discoveries anticipated on theoretical grounds was that of a major planet. The existence of this body was betrayed by irregularities in the movement of the planet Uranus. In 1846 JOHN COUCH ADAMS (1819–92) of Cambridge and U. J. J. LE VERRIER (1811–77) of Paris, working quite independently, indicated the part of the heavens where the perturbing body was to be found. Telescopic search revealed it as foretold and it was given the name Neptune.

A constant desideratum of astronomy has been a determination of the distance of stars. This can be done by measuring the angle that the earth's orbit subtends to a star. The angle is so excessively small that its observation presents great experimental difficulties. These were first overcome by THOMAS HENDERSON (1798–1844). His result, though obtained earlier, was not published till 1839, while that of F. W. BESSEL (1784–1846) appeared in 1838.

(iii) *Astrophysics*

By the first quarter of the nineteenth century there had developed clear ideas of the general structure of the universe and mathematical conceptions of the forms, dimensions, and relations of its constituent

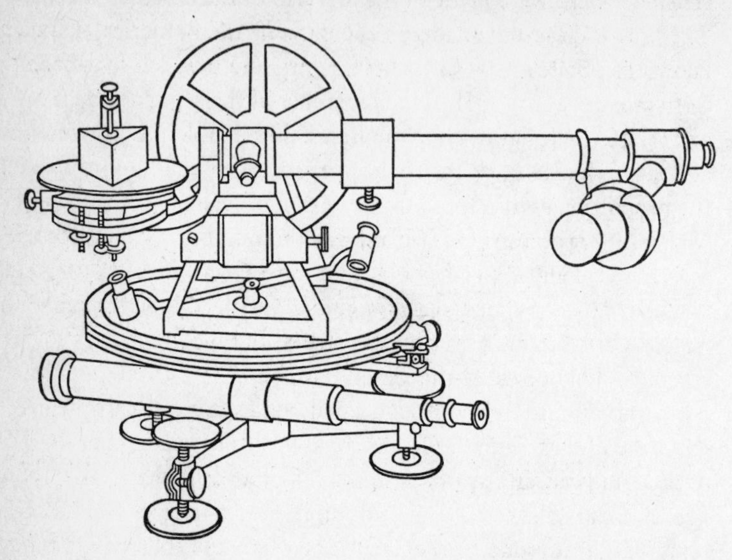

FIG. 132. Fraunhofer's original spectroscope (1814). The light passes through the prism and enters the telescope. He saw in the spectrum black lines which he found constant in position and mapped (Fig. 133 and see pp. 438–41).

members. There was, however, little positive knowledge of their physical and none of their chemical constitution.

The science of astrophysics began with W. H. WOLLASTON (1766–1828). Examining the solar spectrum in 1802, he saw dark streaks crossing the coloured band, which he took to be boundaries of the natural colours. Some twelve years later a self-educated Bavarian instrument-maker, JOSEPH FRAUNHOFER (1787–1826) attached a telescope to the prism and examined the solar spectrum more closely (Fig. 132). He found that the resulting spectrum exhibited numerous black transverse lines of constant position (1814, Fig. 133). Similar lines were visible in all forms of sunlight, whether direct, as from the sun itself, or reflected as from the clouds, moon, or planets. In the spectra from the stars the distribution of lines was different.

In 1859 the two Heidelberg professors, GUSTAV ROBERT KIRCHHOFF (1824–87) and R. W. BUNSEN (1811–99), succeeded in showing that there was an invariable connexion between certain rays of the spectrum and certain kinds of matter (Fig. 172). The assurance of their conclusion was certified by their discovery, through their spectra alone, of two new elements (*Caesium* and *Rubidium*). Kirchhoff went on to demonstrate certain essential characteristics of spectra and so

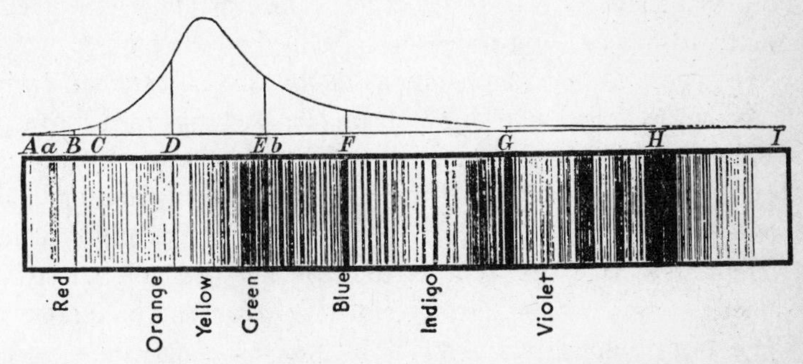

FIG. 133. Fraunhofer's map of the solar spectrum (1814) modified and reduced, to which are added the letters used to identify the more prominent absorption lines and a curve indicating his estimate of the relative luminosity of different parts of solar spectrum.

was able to determine the existence in the sun of many elements (pp. 438 f.).

With the advent of the spectroscope and its application to the heavens, all departments of astronomy became intimately linked. It must suffice to attempt a mere enumeration of some of the results of this modern phase which opened with John Herschel (1792–1871).

The subject of double stars, to which William Herschel (p. 307) had drawn attention, was developed by F. G. W. STRUVE (1793–1864) and his successors at St. Petersburg, working at first with telescopes made by Fraunhofer. A great many multiple stars have been made known. Their numbers render it certain that the forces that have given rise to our universe have a special tendency to the production of these bodies.

No general picture of the universe can be formed unless the laws of the motions of the stars are known. The proper motions of a few stars were known to both Herschels. In 1837 F. W. A. ARGELANDER (1799–1875) knew about 400. The number now known is many thousands. In recent years great stress has been laid on the prevalence among brighter stars of opposite stream-flows towards two regions in the Milky Way. This appearance is now considered to be due to the rotation of the whole stellar system.

Spectroscopic research from Kirchhoff's time has been persistently directed towards the sun. The majority of elements have been identified in the sun. During the year 1868 the solar spectrum was found to include a gas to which the name 'helium' was given. Twenty-seven years later the gas was obtained on our earth (p. 454).

The conception of the conditions of the sun underwent a great change in the period following William Herschel (pp. 307 f.). Much attention was paid to the sun-spots which were shown, as early as 1843, to have a definite period, and later a definite distribution and order of appearance, and a rate of rotation which is different in different solar latitudes. The relation of sun-spots to terrestrial magnetic storms was found to be remarkably constant (p. 444).

The solar prominences observable by the eye during eclipses only, now began to be examined with the spectroscope in full daylight (1868). Investigations have shown that the prominences increase and decrease in harmony with the sun-spots. The prominences originate in a shallow gaseous layer, the *chromosphere*, which is distinguished from the brilliantly incandescent inner layer the *photosphere*. Between the two is a narrow 'reversing layer' detectable only during eclipses and exhibiting special spectroscopic properties (p. 442).

A very important principle associated with the name of CHRISTIAN DOPPLER (1803–53) was introduced in 1842. According to 'Doppler's principle' the movement of a spectrum-yielding body or part of a body can be measured by the shifting of lines in its spectrum. This has rendered possible the estimation of the sun's rotation rate and also of the rate of approach and recession towards or away from us of various stars (pp. 444–5).

3. THE TERRESTRIAL GLOBE

(i) *Measurement of the Earth and Cartography*

The size of the earth was the subject of discussion from an early date. That it was an exact sphere was assumed at least from Aristotelian times. A more accurate method of measuring angular elevation became possible with the invention of the telescope. With its aid an estimation of the length of a degree was undertaken (1669–71) for the Académie des Sciences by JEAN PICARD (1620–82, p. 301). The figure reached was 69·1 miles, which was a large variant from that of 60 miles which had been the estimate generally accepted. The method adopted by Picard was in principle that of Eratosthenes (pp. 76 f.), a star being used instead of the sun. Picard's result was issued in a somewhat inaccessible form (1671). Thus it was at first missed by Newton, who, in ignorance of it, abandoned for some years his calculations, based on earlier measurements, seeking to identify gravity as the force that kept the moon and planets in their orbits (p. 293).

Soon after Picard's determination the Académie organized an astronomical expedition (1671–4) to Cayenne in French Guiana, then occupied by a French commercial company. Cayenne is in latitude 5° N. It was found that, to keep time there, the pendulums of clocks set for Paris in latitude 49° N. had to be shortened. The explanation of this, as we now know, is the bulging of the earth in the region of the equator. Gravitation decreases as we pass southward, since we are also getting farther from the earth's centre, and the pendulum therefore swings slower and has to be shortened if it is to keep time.

The results of the Cayenne expedition were published in 1684. In 1673 Huygens in his *Horologium Oscillatorium* (p. 303) had set forth the relation between the length of a pendulum and time of oscillation. This principle, together with the measurement of Picard, was utilized by Newton for the investigation of the figure of the earth in the *Principia* (1687, p. 294).

Between 1684 and 1714 long series of pendulum measurements were undertaken in France by G. D. Cassini (p. 304) and his son Jacques (1677–1756). The results of these suggested that the form of the earth is that produced by the rotation of an ellipse round its major axis (a *prolate spheroid*).

This conclusion was in discord with that of Huygens and Newton. Thus the form of the earth became a main subject of scientific discussion, and several expeditions went forth to make measurements and to take pendulum observations. Of these, the most important left Paris in 1735 for South America under C. M. DE LA CONDAMINE

(1701–74) to determine the length of a degree of longitude in the neighbourhood of the equator. The expedition laid down a famous and well-measured base, still spoken of as the 'Peru line'. In 1738 it was proved by P. L. M. DE MAUPERTUIS (1698–1759), who had been a member of a similar expedition to northern Sweden, and within the Arctic circle, that the form of the earth was that derived from the rotation of an ellipse round its minor axis (an *oblate spheroid*), flattened at the poles. These results came to be finally accepted and the era of exact geodetic survey began.

If the French excelled during this period in the exactness of their observations, the English made such observations possible by the skill and ingenuity of their instrument-makers. Thus GEORGE GRAHAM (1673–1751) invented the 'dead beat escapement' of clocks (Fig. 134) and arranged the pendulums of his clocks to swing through a very small arc (Fig. 135). He invented also the mercurial pendulum which remains always of the same effective length, since any expansion by heat of the suspending rod is compensated by expansion of mercury in a suspended jar. He constructed astronomical instruments for Halley and Bradley (pp. 306, 322) and geodetic instruments for Maupertuis.

JOHN HARRISON (1692–1776)—'Longitude Harrison'—devised the self-compensating gridiron pendulum (1726) and a maintaining mechanism by which a clock continues to go during winding. He is specially remembered for his chronometer which made possible, for the first time, the exact determination of longitude at sea. A chronometer is essentially a large well-made watch equipped with an heliacal balance-spring and a special type of escapement. The apparatus is suspended in gimbals and poised to remain horizontal despite movements of the ship. Harrison's best instrument kept time to within 54 seconds in 5 months, mostly at sea, a very fine performance for any date. Yet despite his just fame as a chronometer-maker, his instrument was too complicated for maritime use, costing £450, equivalent to a very much larger sum in present values. It was simplified by others and assumed its modern form about 1785.

The instruments of JESSE RAMSDEN (1732–1800) were no less renowned. Best known of them was his instrument known as an 'Equatorial' (1774), which can be adjusted so as to cause a telescope to follow by clockwork the apparent motion of any point in the heavens to which it is directed. Modifications of it are in use in every modern observatory. Of comparable value was his engine for the mathe-

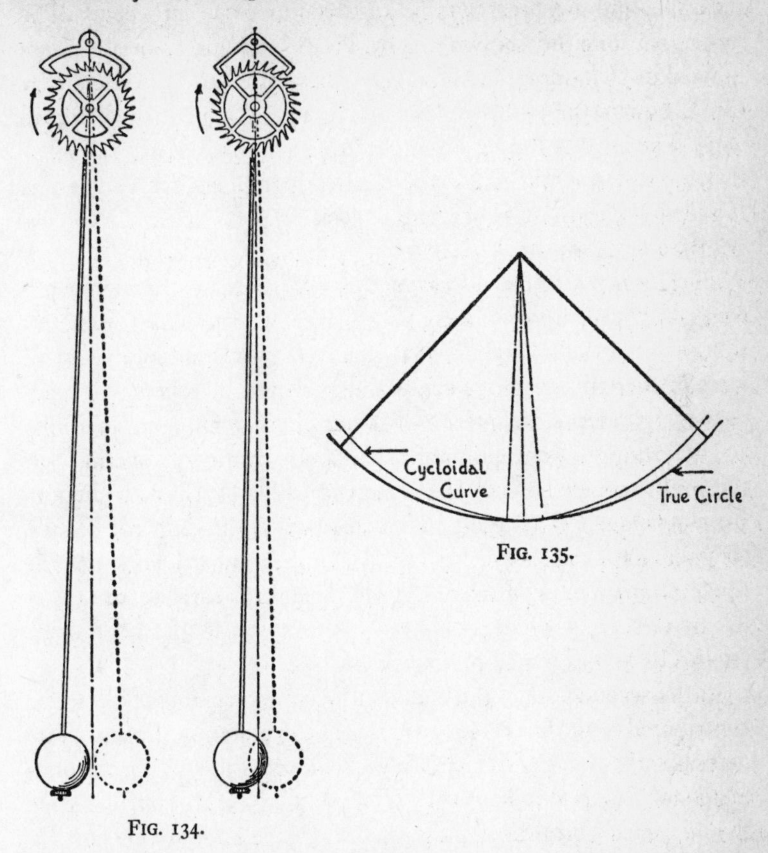

FIG. 134.

FIG. 135.

FIG. 134. Two forms of escapement. *Left*, 'anchor' escapement of William Clement (*c.* 1670) usual in clocks of the later seventeenth and early eighteenth century. The amplitude of the pendulum is small and the pallet striking the teeth causes recoil. *Right*, 'dead-beat' escapement of Graham (*c.* 1715) lessens amplitude further and avoids recoil. The smaller the arc the nearer to a cycloid and to isochronism (see Fig. 135). Graham's escapement remained standard for scientific purposes till the end of the nineteenth century.

FIG. 135. Diagram showing coincidence of circle and cycloid over a very limited arc.

matical gradation of the circle. He also completely transformed the surveying instrument for measuring angles, known from Elizabethan days as the 'theodolite'.

It was a period of great exploratory activity. Exacter determinations of the position of geographical points were constantly being recorded, and a more scientific cartography came into being. The numerous longitudes observed by Picard and his associates were utilized in 1679 for a map of France drawn up for the Académie by G. D. Cassini (p. 304), who also issued a good map of the world in 1694. The interest thus aroused produced a number of firms of map-makers, and several States appointed cartographers. At Venice was founded the earliest geographical society, the *Accademia Cosmographica dei Argonauti*.

The French excelled in cartography for most of the eighteenth century. Especially prominent as a maker of maps was J. B. BOURGUIGNON D'ANVILLE (1697–1783), many of whose admirable maps were in current use until a century ago. He was merciless to legend, preferring to leave the interior of Africa blank to filling it fancifully, and rejecting the conception of an Antarctic continent covering half the southern hemisphere. He portrayed China (1718) according to surveys conducted by Jesuit missionaries under the Emperor Kanghi (reigned 1661–1721). For long the best topographical work was the *Carte géométrique de la France*, based on surveys carried out (1744–83) by C. F. CASSINI (1714–84) and his nephew JACQUES DOMINIQUE (1748–1845), and issued in 1793.

In the second half of the eighteenth century a number of factors contributed to the furtherance of maritime exploration. The accurate determination of longitude at sea was made possible by the chronometers of 'Longitude Harrison' (1692–1776, p. 318). Until his time the only mechanical time-piece that would be accurate enough for the purpose worked on the pendulum principle and this was, of course, inapplicable on a moving base. The conditions of seamen were greatly ameliorated also by the use, as recommended by the British naval surgeon JAMES LIND (1736–1812), of orange and lemon juice as a preventive of scurvy, then the main obstacle to long voyages.

The three voyages of Captain JAMES COOK (1728–79) which occupied the last twelve years of his life will always be memorable It has been said that Cook's real monument is the map of the Pacific. In cartographical achievement he is, however, rivalled by the two French officers J. F. DE GALAUP, Comte de la Pérouse (1741–89), and

J. A. BRUNI D'ENTRECASTEAUX (1739–93), who began the exact record of geographical points in Chinese and Japanese waters and in the Eastern Archipelago. Their names are on the map of Australia.

The labours of explorers of this type mark the opening of the exact scientific stage of topographic development. In 1787, working with a theodolite provided by Jesse Ramsden (p. 318), General WILLIAM ROY (1726–90) measured a base line for the triangulation of the British Isles that was to lead up to the Ordnance Survey. The primary triangulation was not completed till 1858, but the detailed survey was begun in 1791, the first inch-to-the-mile sheet was issued in 1801, and the first six-inch-to-the-mile sheet (that is 1 in 10,560) in 1846.

Other countries have followed along somewhat similar lines but at later dates. Proposals in France to replace the Cassini map were held up by war, and no steps were taken till 1817. The map was brought to final completion only in 1880. Among continental surveys, of special interest as presenting peculiar difficulties, is the beautiful map of Switzerland published in 1842–65 and based on a triangulation completed in 1833. The scale, however, as with all continental maps, is less than that of the Ordnance Survey.

(ii) *Wind, Water, and Terrestrial Magnetism*

Along with the exploration of the globe there developed a desire to reach some generalized conception of its phenomena, its magnetism, the watery atmospheric envelope, the tides, the currents, the winds, and the climates. 'Geophysics', the body of knowledge thus collected, is a quite modern term (1888), but the kind of inquiry that it represents came into prominence in the eighteenth century.

The knowledge of the prevalent winds was brought into relation with the study of the earth as a whole by Halley (pp. 305 f.), who published in 1686 his account of the trade winds and monsoons. The map which accompanies it shows a clear line of demarcation between the variable winds of the temperate zones on the one hand and the more reliable tropic winds on the other, along a line which runs at about 30 degrees both north and south of the equator (Fig. 136). Halley was the first to connect the general circulation of the atmosphere with distribution of sun's heat over the earth's surface. In a later version of this map (1700) he added observations of the deviations of the magnetic compass, indicating the lines of equal variation (see p. 305).

GEORGE HADLEY (1685–1768) enunciated in 1735 the current theory of trade winds as the resultant of the rotation of the earth and the displacement of air by tropical heat (Fig. 136). Later the same view was taken by Dalton (1793, p. 343). The first general work on winds was produced in 1742 by the French mathematician JEAN LE

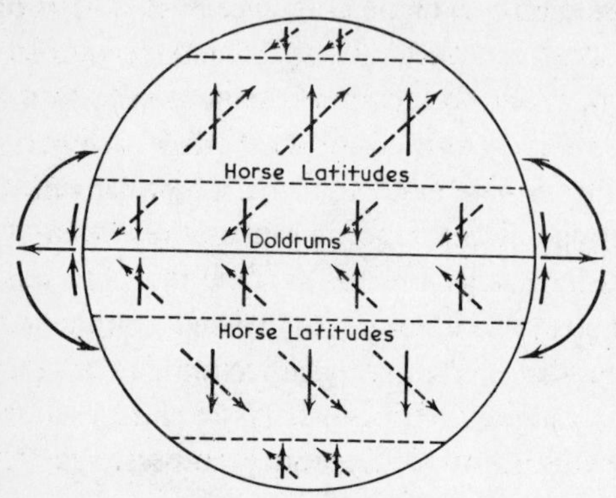

FIG. 136. Trade winds.

ROND D'ALEMBERT (1717–83). Of meteorological advances during the century, following the appearance of this work, the most significant were perhaps the investigations on the watery content of the atmosphere (1783) by H. B. DE SAUSSURE (1740–99) of Geneva. There were also balloon ascents to ascertain the properties of air at high altitudes notably by Gay-Lussac (1804, p. 344), the introduction of the 'wind scale' (1805) by Admiral Beaufort (1774–1857), and the theory of dew set out (1814) by the Anglo-American W. C. Wells (1757–1817).

Scientific weather-forecasting may be said to begin with WILLIAM REID (1791–1858), a Scottish military engineer officer. From 1831, when stationed in the West Indies, he collected information about the paths of the hurricanes in the Caribbean Sea. His *Attempt to develop the Law of Storms by means of Facts* (1838) went through many editions, translations, and enlargements. It was the basis of the work of the American naval officer MATTHEW FONTAINE MAURY (1806–73). From 1839 onward he occupied himself in extracting from log-books great numbers of observations of winds, currents, temperature, and so forth. By collating these he was able to draw up marine charts which led to such shortening of passages that an international conference was called in 1853 to consider further organization of such observations. Maury's *Physical Geography of the Sea* (1855) is the foundation work of modern knowledge on the subject. Largely as the result of his work, meteorological offices were established by several governments, and the international meteorological services initiated. In England the first director of the Meteorological Office, Admiral ROBERT FITZROY (1805–65), was appointed in 1855. Darwin had sailed with him twenty years previously in the *Beagle*, and he is still remembered by the 'Fitzroy barometer'.

In the early sixties the department began to collect observations by telegraph from land-stations, to issue storm warnings to the ports, and to publish weather-forecasts. The British Association was at this time maintaining 'Kew Observatory' for studies in terrestrial magnetism, atmospheric electricity, meteorology including solar radiation, and verification of instruments. Self-reading instruments, using photography for recording pressure and temperature, were introduced there. FRANCIS GALTON (1822–1911) was very active in promoting this work. His *Meteorographica* (1863) contains the basis of modern weather-maps. He is remembered also for his device of identification by finger-prints and his interest in eugenics.

Britain, from her geographical position and trading interests, was a natural home for meteorology. The work on it was centralized in the seventies by a committee appointed by the Government, the Royal Society, and the British Association. This body established a number of weather-observatories in the British Isles. It issued meteorological charts, current charts, and a *Quarterly Weather Report*. Meteorology was sufficiently a science to be represented by international gatherings from 1872. Investigation of the upper air was attempted first by balloons and then by watching the smoke of shell-bursts. In the eighties and nineties the general methods of measuring and recording wind direction and wind-pressures, of making and distributing weather maps, and of making forecasts reached a modern form. Meteorology grew to be an accepted discipline during the life of Sir WILLIAM NAPIER SHAW (1854–1915), whose publications on the subject began in 1881, who became a member of the meteorological council in 1897 and Director of the Meteorological Office in 1905.

Of all aspects of geophysics, the theme of the tides has perhaps attracted the greatest amount of scientific ability. Kepler and Galileo devoted attention to the subject. Newton, in the *Principia* (1687), placed the theory of the tides on a gravitational basis.

An adequate exposition of the tides is a very difficult task, nor is the tidal theory of Newton adequate for the prediction of the times or the heights of tide at any required place. Newton, however, did give a satisfactory explanation of many of the characteristics of tides. The Newtonian view was expounded by Halley for the benefit of King James II, and this exposition has since become traditional in textbooks. It is illustrated by a diagram to be found on the first plate of nearly every school atlas. It is misleading since the problem is represented as one of statics when it is, in fact, one of dynamics.

In the nineteenth century no one contributed more to tidal theory than Sir GEORGE DARWIN (1845–1912), son of Charles Darwin, especially in a series of papers on tidal friction as a factor in determining the relation of earth and moon. He showed that, because of viscosity, the tides will always have their high point in advance of the positions they would hold if the whole earth were perfectly fluid. Thus a force is ever checking earth's rotation and increasing the distance between moon and earth with a consequent lengthening of the month which was equal to the day 54 million years ago. His work *The Tides* (1898) is a masterpiece of popular exposition based very largely on his own researches.

The subject of terrestrial magnetism has been especially studied because of its importance to navigation. A mass of data was collected, though there were few general ideas to connect them until the nineteenth century was well advanced. That the magnetic compass does not normally point to the true north is said to have been discovered by Columbus during his first voyage to America in 1492. The degree by which it departs from this line is the *variation*. That the compass suspended about a horizontal needle in the magnetic meridian will also dip was discovered at the end of the sixteenth century. Gilbert (1600, pp. 221 f.) knew that both variation and dip were different in different places. In the early years of the seventeenth century it was found that the variation changed in the course of years in the same place. George Graham (p. 318) showed in 1724 that there was also a diurnal change in the variation. Much work was done by Halley on the difference in the degree of variation in different parts of the world. In 1700 he drew up an interesting chart in which the distribution of equal degrees of variation in the earth's surface is represented by lines, *isogonic lines* as we now call them. The method, there used for the first time, has since been adopted for innumerable other terrestrial variations such as isoclinals (lines of equal magnetic dip), isomagnetics (lines of equal magnetic force), isobars (lines of equal barometric pressure), isotherms (lines of equal temperature, p. 331), and the like.

Between 1756 and 1759 numerous observations by JOHN CANTON (1718–72) showed that on certain days the movements of the compass were conspicuously irregular and that the Aurora borealis was then often visible. These phenomena, it was soon realized, were related to the occurrence of sun-spots.

Another landmark in the history of terrestrial magnetism was the discovery, towards the end of the eighteenth century, that the intensity of magnetic force varies at different parts of the earth. The first published observations on this were made in equatorial America (1798–1803) by Humboldt (p. 331). In 1827 Arago (p. 359) showed that this intensity also exhibits diurnal variation. In 1834 the mathematician K. F. GAUSS (1777–1855) instituted at Göttingen the first special observatory for terrestrial magnetism. He greatly improved the type of instrument for magnetic observations. In 1840 a number of magnetic laboratories were established in various parts of the British Empire under the general superintendence of EDWARD SABINE (1788–1883), who had long been occupied on the subject. His numerous publications on terrestrial magnetism issued between 1823 and 1871 are still consulted.

(iii) *Early Views of Earth History*

That something of the history of the earth might be learned by a study of its crust was believed from of old. Much positive mineralogical knowledge accumulated from the mining industry. Among the most puzzling phenomena presented by the crust of the earth was that of fossils. The Dane NIELS STENO (1648–86), who spent some years in Italy, discussed the formation, displacement, and destruction of the stratified rocks in Tuscany (1669) and recognized the organic origin of fossils. Steno added broad powers of discernment to extensive scientific equipment and creative imagination. He sought to reconstruct the eventful geological history of Tuscany by portraying in a series of diagrams what he believed to be the fates of its main strata. These figures are the first attempt to represent geological sections (Fig. 137). They imply an enormous extension of the then customary time-scale, as he timidly and vaguely recognizes in his text. A number of Italian, English, and French writers concurred with Steno, and during the first three-quarters of the eighteenth century there was an extensive accumulation of geological data and many theories were proposed to explain them.

The first comprehensive general account of the history of the earth, which included a consideration of the nature of fossils, was put forward by GEORGES LOUIS LECLERC, Comte de Buffon (1707–88, pp. 503 f.) in his *Époques de la Nature* (1778). In forming his theory he laid stress upon certain data, not all of which can now be interpreted as he would have had them. He held in mind primarily (*a*) the oblate spheroid form of the earth (p. 318); (*b*) the contrast between the small amount of heat received from the sun and the large supply possessed by the earth; (*c*) the effect of the earth's internal heat in altering the rocks; and (*d*) the presence of fossils in all sorts of situations, even mountain tops. In association with the last he noted that limestone in north Europe, Asia, and America often consists largely of the remains of marine organisms; and that the remains of large terrestrial animals, more or less similar to living forms, often occur near the surface, showing that they were recently living, whereas the deeper-lying remains of marine creatures in the same region belong to extinct forms or to forms related only to the inhabitants of far distant seas. He conceived that the earth (and other planets) arose from the collision of a comet with the sun. Thus arose a molten spheroid, the history of which can be divided into seven epochs, thus:

1st epoch. Incandescent to molten. 3,000 years.
2nd epoch. Gradual consolidation. Rents in crust allow influx of molten metallic ores. 35,000 years.

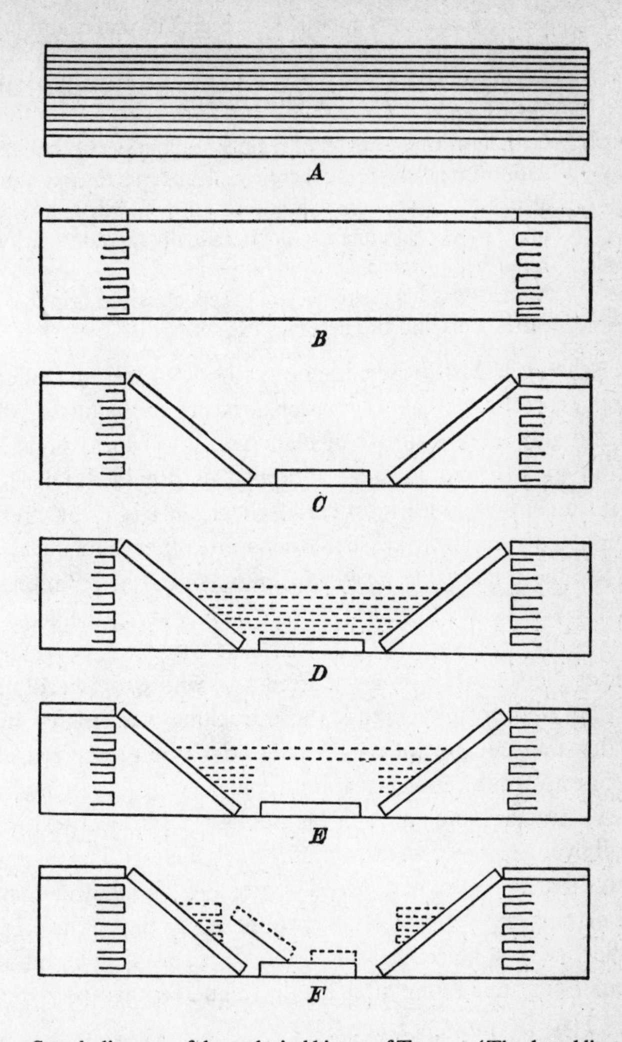

FIG. 137. Steno's diagrams of the geological history of Tuscany. 'The dotted lines represent sand strata with which various quantities of clay and rocks may be mixed. The rest represent strata of rock, though other strata of softer substance may be mixed with them. Here I review briefly the order of change. A, The rocky strata are still whole. B, Huge cavities have been eaten out by water or fire while the upper strata remain unbroken. C, Mountains and valleys have been caused by breaking up of upper strata. D, New strata have been laid by the sea in the valley. E, A portion of the strata in the new beds has been destroyed while the upper strata remain unbroken. F, Hill and valleys produced by the breaking of the upper strata.' Slightly modified from Steno. Florence, 1669.

3rd epoch. Atmospheric vapours precipitated as the primitive universal ocean. Continents appear. Life begins in waters and marine sediment accumulates. 15,000–20,000 years.

4th epoch. Access of internal heat. Period of violent volcanic activity. 5,000 years.

5th epoch. Calm restored. Equatorial regions still too hot for habitation. Life over polar areas where dwell huge terrestrial animals, elephants, mastodons, rhinoceroses, &c., which now came into existence. Fauna and flora gradually migrate southward.

6th epoch. Land mass broken up. Man appears.

7th epoch. Man asserts his supremacy. This epoch will continue till the earth cools and life becomes extinct.

The scheme is historically important both as the first effective attempt to correlate actual observations bearing on the history of the earth, and also as an estimate of many geological formations as of very slow growth and of great antiquity. It provided a basis for inquiry. In common with most early schemes it laid great stress on volcanic activity, earthquakes, explosions, and other dramatic events.

Despite the remarkable insight of the accomplished Buffon, and the attractiveness and popularity of his literary style, the geological dictator of the age was ABRAHAM GOTTLOB WERNER (1750–1817), a teacher at the school of mines at Freiburg, who wrote hardly anything at all, did not travel, and whose teaching was vitiated by his belief that the sequence of rock masses which he recognized in his native Saxony was of universal application. Werner was an unusually successful teacher, and through his pupils the physical features of rocks all over the world became more widely known. His main doctrine was that rocks are of aqueous origin, and his followers, known as Wernerians or 'Neptunists', were opposed by those who stressed the influence of subterranean heat, the 'Vulcanists'. The influence of Werner continued long after his death and reached the youthful Charles Darwin (pp. 506 f.).

Very important in the history of geology is the influence of the French naturalist Cuvier (p. 381). He realized that the evidence of the rocks reveals a succession of animal populations. He perceived that vast numbers of species, many no longer existing, appeared upon the earth at different periods. Following Linnaeus, he was a firm believer in the fixity and unalterability of species, though his contemporary Lamarck (p. 505) was engaged in putting forward the opposite view. Cuvier had, however, to account for the extinction of some forms of life, and for what seemed the creation, or at least the appearance, of new forms. His explanation of these remarkable facts was that the earth has been the scene of a series of great *catastrophies.* He believed that of the last of these catastrophies we have an historic record. It is the flood recorded in the Book of Genesis. He expressly denied the existence of fossil man of great antiquity.

(iv) *Stratigraphy*

The work of JAMES HUTTON (1726–97) initiates a more modern attitude. He travelled widely in Britain to study rocks, and satisfied himself that it is mostly in stratifications that fossils occur. He saw clearly that the imposition of successive horizontal layers is inexplicable as a result of a single great flood but suggests rather a quiet orderly deposit over a long period. In his *Theory of the Earth* (1795) he interpreted the strata as having once been the beds of seas, lakes, marshes, &c.

It was soon recognized that rocks often contain fragments from lower layers, nor could the fact be missed that stratified series are often tilted, bent, or broken. Many, encouraged by Cuvier's doctrine of 'catastrophies', ascribed these irregularities to violent upheavals. In this connexion it is interesting to observe that the *Essai sur la géographie minéralogique des environs de Paris* (1811) of ALEXANDRE BRONGNIART (1770–1847), though written in collaboration with Cuvier, inclines more to the views of Hutton.

WILLIAM SMITH (1769–1839), a civil engineer, obtained an insight into the nature of strata while cutting canals. He published the first coloured geological map (1815). His *Stratigraphical System of Organised Fossils* (1817) showed that certain layers have each their characteristic series of fossils. Some members of a series are wont to occur also in the layer below, others in the layer above, others in all three. Therefore changes in the flora and fauna which these fossils represent could not have been sudden. He saw, too, that the farther back we go, the less like are the fossils to forms still living.

A third British geologist, CHARLES LYELL (1797–1875), finally exorcized the catastrophic demon. He took to the study of geology while at Oxford, travelled considerably, and was influenced both by William Smith and by Lamarck. He saw that the relative ages of the later deposits could be determined by the proportion they yielded of living and of extinct molluscan shells. In his great *Principles of Geology* (1830–3) he showed that rocks are now being laid down by seas and rivers and are still being broken up by glaciers, rain, sandstorms, &c.:

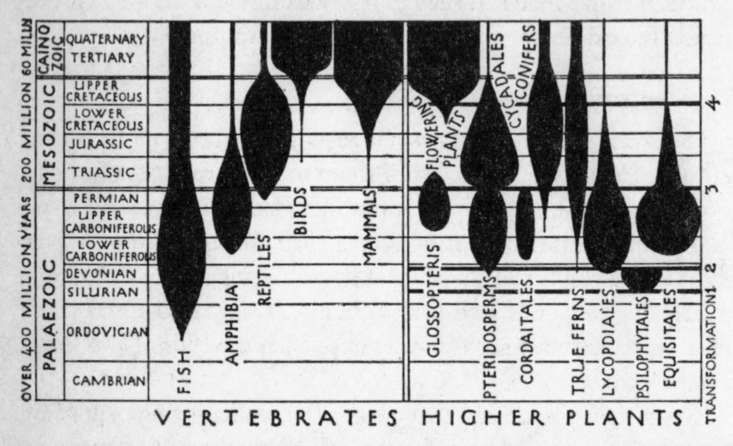

FIG. 138. Diagram of geological succession of higher organisms. The varying width of the black bands represents the relative dominance of the various classes at the different geological periods. The time estimates to the left must be taken as the roughest of approximations. The numbers to the extreme right and the black horizontal lines corresponding to them indicate the four great floral transformations.

that geologically ancient conditions were in essence similar to those of our time. Few books have exercised more influence on biological thought. Darwin's early observations were made in the light of Lyell's work. The conception of the general geological sequence of groups of living things, as conceived in the fifties, was not very far from that of about 1900 (Fig. 138).

Geology became an essentially British science as the names of the formations reveal. Lyell is responsible for *Pliocene* (Greek, 'more recent'), *Miocene* ('less recent'), and *Eocene* ('dawn of recent'); Sedgwick, the Cambridge geologist with whom Darwin went on geological excursions, invented *Devonian* (from its predominance in Devonshire), *Cambrian* (Cambria = Wales), *Palaeozoic* (Greek, 'ancient life') and *Cainozoic* ('new life'). Between the last two formations John Phillips of Oxford (1800–74) interpolated *Mesozoic* ('intermediate life'). Other British contemporaries are responsible for *Carboniferous* (or 'coal-bearing'), *Ordovician* and *Silurian* (the Ordovices and Silures are British tribes mentioned by Caesar), *Permian* (from the province of Perm in east Russia), and *Cretaceous* (Latin, 'chalky'). On the other hand, *Triassic* (Latin *trias*, the number 'three') and *Jurassic* (from the Jura mountains) were titles given by German geologists at the beginning of the nineteenth century. The term Tertiary is older and was used by eighteenth-century Italian writers. The tertiary formations were held to be the third of a series of which the *Secondary* correspond roughly to the *Mesozoic* and *Palaeozoic*, and the *Primary* to the non-fossil-bearing rocks. The word *Geology* itself was effectively introduced (1779) by H. B. de Saussure (1740–99) of Geneva, founder of modern mountaineering.

Of all writers on geophysics none has treated the subject so comprehensively and philosophically as ALEXANDER VON HUMBOLDT (1769–1859, p. 401). His life was largely spent in travel and exploration of the most varied kind, and his status as diplomatic agent in Paris brought him contact with nearly all the leading scientific men of his day. Among his positive additions to science is the introduction of isothermal lines (1817), and he was the first to make a general study of temperature and pressure over the globe which has been essential to the modern science of meteorology. He was the first to investigate the rate of decrease of mean temperature with increased altitude. He made many studies of volcanoes and showed that they occur in linear groups, presumably corresponding to subterranean fissures. He showed that many rocks thought to be of aqueous were really of igneous origin. He discovered that the magnetic force of the earth decreases from the poles to the equator (1804). He made the preliminary steps to a real geography of plants, studying them in relation to the physical conditions in which they grow.

Humboldt is seen at his best in the magnificent summary of his own life-work, his *Kosmos*, of which the publication was begun in 1845 and completed posthumously in 1862. This book has been said to combine the large and vague ideas, typical of eighteenth-century thought, with the exact and positive science of the nineteenth. It is a truly transitional work, but still forms an excellent introduction to the study of geophysics.

During the first half of the nineteenth century, as geology grew into an independent science, the structure of the earth was studied from the point of view of the distribution and arrangement of its rocks (stratigraphy), from the point of view of the structure and composition of its rocks (petrography), and from the point of view of the nature and affinities of its fossils (palaeontology). Perhaps no country in the world presents so much geological variety within so small an area as does England. It is thus not inexplicable that geology became an especially British science. 'The Geological Survey of England and Wales' was begun by THOMAS DE LA BECHE (1796–1855) in 1832. It was far earlier in inception and execution than any comparable work produced in any other country.

A series of other English investigators gave to geology its rational framework for the detailed research of the next century. Thirty years of work by G. POULETT SCROPE (1787–1876), beginning with his *Considerations on Volcanos* (1825), marked the end of the Wernerian view (p. 328). He laid the foundations of the theory of volcanic action and drew attention to the peculiar distribution of volcanoes. RODERICK MURCHISON (1792–1871) in his great *Silurian System* (1839) expounded the chronological correspondence of rocks, introduced much of the nomenclature now in use, and explained the nature and incidence of many scenic details. His views were shown to be applicable over a wide area by his geological exploration of Russia (1841–5). Behind the band of British geologists stood ADAM SEDGWICK (1785–1873), who worked with them all and among them with his pupil Charles Darwin (pp. 507 f.).

4. TRANSFORMATIONS OF MATTER

(i) *Rise of Quantitative Method*

A belief in the indestructibility and uncreatability of matter is, in some degree, implicit in many operations outside the scientific sphere (p. 268). In the seventeenth century the belief sometimes became explicit. Thus Francis Bacon wrote: 'It is sufficiently clear that all things are changed, and nothing really perishes, and that the sum of matter remains absolutely the same' (*Cogitationes de natura rerum*, published posthumously, 1653), and there are comparable passages in the writings of Boyle (pp. 271 f.). The doctrine was given express form by Newton.

The law on which gravity acts, that of inverse squares, implies that the *weight* of a body is not constant, but varies according to its relation with other bodies. But Newton's second law of motion, that 'change of momentum[1] is proportional to the impressed force', implies that quantities of matter, that is to say *masses*, are equal if they suffer equal changes of momentum under the action of equal forces, and that, conversely, forces are equal if they produce the same changes of momentum in the same body. Thus Newton distinguished clearly between *mass* and *weight*. The *mass* of a body is proportional to the force that produces a given acceleration to it. This force, in the case of a freely falling body, is the weight. Since all bodies fall at the same place with the same acceleration, their masses are proportional to their weights at the same place.

This exact and express doctrine of the *constancy of weight at the same place* (provided that other attracting bodies are unmoved) was a condition for the development of conceptions concerning the nature of physical changes. Without that doctrine the belief in any sudden, inexplicable, or magical appearance is possible. With it all changes in the state of matter can, in theory, be expressed in terms of number, weight, and measure. The changes that are specially investigated on the basis of weight are those known as 'chemical'. Thus the Newtonian conception gave a special impetus to the rationalization of chemistry and provided, in effect, the doctrine of the indestructibility and uncreatability of matter.

The investigation of chemical processes in the seventeenth century had yielded, by the dawn of the eighteenth, a vast accumulation of facts for which no satisfactory system of classification had been suggested. Antitheses, as *acid and alkali*, were emphasized and tests for them were devised. Categories were invented and defined, such as *salts* (that is soluble, sapid, and crystalline substances), *earths* (that is friable, fire-resisting, and tasteless substances), and *calces* (that is

[1] Newton's word is *motion*, not momentum, but he means what we mean by the latter word.

powdery products of heated minerals). JAN BAPTIST VAN HELMONT (1577–1644) had indicated the existence of various aeriform substances, for which he devised the name *gas* (1644, p. 269),

FIG. 139. Hales's apparatus for collecting gas over water, 1727.

which could be condensed, as he supposed, into solid bodies and released therefrom by chemical change. He had, however, no method of collecting gas. Chemical theory, though it had emerged from the alchemical stage, was a confused mass of doctrine and tradition.

The Rev. STEPHEN HALES (1677–1761) devised an apparatus for collecting gases by leading them, from the retorts in which they were produced by heating, through a pipe to a vessel filled with and inverted over water, in the so-called 'pneumatic trough' (Fig. 139).

He was able to measure the volumes of gases produced from weighed amounts of solids. He made, however, no further chemical examination of these gases because he supposed the product to be, in all cases, 'air' which had functioned as a kind of cement binding together the particles of the solids that he had heated.

Chemical technique was, in other respects, advanced and refined. This process was aided, from about 1670, by apparatus made from the newly introduced transparent flint glass in place of the older more opaque vessels. The knowledge of the age was admirably summarized by the distinguished Dutch physician HERMANN BOERHAAVE (1668–1738). His *Elements of Chemistry* (1732) is among the very few great works written expressly as a students' textbook. Though exhibiting few new departures, it is firmly based on personal experience and is exceptionally lucid. Boerhaave held that all chemical events are ultimately reducible to relatively few and simple categories, and he believed 'vital' processes to be expressible in chemical terms. Boerhaave's attitude gave to experimental chemistry a hopeful outlook which supported it for more than a generation, despite the paucity of important general laws.

The most notable chemical development of the earlier eighteenth century was the idea of 'affinity'. In 1718 the French physician ÉTIENNE FRANÇOIS GEOFFROY (1672–1731), influenced by Boerhaave, drew up tables in which acids were arranged in the order of their affinity for certain bases, and metals were arranged in the order of their affinity for sulphur. The relative degrees of affinity were estimated by ascertaining whether one base turned out another base or one metal another metal from a given compound. This idea of Geoffroy was further pursued by Black and others and notably by Bergman and Berthollet (p. 342).

(ii) *Intensive Study of Chemical Reaction*

The Scottish investigator JOSEPH BLACK (1728–99) published in 1756 his *Experiments upon Magnesia alba, Quicklime, and some other Alcaline Substances*. Perhaps no brief chemical essay has ever been so weighted with significant novelty. Black was a cautious investigator and his success was due to the accuracy of his measurements. He knew that chalk, by being heated and thus turned into quicklime (equation 1), ceases to effervesce with acids but gains the power of absorbing water (equation 2). As we would now formulate it:

$CaCO_3 = CaO + CO_2$ (Chalk into quicklime. Gas evolved.) (1)
$CaO + H_2O = Ca(OH)_2$ (Slaking of quicklime. Water absorbed.) (2)

Moreover, Black showed that, in the process of heating, the chalk loses weight, a loss which, by applying the methods of Hales (p. 333), he attributed to the removal of air in the process. And it had long been known that if the slaked lime be treated with a mild alkali, e.g. carbonate of soda, it is changed back into the state in which it was before heating, in fact, into chalk, while the mild alkali is converted into a caustic alkali. The process would now be represented thus:

$$Ca(OH)_2 + Na_2CO_3 = CaCO_3 + 2NaOH. \quad (3)$$

Moreover, he showed that a definite amount of chalk, whether heated into quicklime or not, neutralizes an equal weight of acid, the only difference being that the neutralization takes place with effervescence and loss of weight if the chalk is unheated, and without effervescence or loss of weight if the chalk is first heated into quicklime. Thus:

Unheated $CaCO_3$ (chalk) $+ 2HCl = CaCl_2 + H_2O + CO_2$. (4)
Heated CaO (quicklime) $+ 2HCl = CaCl_2 + H_2O$. (5)

The gas given off by chalk in (1), transferred from one alkali to the other in (3) and given off in the effervescence produced by the reaction (4), he named 'fixed air', thus differentiating it from the ordinary air of the atmosphere more clearly than van Helmont (p. 269) had been able to do in his tentative and chemically incomplete work. We now call it 'carbon dioxide'. The conversion of quicklime into ordinary chalk by exposure to air:

$$CaO + CO_2 = CaCO_3, \quad (6)$$

proved that carbon dioxide is a normal constituent of the air.

Black's work is of very great importance as the first intensive and detailed study of a chemical reaction. His especial triumph consisted in showing that the chemical changes occurring in this series of reactions could, without isolating the 'fixed air', be detected by subjecting them at every stage to the arbitrament of the balance. Black had thus discovered a gas different from air, which could exist in either the free or combined state, could be transferred from combination with one substance to another, and had many properties peculiar to itself. It had not till then been generally and clearly realized that there were any kinds of gases distinct from air. Attention was now drawn to this fact. The development of a technique for the isolation and study of gases and the discovery of the characters and laws of combination of gases was the main task of chemical endeavour of the later eighteenth and early nineteenth centuries.

(iii) *Gases*

In 1766 the eccentric philosopher HENRY CAVENDISH (1731–1810), as exact an experimenter as Black, sent his first paper to the Royal Society. It bore the title *On Factitious Air*, by which he implied gas produced artificially in the laboratory as distinct from the 'natural' air. He found that a definite, peculiar, and highly inflammable gas, which he called 'inflammable air'—'hydrogen', as we now call it—is produced by the action of acids on certain metals. Continuing his investigations on an exact quantitative basis he published his *Experiments on Air* (1784). These demonstrated that the only product of the combustion of 'inflammable air' (hydrogen) and 'dephlogisticated air' (oxygen) is water. He gave an approximately correct estimate of the proportions of the two in water.

Cavendish ascertained the amount of hydrogen evolved by action of acids on different metals. Adjusting his figures according to modern findings, we may say that he found that one part by weight of hydrogen was displaced by twenty-four parts of iron, twenty-eight of zinc, or fifty of tin. These numbers correspond to the 'equivalents' of these elements, and were used by him in 1766 to describe the different weights of different bases that neutralized a fixed amount of a given acid. He was the first to determine the weights of equal volumes of gases, a very fruitful line of research.

The chemical activities of the Unitarian divine, JOSEPH PRIESTLEY (1733–1804, p. 406), were contemporary with those of Cavendish. He greatly improved the technique for studying gases by collecting them over mercury instead of water (Fig. 140). A series of important observations was made by him in the seventies and eighties. He showed that green plants would make respired air again respirable, and that they gave off a respirable gas. He prepared and studied a number of gases (ammonia, hydrogen chloride, sulphur dioxide, nitric and nitrous oxides, nitrogen peroxide), he investigated nitrogen

FIG. 140. Priestley's apparatus for collecting gas.

and silicon tetrafluoride, and he isolated oxygen (1774–5) by heating certain oxides. He was hampered by his obstinate adherence to the old phlogiston theory. Phlogiston was a hypothetical substance supposed to exist in all combustible bodies and to be disengaged during combustion. It was 'the matter of fire'.[1]

Contemporary also with Cavendish and Priestley was CARL WILHELM SCHEELE (1742–86), a Swedish apothecary and one of the greatest of chemical experimenters and discoverers. His *Treatise on Air*

[1] This false idea, formulated by J. J. Becher (1635–82) was given currency by Stahl (p. 281) from 1702. It died with the acceptance of the work of Lavoisier (p. 339).

and Fire (1777) proved that air consisted of two different gases now known as oxygen and nitrogen. Most of this research had been made, though not published, before 1773. Thus Scheele's recognition and isolation of oxygen preceded Priestley's. Scheele's numerous chemical discoveries include, not only oxygen, but also chlorine, manganese, baryta, silicon tetrafluoride, hydrofluoric acid, various inorganic acids and the first extensive range of organic acids, glycerol, arseniuretted hydrogen, copper arsenite (still known as 'Scheele's green'), and many other substances. His *Treatise* and his many memoirs mark him as a rigorous experimenter and a concise writer.

(iv) *The Elements*

Black, Cavendish, Priestley, and Scheele assumed that matter was completely 'conserved', that is neither came into being nor passed out of being during their experiments. Further, they assumed weight to be the measure of the amount of matter. But the old view of the four elements had not quite gone out of currency. It was, in fact, widely mooted that prolonged boiling converted water into earth. This question was taken up and finally resolved by the great French chemist ANTOINE LAURENT LAVOISIER (1743–94). He began the investigation with a simple but extremely carefully conducted series of experiments. By exact weighing he showed (1770) that if ordinary water be boiled in a suitably designed vessel, in such a way that the steam produced is condensed and it and the residue weighed, then the weight of the solid particles that remain behind corresponds to the weight lost by the water. Thus nothing is lost and nothing gained.

Lavoisier next investigated the phenomena of calcination of metals. This process, it had long been known, results in the increase in weight of the calcined metal, an increase which Lavoisier was able to show as due to something taken from the air (1774–8). This was a serious blow to the phlogiston theory (p. 281). He proceeded to an extensive and quantitative investigation of the changes occurring during breathing, burning, and other forms of combustion (1772–83). In the course of this he discovered the true nature of respired air, and showed how both carbon dioxide and water are products of the normal act of breathing.

If clear grasp of its implication be accepted as the test of a discovery, Lavoisier was the discoverer of oxygen. We owe the word *oxygen* to him. He proved that in all cases of combustion there is a combination of oxygen with the substance burned. He repeated the experiments of Cavendish on exploding 'inflammable air' (hydrogen) and 'dephlogisticated air' (oxygen), and thence concluded that water was a compound of these two gases (1784). These experiments mark the end of the phlogiston theory. Men of science had now in their hands a technique by which the laws of chemical combination could be investigated.

Among Lavoisier's major contributions to science was his establishment, once and for all, of the conception of chemical 'elements' in the modern sense—'simple radicles' was the title attached to them by one of his French contemporaries. These 'simple radicles', following Boyle, he defines as substances which cannot be further decomposed. He divides them into four groups: (*a*) the gases oxygen, nitrogen, and hydrogen, and the 'imponderables' light and 'caloric'; (*b*) elements such as sulphur, phosphorus, and carbon which, on oxidation, yield acids; (*c*) metals, of which he distinguished seventeen; (*d*) the 'earths', lime, magnesia, baryta, alumina, and silica. These last had not yet been decomposed. The same might be said of the 'alkalis', potash and soda, but Lavoisier was so certain that the alkalis were compound substances, produced by the union of oxygen with other 'simple radicles', as yet undiscovered, that he refused to include them among the 'simple radicles'.

Lavoisier was able to recognize correctly twenty-three elements in the modern sense, though his actual list was considerably longer. Together with de Morveau and Berthollet, in their joint work, *A New Chemical Nomenclature* (1787), he introduced a new system of naming substances according to their chemical composition, a reform that contributed greatly to the progress of chemistry by its rejection of the fanciful and often ridiculous alchemical names and the substitution of many now in use.

Lavoisier is generally regarded as the founder of the modern phase of chemistry, which he set forth in his classic *Elementary Treatise on Chemistry* (1789). His writings were widely studied. His experiments were models of painstaking ingenuity. Perhaps his numerous and varied achievements may be summed up in the statement that he gave coherence and clarity to the conception of the conservation of matter. All his work was based on the explicit assertion of the principle that, within experimental limits, the same weight of simple bodies can be drawn from compound bodies as has been put into them, no more and no less, and that compound bodies represent the combined weight of the simple bodies of which they are composed. This view became, with Lavoisier, explicit and axiomatic.

(v) *Atomism*

As the eighteenth century turned into the nineteenth, the question of the innate constitution of matter was again raised. In the seventeenth century 'Epicureanism' based on atomic views had become a philosophic vogue. It was opposed to the current Cartesianism which it survived. Among early exponents of the atomic view were Gassendi (p. 273), whose main work appeared in 1649, and Boyle (pp. 271–3), who treated the subject at intervals between 1661 and his death in 1691. Huygens also supported the atomic view. Newton, in his calculations of the motions of the planets, found it necessary to assume interstellar space to be a vacuum. He extended this conception to terrestrial matter involving the view of atoms surrounded by vacuum (*Principia*, 1687). It is, however, difficult to find any definite formulation regarding the exact nature of his 'corpuscles' or 'particles' in his works. But from his time onward, despite the opposition of Leibniz (p. 309), the constitution of matter was generally considered as atomic by physical investigators. The view was popularized and widely disseminated by Voltaire (p. 294).

The older investigators had great difficulty in obtaining their substances in a pure state. Indeed, chemical purity is an idea of very gradual growth, and is perhaps hardly consistent with the older doctrine of the 'four elements'. The work of Black, Cavendish, and Lavoisier, however, drew general attention to the high degree of exactness possible in chemical operations. This conception was pressed by Lavoisier's fellow countryman JOSEPH LOUIS PROUST (1755–1826), who emphasized the constant composition of chemical compounds. With improved methods for preparation of pure substances he showed that a definite compound, however formed, whether in

Nature or by the hand of man in the laboratory, always contains the same 'simple bodies' (i.e. elements) combined in the same proportions by weight. This fact is expressed as the so-called 'Law of Definite Proportions'. Working on this law were several chemists. Notable among them was E. G. FISCHER (1754–1831), who prepared a table of equivalents (1802) from the figures of J. B. RICHTER (1762–1807), to correspond to the law of equivalent proportions.

Proust's conclusions were disputed by CLAUDE LOUIS BERTHOLLET (1748–1822), who in his *Essay on Chemical Statics* (1803) had set forth his views on chemical affinity and had criticized the development of Geoffroy's affinity table (p. 335) by TOBERN OLAF BERGMAN (1735–84). Bergman, recognizing (1773) that affinity tables should be double, one table showing the affinities for reactions in solution (the 'wet way') and the other showing the affinities when the substances were heated together (the 'dry way'), had drawn up large duplicate tables in his *Elective Attractions* (1775–83).

Bergman, moreover, recognized that in some reactions the chemical change could be carried to completion only if the amount of the reacting substance added exceeded that demanded by the amount of the substance acted upon. In more familiar phraseology he showed that it was necessary to add more than the amount 'chemically equivalent'. Berthollet clearly demonstrated that the relative amounts of the substances concerned in a chemical reaction, together with such factors as insolubility and volatility, affected the completeness of the reaction; that increasing proportions of one reagent caused the reaction to proceed still farther in one direction; and that chemical reactions in general were incomplete, the substance upon which two other substances acted with opposing forces being divided between them in proportion both to their affinities for that substance and to the quantities of those substances present.

From these theoretically sound principles, unfortunately neglected for many years, but later to become the basis of modern chemical dynamics, Berthollet erroneously concluded, against Proust, that chemical compounds were produced in analogous ways, and that their constituents were therefore combined; not in fixed and constant proportions, but in proportions that varied with the conditions under which the compounds were formed. Proust's conclusions were, however, accepted by chemists, and his law presently received a new and wider interpretation as a result of the atomic speculations of Dalton.

JOHN DALTON (1766–1844), a Quaker teacher of Manchester, had greater philosophic insight than Proust. Dalton's first important contribution to science was his rule that all gases expand equally with equal increments of temperature (1801). This law was about that time more explicitly formulated by the French chemist JOSEPH LOUIS GAY-LUSSAC (1778–1850), and his name is not unjustly associated with it.[1] His own 'law of partial pressure' (1801) Dalton decided might be explained on the atomic basis, 'a conclusion', he assures us, 'which seems universally adopted'.

Dalton's line of thought on the constitution of matter had come to him first through his interest in meteorology. His analyses of air showed that it was always composed of the same proportion of oxygen and nitrogen, with small quantities of water vapour and carbon dioxide. He knew that these gases are not in combination and have different densities. Why then does the heaviest not sink to the bottom and the lightest rise to the top? These facts might be explained if they were all composed of minute particles of different sizes in movement as suggested by philosophers of antiquity such as Lucretius (p. 104). Adding to the ancient atomic conception the new view that matter was composed of a large number of elementary, homogeneous, and distinct substances, themselves composed of indivisible, indestructible, uncreatable atoms, it must also be assumed that all the atoms of any particular element are like each other but different from the atoms of other elements.

This view fitted well to Proust's recently formulated 'Law of Definite Proportions' (p. 341). In applying his theory to the facts of chemistry, Dalton started with the assumption that chemical combination takes place in the simplest possible way, one atom of one element combining with one atom of another, water being composed of H and O in a 1:1 ratio, and ammonia of N and H also in a 1:1 ratio. He assumed also that when two elements form more than one compound, higher ratios are possible, as for instance with the oxides of carbon (CO and CO_2).

[1] Gay-Lussac himself indicated that J. A. C. Charles (1746–1823) had preceded him in this discovery but had published no results.

Dalton had been working on his theory since the beginning of the century and gave it formal enunciation in 1808. The first number of his *New System of Chemical Philosophy* (1808–27) which appeared in that year has gained general acceptance as a classic. In it he pointed out that, though atoms must be far too small to measure or weigh directly, yet nevertheless it should be possible to determine the relative weights of atoms of different elements. For this we need only know the relative number of atoms combining to form a compound, and the relative weights in which the constituent elements combined to form that compound.

Dalton had very little real experimental guidance as to the number of atoms that form compounds. Thus he wrongly assumed that, in water, hydrogen and oxygen are combined in the ratio of 1 atom to 1 atom, instead of in the ratio 2 to 1. He then introduced experimental error in estimating the relative weight of the hydrogen and oxygen in water as 1 to 7 (instead of 1 to 8). Thus he ascribed to oxygen the relative atomic weight of 7 instead of 16.

(vi) *Molecular Theory*

The publication of the atomic theory attracted attention in France. GAY-LUSSAC (1778–1850, pp. 322, 343) was working on similar lines. He was interested in the combination of gases and showed that, when gases combine, their relative volumes bear a very simple numerical relation to each other and to the volume of their product, if gaseous (1808). Thus one volume of oxygen combines with two volumes of hydrogen to form two volumes of water vapour; one volume of nitrogen combines with three volumes of hydrogen to form two volumes of ammonia gas, and so on.

The atomic theory and the findings of Gay-Lussac were clearly linked together in the exposition of the Italian AMEDEO AVOGADRO (1776–1856). Avogadro pointed out (1811) that if there is a simple numerical relation between combining volumes of gases and if they combine into uniform atomic groups, then there must be some simple connexion between the actual numbers of these atomic groups in equal volumes of combining gases. The simplest relation—and that which has been shown to be the real one—is that equal volumes of all gases contain in similar conditions the same number of atomic groups. Avogadro assumed that the atomic groups, as conceived by Dalton, were not indivisible but in the simplest case consisted of two parts, separable during chemical reaction. The divisible groups he named *molecules* (Latin = 'little masses'). Avogadro also assumed that these molecules, and not the individual atoms, were equally distributed throughout space in the case of all gases (1811). Both assumptions, he observed, fitted Gay-Lussac's law.

Avogadro's hypothesis, that 'equal volumes of all gases under the same conditions of temperature and pressure contain the same number of molecules', was, to the confusion of their subject, unfortunately not received by chemists, owing, firstly, to the small number of cases to which it could then be applied, and, secondly, to the fact that several of those cases gave anomalous results not understood until much later. It was not until 1858, after Avogadro's death, that authoritative attention was called to it by another Italian chemist, STANISLAO CANNIZZARO (1826–1910). This long eclipse of an important law rendered the results of physical chemistry far less profitable than they might have been for nearly half a century.

During this period there was enunciated a hypothesis that has had a somewhat similar history. In an anonymous paper published in 1815, the English physician WILLIAM PROUT (1785–1850) called attention to the closeness with which the atomic weights of the elements, expressed in terms of relation to hydrogen, approximated to whole numbers. Hydrogen, therefore, he regarded as the universal substance. In more modern times there was a general movement towards Prout's hypothesis of a *materia prima*, and his conception of atomic weights approximating to whole numbers has assumed a new significance in the twentieth century.

Much of the chemical activity of the first half of the nineteenth century naturally went to the exact determination of atomic and molecular weights. Notably the Swede JÖNS JAKOB BERZELIUS (1779–1848) devoted himself to this task from 1811 onwards, ascertaining the molecular weights of thousands of substances. He also did important work as the founder of electrochemical theory. He developed the conception that a group of atoms or *radicle* can form an unchanging constituent through a series of compounds, behaving as though it were an element. He rendered a great service in establishing chemical nomenclature and developed the convenient mode of formulating elements by the capital initial letters of their Latin names, adding to them the numbers of the various atoms present in a compound.

Many of the most fruitful lines of Lavoisier's work were continued by Sir HUMPHRY DAVY (1778–1829). Notably he succeeded by means of the electric current (pp. 357 f.) in resolving the alkalis, potash and soda, and the alkaline earths, baryta, strontia, lime, and magnesia, into their elements. Those elements were oxygen on the one hand, and a series of metals which he called potassium, sodium, barium, strontium, calcium, and magnesium, deriving these names from the old terms for the substances in which the respective elements were contained (1807–8). He also showed that the gas chlorine, prepared by the Swedish chemist Scheele in 1774 and thought to contain oxygen, was of elementary character (1810, p. 338).

Davy was especially fortunate in the practical application of much of his work. His 'safety-lamp' still bears his name, and deservedly so, for his detailed and important researches on flame and explosions made it practicable, though the principle on which it is based was discovered by George Stephenson, the engineer. He performed a great service to agriculture by codifying, for the first time, the mass of chemical knowledge applicable to it. His *Elements of Agricultural Chemistry* (1813) contains the first use in the English language of the word *Element* defined in the modern chemical sense:

> All the varieties of material substances may be resolved into a comparatively small number of bodies, which, as they are not capable of being decomposed, are considered in the present state of chemical knowledge as *elements.*

At that date Davy recognized forty-seven of these elements.

An impressive figure in the scientific world of the thirties and forties was JUSTUS VON LIEBIG (1803–73), professor of chemistry first in Giessen and then in Munich. He applied to organic substances the exact methods that had been developed in the previous decades. Over his laboratory was inscribed 'God has ordained all things by measure, number and weight'.[1] His great achievement was his application of exact chemical knowledge to the processes and products of vital activity. (For Liebig's physiological work see pp. 407 ff.)

[1] This is a quotation from the apocryphal *Book of Wisdom*, xi. 21.

5. TRANSFORMATIONS OF FORCES

(i) *The Imponderables*

The seventeenth century—the age of Galileo—and the eighteenth —the age of Newton—established a view of a universe maintained by a balance of *forces* acting on *bodies*. There was still much vagueness as to the limits of the two. Thus, 'phlogiston', which was supposed to go forth from a body on combustion, and 'ether' which was at once agent and medium of light, no less than the electric and magnetic 'fluids', remained ambiguous conceptions to the very end of the eighteenth century and even into the nineteenth. Some of this group of imagined entities, heat, ether, the electric and magnetic fluids, were regarded as weightless substances: 'imponderables'. The confusion of language created by the 'imponderables' long persisted. 'It is the imponderables—heat, electricity, love—that rule the world', wrote Oliver Wendell Holmes (1809–94)—himself a man of science—as late as 1858 (*The Autocrat of the Breakfast Table*).

Among the imponderables a place of special importance was occupied by the supposed substance of heat: 'caloric'. During the earlier eighteenth century two views of the nature of heat were current. On the one hand, it was generally conceived as a fluid held in greater or less quantity within the pores of all bodies. Thus when a metal grows hot on being hammered, the heat becomes more perceptible because the caloric, it was thought, was squeezed out by the pressure. The material and fluid nature of heat was a generally accepted idea which was not greatly disturbed by the victorious advance of the Newtonian philosophy.

On the other hand, the suggestion was made by Galileo (1624), Boyle (1664), Hooke (1665), and Huygens (1690) that all basic physical phenomena—heat, light, chemical action, electricity, magnetism —were susceptible of mechanical explanation. It was believed that all were due to the movements on the part of small particles of the affected bodies, varying in form, velocity, order of arrangement, attractive power, &c.

Certain relations between forces of different kinds were, of course, evident to every observer. This was the case, for example, with the general interconnexions of light with heat and again with electricity, and especially of heat with work. The production of fire by friction was a device of the highest antiquity; frictional electricity was well known; the steam-pump was becoming familiar; the production of heat, light, and sound in a variety of chemical and physical operations was also naturally very familiar. Nevertheless no exact relation between these various phenomena was yet recognized.

(ii) *Temperature Measurement*

Methods of estimating temperature were greatly improved even before the elucidation of what may seem now the obvious distinction between heat and temperature. An air thermometer or rather thermoscope had been invented by Galileo about 1592 (Fig. 109), and an open-ended water thermoscope was described by Jean Rey in 1632. A distinct advance, making the passage from the thermoscope to thermometer, was the sealed alcohol indicator, invented about 1641, probably by Ferdinand II, Grand Duke of Tuscany. It was used for the experiments of the Italian Accademia del Cimento during its brief life (1657–67, Fig. 141). All these instruments were provided with arbitrary scales.

At the very beginning of the eighteenth century (1701) Newton suggested an oil thermometer with a rational thermometric scale, in which the temperature of freezing water was taken as 0° and that of the human body in health as 12°. By assuming that the rate of cooling of a hot body is proportional to the 'whole heat' [= temperature] of that body, he was able to estimate higher temperatures, such as 'red-heat', by observing the times taken by hot bodies to cool down to temperatures measurable on his thermometer. The proportionality here assumed has since become known as 'Newton's Law of Cooling'. This, more exactly, is that, for small ranges of temperature, the rate of cooling of a hot body is proportional to the difference in temperature between that body and the medium by which it is surrounded.

The mercury thermometer was introduced and thermometric standards fixed about 1715 by D. G. FAHRENHEIT (1686–1736) and described in a communication to the Royal Society in 1724. A maximum and minimum thermometer was constructed in 1757 by CHARLES CAVENDISH (1703–83), whose son Henry (pp. 337 f.) explored the thermometric conduct of mercury in 1783. This instrument was im-

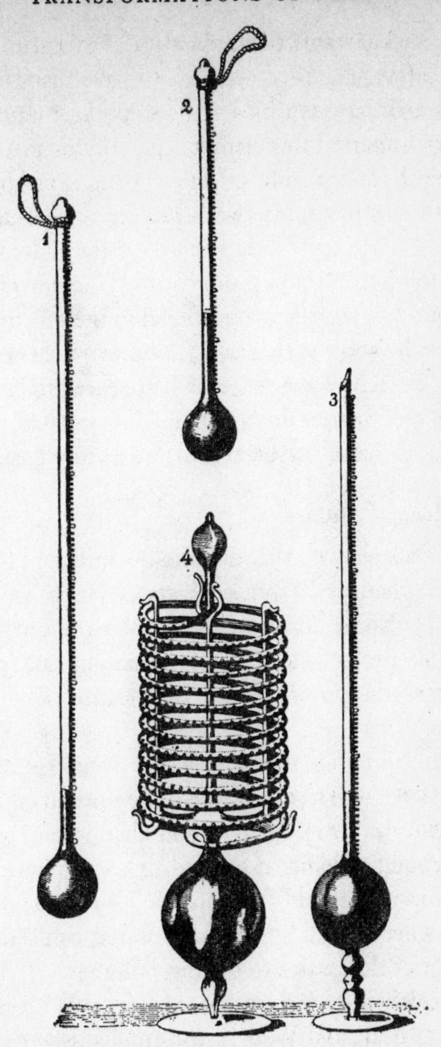

FIG. 141. Sealed alcohol thermometer of the Accademia del Cimento at Florence *c.* 1664. They were graduated by small spheres of coloured glass fused to the stem.

proved in the later years of the eighteenth century and assumed its modern form in the hands of DANIEL RUTHERFORD (1749–1819) in 1794.

The invention of a satisfactory instrument for the measurement of temperature, with fixed points giving concordant readings in all circumstances to investigators in different places, had, as its most immediate important result, the foundation of the quantitative science of heat by JOSEPH BLACK (1728–99). About 1760 Black introduced the method of measuring quantities of heat by the number of degrees of temperature imparted to a definite quantity of matter, a method destined to have far-reaching effects. At the same time Black set forth clearly the distinction between *heat* and *temperature*, or *quantity of heat* and *intensity of heat*. Rejecting the older view that the quantities of heat necessary to produce equal increments of temperature in different bodies were proportional to the quantities of matter in these bodies, he showed that every kind of substance had its own characteristic 'capacity for heat', which appeared to bear no relation to the quantity of matter in the body investigated. Black's term 'capacity for heat' has since been replaced by the term *specific heat*.

(iii) *Heat a Mode of Motion*

In 1761–4 Black showed that definite quantities of heat disappear during certain changes of physical states, such as melting and evaporation. He also demonstrated that the same quantities of heat reappear during the reverse changes, freezing and condensation. Black called this disappearing and reappearing factor the 'latent heat'.

Black's discovery of latent heat was shortly afterwards applied by the engineer, JAMES WATT (1736–1819), then occupied in improving the steam-engine (p. 351). Watt found that water, on conversion into steam at boiling-point, expanded at atmospheric pressure to about 1,800 times its liquid volume. He also found that steam at boiling-point, when passed into ice-cold water, could raise about six times its weight of that water to boiling-point (1764). This puzzling result Black explained to him in accordance with his discoveries of 1761–4 on latent heat; and Watt (1765) applied Black's discovery in his contrivance of the separate condenser, the greatest of all his many improvements of the steam-engine (Fig. 142). This simple principle is still in use and has made possible many subsequent developments.

The conception of the nature of heat, from being a subject of speculation, was now on what seemed an exact basis, susceptible of practical application. Heat was held to be an elastic, uncreatable, indestructible, measurable fluid. To emphasize this new outlook, Lavoisier and the French Academicians introduced for it the name 'calorique' (1787).

The theory of caloric was, however, soon being undermined by the adventurous American, BENJAMIN THOMPSON, Count Rumford (1753–1814). Employing a balance, sensitive to one part in 1,000,000,

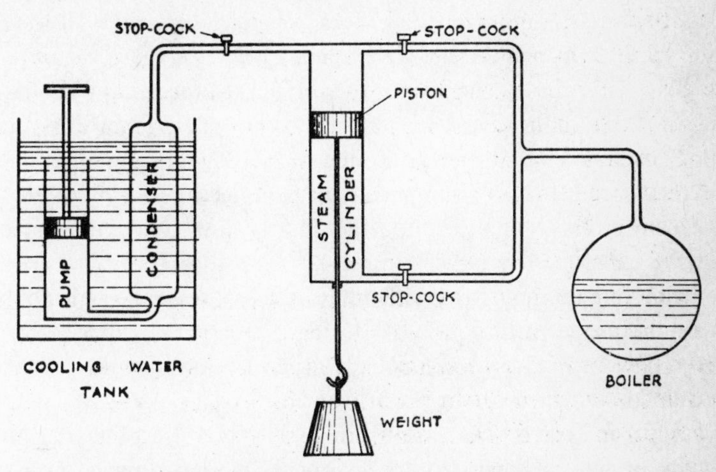

FIG. 142. Diagram of Watt's model illustrating condensing principle for steam-engine, 1765. In older engines the cylinder itself had been cooled at each stroke, after entry of steam. Watt attached a condenser and 'air-pump' to empty the cylinder, which could then be kept permanently at steam heat while the vacuum produced by condensation did its share of the work and thus added to the efficiency of the engine.

he showed (1799) that there was no measurable alteration of weight in a mass of water on conversion into ice or on the reconversion of the ice into water, despite a heat-change of an order that would raise $9\frac{3}{4}$ oz. of gold from freezing-point to a red heat. Heat, therefore, if a fluid, 'must be something so infinitely rare, even in its most condensed state, as to baffle all our attempts to discover its gravity'. Therefore to Rumford it did not appear likely that heat was a substance distinct from, and accumulated in, the heated body. If, however, heat were 'nothing more than an intestine vibratory motion of the constituent parts of heated bodies', then no change of the weights was to be expected on heating, since only the internal motions, not their mass, would be affected.

In 1798 Rumford had published his *Inquiry concerning the Source of the Heat which is excited by Friction.* In boring cannon he estimated the heat produced by measuring the rise in temperature of a mass of water contained in a box suitably arranged around the boring-point. The heat generated by the friction of the borer and the cannon appeared to be inexhaustible, and he reasoned 'that any thing which any *insulated* body, or system of bodies, can continue to furnish *without limitation*, cannot possibly be a *material substance*'. Heat was therefore, he concluded, 'a kind of motion'.

Soon after these experiments there appeared the first publication (1799) of Humphry Davy (1778–1829, p. 346), describing work that he had carried out at the age of nineteen. It contains the often misquoted account of an attempt to melt two pieces of ice by the heat developed on rubbing them together in a vacuum. The arrangement of the experiment was very imperfect and, since Davy's recorded results were thermodynamically impossible, there can be little doubt that the proper technique was lacking. Perhaps the experiment is even now beyond the powers of any experimenter. The results were assumed, and throughout his brilliant career Davy held fast to his youthful and correct conclusion—unjustified or at least unconfirmed by his premisses—that heat was a vibratory motion of the corpuscles of bodies. It is a remarkable case of that feeling or instinct for the correct solution that is the special gift of some talented investigators.

Count Rumford had come very near to a more demonstrable treatment of the transformation and conservation of energy, for he was not far from revealing the nature of the relation between heat and mechanical effort. He observed in his experiments on the boring of cannon that two horses, working steadily against frictional resistance, produced heat at a steady rate. He even compared the heat thus produced with the heat that would result from the combustion of the food consumed by the horses. Yet since he had no exact and transferable conception of *work* as a measure of mechanical action, he could not develop a complete doctrine of the transformation of one form of energy into another.

The development of the steam-engine by Watt, and its use in the pumping of Cornish mines was, about this time, much in men's minds. When the firm of Boulton and Watt first began to manufacture their engines, the terms of sale devised by Watt involved the annual payment by the buyer, over a period of years, of one-third of the value of the savings in fuel effected by the new engine where it replaced an older type. But since the new engines were often for use in new mines, or were to do more work than those they replaced, or were required to pump from greater depths, a method of comparing engines was needed. Thus the determination of the *duty* of an engine was introduced (1778) as a quantitative relation between output of work and consumption of fuel. The 'duty' was the number of pounds of water raised by the engine through a vertical height of one foot per bushel of coal consumed. From this could be calculated the *power* of an engine, i.e. its *rate of doing work*. A standard of power was introduced by Watt in 1782–3 from calculations of the rate of working of a mill horse, and the term *horse-power* was applied to define a rate of doing work equivalent to the raising of 33,000 pounds one foot per minute. It was not, however, until the middle of the nineteenth century that the general convertibility of heat into work was finally recognized (p. 375 f.).

(iv) *Static Electricity*

In the field of electricity, until the end of the eighteenth century, only the static form was recognized. The process of electrical conduction was demonstrated in 1731 and it was shown that, while some bodies would conduct electricity, others would not. Thus 'insulation' became possible. It was shown also that all bodies are capable of electrification.

Early attention was drawn to electrical attraction and repulsion. To explain them a theory of *two fluids* was introduced (1730) by the French experimenter C. F. DU FAY (1698–1739). These fluids were supposed to be separated by friction and to neutralize each other when in combination.

The striking way in which an electric charge may be fixed by two conductors separated by a non-conductor, as in the familiar 'Leyden jar', was discovered at that town in 1746 by two Dutch experimenters. About this time BENJAMIN FRANKLIN (1706–90) began to take an interest in electricity and soon observed that electric charges could be drawn off with peculiar facility by metal points. He supposed that 'electric fire is a common element' existing in all bodies. If a body had more than its normal share it was called *plus*, if less *minus* (1747). This was the 'one fluid theory' which held the field until the time of Faraday (pp. 361 f.). Franklin explained lightning as of electrical origin, suggested lightning conductors (1749), and put the idea to a practical test (1752). Through the survey by Priestley of the general state of electrical knowledge in his *History and Present State of Electricity* (1767) such phenomena became generally recognized. A number of types of frictional electrical machines were introduced and the subject attracted much attention. Electrical investigation had hitherto been almost entirely qualitative. In 1767, from the observation that there was no charge on the inner surface of a hollow electrified metal body, Priestley had suggested that the law of electrical attraction was the same as that of gravitational attraction, namely that of the inverse square of the distance (p. 293). Cavendish gave an experimental proof of this in 1771. Unfortunately, however, he did not publish his experimental verification, and it remained unknown till 1879.

The first method of measurement applicable to electricity was the action of an electrified object on light suspended bodies such as threads, metal foil, or pith-balls. An early attempt at quantitative expression was made in 1786 with a gold-leaf electroscope by measuring the angular divergence of the leaves when charged. But the first effective verification of the law of attraction was made by the French engineer CHARLES AUGUSTUS COULOMB (1736–1806), who adapted to electricity Hooke's principle 'ut tensio sic vis'. Using hairs and wires he constructed a 'torsion balance' (1785). The principle was to measure the amount of torsion required to bring a charged pith-ball within various distances of another pith-ball, equally charged with electricity of the same sign and therefore repelling it (Fig. 143). This method was peculiarly adapted for the investigation of the distribution of electricity on surfaces and of the laws of electrical and magnetic action. Coulomb was the founder of the mathematical theory of these subjects, and by the use of his 'balance' was able to prove that Newton's law of inverse squares (p. 293) holds good for electric and magnetic attraction and repulsion.

In the later eighteenth century there was considerable interest in the shock-producing fishes, the skate-like *Torpedo*, and the electric eel or *Gymnotus*. Accounts of them were given by John Hunter (1773–5), Ingenhousz (1773), and Cavendish (1776), and it was realized that their shocks were of an electrical nature. The attention thus drawn to electricity in the animal body led LUIGI GALVANI (1737–98) of Bologna to investigate the susceptibility of nerves to its action.

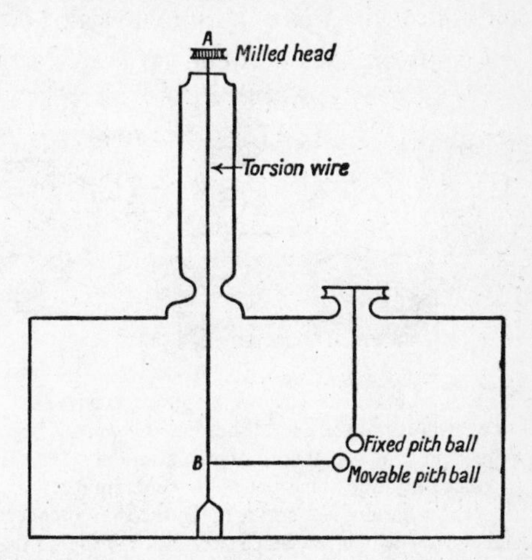

FIG. 143. Coulomb's Torsion Balance. Within a closed chamber two charged balls are insulated. One is fixed to the framework, the other attached at *B* to a wire that can be turned by a milled head *A*. The degree of torsion needed to bring them together is a measure of the force of their mutual repulsion.

He showed that muscular contraction could be produced by electrical action and conversely that electric phenomena could be produced by the muscular contraction (1791, Fig. 144).

(v) *First Study of Current Electricity*

Many thought that this 'animal electricity' was of its own peculiar kind and it was dubbed 'galvanism'. ALESSANDRO VOLTA (1745–1827) of Pavia, working on the results of Galvani, found that electric discharge through a nerve or sense organ not only produced muscular contraction but also sensation. If one end of a bent rod with limbs of different metals were held in the mouth a sensation of light was produced when the other end touched the eye. A silver and a gold coin held against the tongue gave a saltish taste when the coins touched at their edges or were connected by a wire. The essential thing was contact of different metals. Volta showed that a muscle can be thrown into continuous contraction by repeated electrical stimulation, but he was also able to demonstrate (1800) that the animal relationship of 'galvanism' is in no way essential, as had previously been thought. Volta's device of the 'voltaic pile', in which the electric discharge of coins of the original experiment was replaced by a whole series of pairs of coins or disks between cards soaked in brine, soon

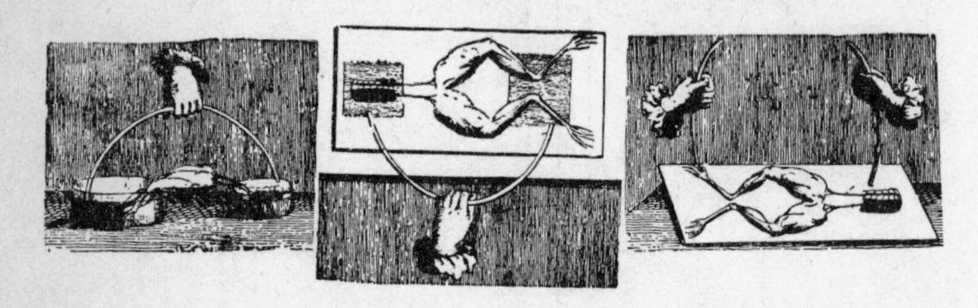

FIG. 144. Galvani's experiments on effects of metallic contacts on nerves and muscles of frogs' legs (1791). To left a metal rod establishes electric contact between water in two dishes. In one lies the end of the spinal cord of a frog, in the other the attached legs. In the middle there is contact by a metal bar between two damp mats, on one of which lies the spinal cord and on the other the legs and feet. To the right there is a similar preparation with a broken contact which can be completed by bringing the rods together.

developed into his famous 'couronne de tasses' (1800, Fig. 145) and was the foundation of electrochemistry. The invention drew immediate and widespread attention. It was the first instrument for producing an electric current.

In England water was decomposed by current in the very year of Volta's publication. It was a generally held view that the chemical changes in the pile were the source of the electric current. Thus chemical affinity began to be correlated with electricity, which Franklin and others after him had come to regard as related to 'fire' or heat. The 'crown of cups', each cup containing two plates of different metals steeped in salt water or a dilute acid, is the direct ancestor of the various forms of electric 'cell'.

The voltaic pile or the crown of cups provided an entirely new means for the decomposition of certain substances. In the decomposition of water by the current so produced, very great interest was aroused by the sight of oxygen and hydrogen bubbling off from the separate plates. Humphry Davy (p. 346) was among the first to develop this fruitful mode of analysis, from which he had very great hopes, believing that it must 'carry with it perfectly new views of corpuscular action'. He himself showed by its means that in the decomposition of water the volume of hydrogen is double that of

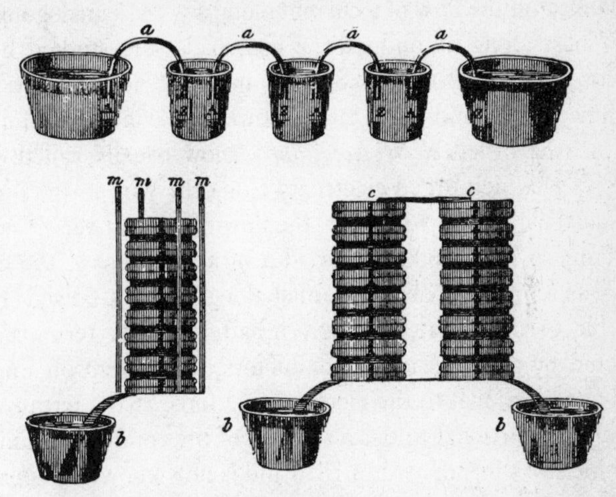

FIG. 145. Volta's Pile and, above, his Crown of Cups (1800). The pile is a series of paired disks of silver and zinc, sandwiched between paper strips soaked in salt water. They are supported by glass rods *m m*. From the lowest disk a metal strip goes to vessel *b*. A current will pass from the top disk to the vessel if the two are linked by a conductor. Two piles may be linked together by a metal strip, as at *c c*, and the effect doubled.

The 'crown of cups' is a series of vessels of salt water or dilute acid in which are pairs of plates of different metals, connected by metal strips *a a*. The action is as with the pile.

oxygen. Before many years electrical decomposition in his hands had yielded a whole series of new elements, notably sodium and potassium (1807–8).

The nature of the process of electrical decomposition and the cause of migration of its products to the two poles of the electric cell gave rise to much speculation. Davy developed or adapted a theory that the electric pile breaks the particles near it into two factors. Thus in decomposition with a zinc-copper couple the copper repels and the zinc attracts the oxygen. Oxygen being given off, the hydrogen is thereby set free and attracts oxygen from the nearest particle. Thus again hydrogen is released and again attracts the nearest oxygen. A chain of decomposition is formed resulting in the discharge of hydrogen at the zinc pole and oxygen at the copper.

Davy also investigated the flow of electric current along metallic conductors and obtained results which follow naturally from a more fundamental relation discovered by GEORG SIMON OHM (1787–1854). Ohm looked on the flow of a current along a wire as analogous to the flow of heat along a conductor, which had been studied by JEAN BAPTISTE JOSEPH FOURIER (1758–1830) in 1822. He therefore sought something corresponding to temperature, and conceived of a quantity which he called *electroscopic force* (now usually called *electromotive force*) which urged electricity along the wire in the direction of the current. Just as with heat, the flow of current varied with the conductor, so he introduced the idea of a *resistance* to the current which was a property of the channel along which it flowed. He was then able to show that, for a given battery whose terminals were connected by a succession of conductors, the current obtained was directly proportional to the electroscopic force at the terminals and inversely proportional to the resistance of the conductor. This relation, which was formulated in 1826 and is now known as *Ohm's law*, forms one of the foundation stones of modern electrical science. Its importance, however, was long unappreciated, but before Ohm's death recognition came first in England where he received the Copley medal of the Royal Society in 1841.

The process of electrical decomposition was given quantitative expression by Faraday (1833, pp. 361 f.). Its two primary laws, still known by his name, are:

(*a*) The mass of the product liberated by electrical decomposition is proportional to the quantity of electricity passed.

(*b*) When the same current is passed through solutions of different substances, the masses of the liberated products are proportional to the chemical equivalents of those products.

Thus a definite relation was established between electrical and chemical action.

(vi) *Electromagnetism*

Soon after the completion of Davy's electro-chemical researches a new orientation of electrical science set in. The year 1820 was especially eventful. In that year the Dane, HANS CHRISTIAN OERSTED (1777–1851), demonstrated exactly the long-suspected connexion of electricity with magnetism. He found that if a wire carrying an electric current was placed near and parallel to a magnetic needle it

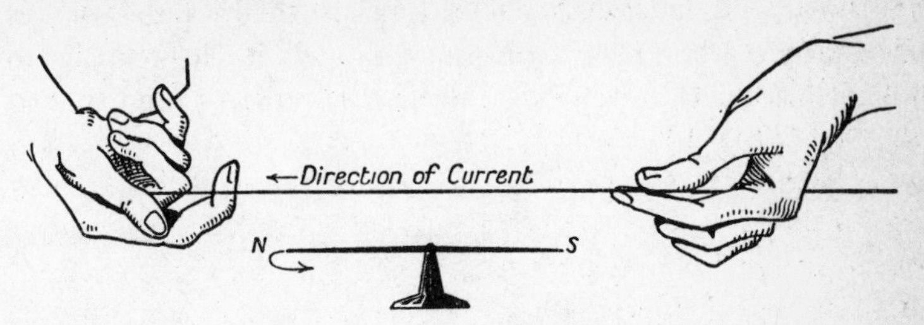

FIG. 146. Oersted's experiment on the effect of an electric current on a magnetic needle.

deflected it (Fig. 146), but not if the wire carrying the current was at right angles to the needle. The direction in which the needle turns depends on whether the wire carrying the current is above or below the needle, and on the direction of the current.

The significance of this linking of electricity with magnetism was at once recognized by the French investigator, FRANÇOIS ARAGO

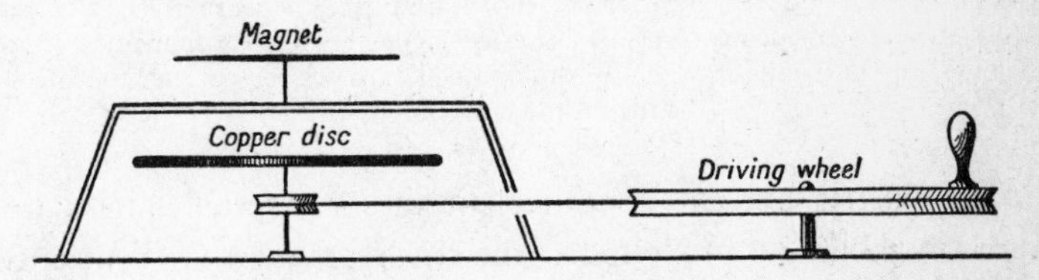

FIG. 147. Arago's experiment of rotating a copper disk below a magnetic needle.

(1786–1853), who showed (1820) that a spiral of copper wire, through which a current was passed, attracted previously unmagnetized iron filings, which clung to the wire as long as the current flowed, but dropped off when the circuit was broken. Such a coil, in fact, acts like a magnet. In 1824 he found that rotation of a copper disk produced rotation of a magnetic needle supported above it (Fig. 147). This phenomenon was rendered intelligible by Faraday in 1831 (pp. 365–6).

ANDRE MARIE AMPÈRE (1775–1836), very soon after Oersted's publication, revealed the laws governing the deflection of the magnetic needle by the electric current and the mutual attractions and repulsions of electric currents. He showed that two parallel wires carrying currents attract each other if the currents flow in the same direction, and repel each other if the currents flow in opposite directions and he showed, as Arago had already done, that a cylindrical coil behaves like a magnet when a current is passed through it. He proceeded to a mathematical analysis of these phenomena (1822–7) and showed

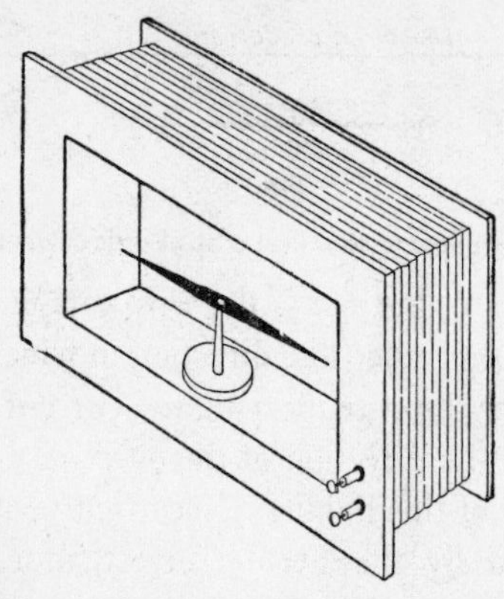

FIG. 148. The simplest form of galvanometer or apparatus for measuring electric current. It consists of a magnetic needle set in a non-conducting rectangular framework around which are wound many turns of wire through which passes the current, the effect of which is to be measured.

that an electric current is equivalent in its external effects to a magnetic shell. He propounded the theory that magnetism is the result of molecular electric currents. His memory is perpetuated in the well-known 'Ampère's Rule', formulated by him for determining the deflection of a magnet by an electric current, and in the *ampere*, the practical unit of electric current, which is named after him.

The work of these investigators, especially of Ampère, provided a means of detecting a current and of measuring it on some arbitrarily chosen scale by means of its magnetic effect. Instruments devised for this purpose, *galvanometers*, appeared in 1821 from the hands of several inventors. In their simplest form, they consist of a coil of many turns of wire carrying the current and deflecting a magnetic needle balanced on a pivot at the centre of the coil (Fig. 148).

(vii) *The Dynamo*

A main achievement of MICHAEL FARADAY (1791–1867), one of the greatest of scientific geniuses, was the demonstration that an electric current can be used as a source of power. From the experiments of Oersted and from his own Faraday realized that a current traversing a wire creates a magnetic 'field of force'. Any part of this may be pre-

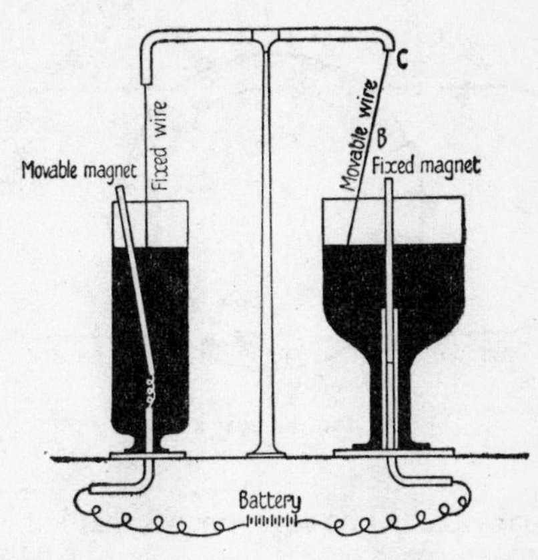

FIG. 149. Faraday's apparatus for demonstrating how an electric current can be disposed so as to produce a continuous rotational movement, the principle of the dynamo.

sented graphically, in any plane at right angles to the wire, by a series of circles concentric to the wire (Fig. 152). Faraday thought that this force might cause a magnet to move round the wire. Moreover he argued that if a magnetic pole can be made to rotate round a current it should be possible to cause a wire carrying current to rotate round a magnetic pole.

A circuit, consisting of two vessels of mercury and connecting wires, was arranged by Faraday so that in one vessel there was a fixed magnet and a wire free to rotate, while in the other the wire was fixed and the magnet movable (Fig. 149). Electric current passed from the wire through the mercury in the left-hand cup to a copper rod running into the base of the vessel. The magnet in this cup was fastened to the copper rod by a thread. In the right-hand vessel the fixed magnet was placed in a socket in the stem of the vessel, and the wire which dipped into the mercury was able to move freely. As soon as the circuit was completed, the magnet in the first vessel and the wire in the second commenced to rotate, and continued to do so while the current was passing. Faraday had thus transformed electrical current into continuous movement (1821).

Faraday knew of the demonstration by Ampère that a cylindrical coil of wire behaves like a magnet when a current is passed through it. The converse that a magnet could produce a current was shown

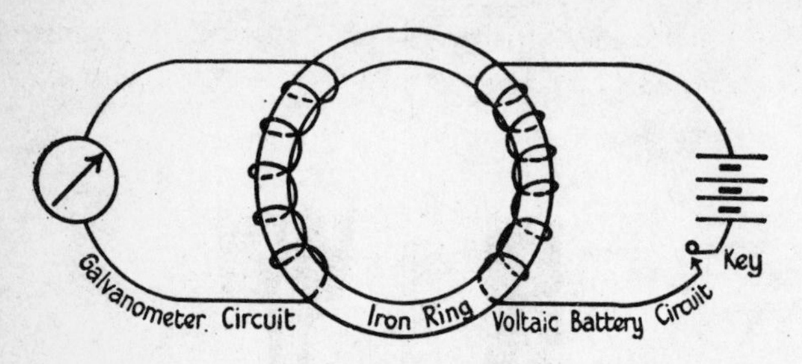

FIG. 150. Faraday's ring.

by him to be equally true. The experiments that led him to this conclusion have become classics.

Around an iron ring he wound two separate coils of wire. One was connected with a voltaic battery, the other with a galvanometer. A key made it possible to break or make circuit. On making or breaking the current in the voltaic circuit, the galvanometer showed that a current also flowed for an instant in its circuit, but the currents on making and breaking were in the opposite direction (1831, Fig. 150).

But if a circuit can act as a magnet, as Arago had shown, cannot a magnet produce this same result with an iron ring? Is not the battery unnecessary? The testing of this point was a critical experiment for the whole future of electrical science. Faraday wound a coil of wire round a bar of iron and completed the circuit so as to include a galvanometer. He then placed the bar between the north pole of one bar magnet and the south pole of another, the other ends of the magnets being in contact. Whenever contact between the magnets was made or broken the galvanometer indicated the momentary passage of a current (Fig. 151).

On this discovery a wit of the time wrote:

> Around the magnet Faraday
> Was sure that Volta's lightnings play:
> But how to draw them from the wire?
> He took a lesson from the heart:
> 'Tis when we meet, 'tis when we part,
> Breaks forth the celestial fire.

Faraday had dispensed with a battery. Could he, by retaining the battery, dispense with a magnet, substituting for it a current? Using

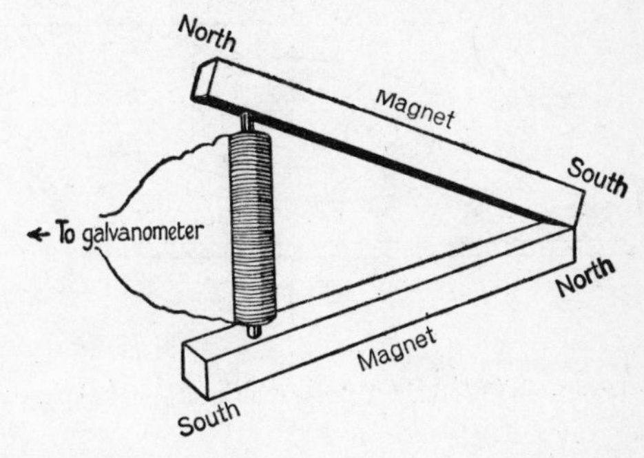

FIG. 151. Production of momentary electric current by magnetic 'make' and 'break'.

a wooden bobbin for the iron ring, Faraday wound a coil of wire round it, and connected it to a voltaic coil. Round this 'primary' coil was wound another and much longer coil, the 'secondary', its ends being joined to a galvanometer. As before, both on make and break momentary currents were indicated by the galvanometer. Faraday had revealed the process of 'inducing' a current, and with the knowledge of induction currents a new era in the application of electricity had opened (1831).

It was now clear that the essential factor in the production of the magneto-electric effects was change, movement of the magnet or of the coil, or making and breaking of the current or the contact. Magneto-electric effects are related somehow to 'fields of force' which fade out as we pass farther from the site of the change. These fields of force can be arranged or mapped in lines as indicated by the behaviour of iron filings placed on cards within their area.

(viii) *The Magneto-Electric Field*

In seeking a general explanation of these phenomena Faraday was thinking much about the lines of magnetic force which came to play a very important part in electrical sciences. They were by no means a new conception. Gilbert (p. 221) had a clear idea of them, Descartes had seen in them evidence for his hypothetical vortices (p. 262), and

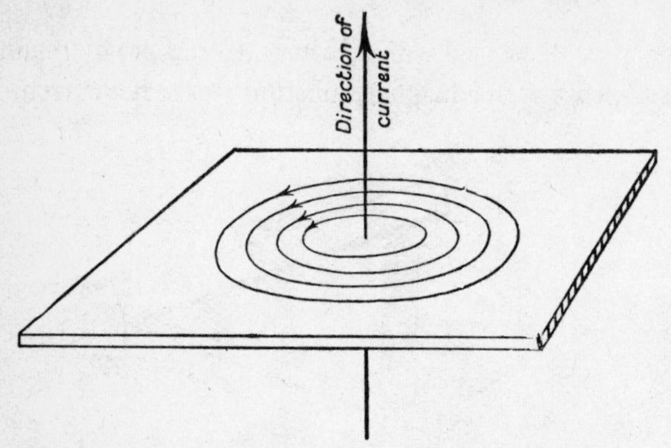

FIG. 152. Lines of force due to current in a straight conductor.

certain eighteenth-century physicists had mapped them, but it was reserved for Faraday to indicate their significance. Throughout the rest of his career he continued to speculate and experiment on these lines of force which are now a familiar scientific conception.

The general character of the lines of force due to a current can be easily demonstrated either by manipulating a small compass needle in the neighbourhood of a current or by running an electric wire carrying a current through a card on which iron filings are spread. These filings take the position of curves in the neighbourhood of the wire and the lines of force can similarly be represented as concentric circles at right angles to the current (Fig. 152).

Faraday had already succeeded in making a magnet rotate round a wire carrying a current, and a wire carrying a current rotate round a magnet. Such movements are related to the distribution of the lines of force due to current or magnet setting up certain stresses in the medium. This wire or magnet is continually urged away from the strong part of the field. Ampère had shown that parallel wires carrying currents attract one another if the currents are in the same direction and repel one another if the currents are in opposite directions. This fact Faraday was able easily to fit into his conception of 'lines of

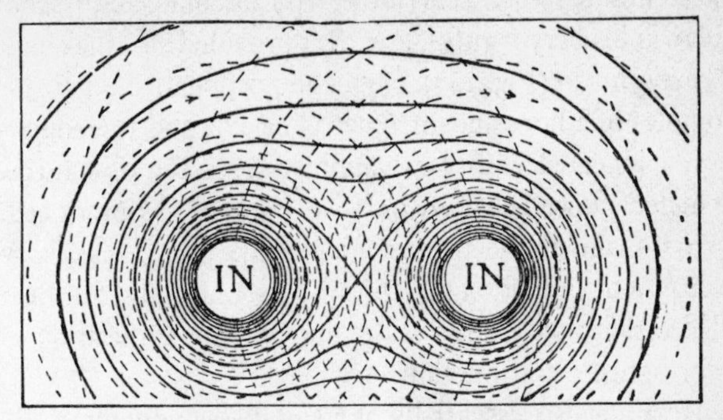

FIG. 153. Field due to currents in the same direction.

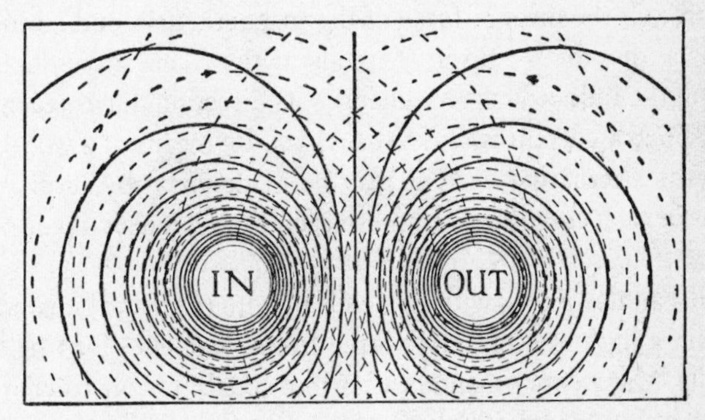

FIG. 154. Field due to currents in the opposite direction.

force'. If currents run in the same direction in two neighbouring wires, the resultant field of lines of force will be such that they will be driven from the strong parts of the field to the weaker and so drawn together (Fig. 153). If the currents run in opposite directions, they will again be driven to the weaker parts of the field and so driven apart (Fig. 154). The idea of 'field of force' was very fruitful.

Arago's demonstration of the effect of a rotating copper disk on a magnet suspended over it (p. 359) was now explicable in terms of lines of force. As the disk moves, it cuts through the lines of force of the magnet. Induced currents are therefore set up. The movement of the magnet is simply the result of the mutual action of the magnet and of the magnetic fields due to the induced current. By visualizing the lines of force as endowed with certain physical properties, it is possible to link together many otherwise disconnected phenomena.

Faraday in his very fruitful year 1831 provided also the converse to Arago's experiment (Fig. 147). He made a copper disk rotate between the two poles of a horseshoe magnet. The axis and the edges of the disk were connected with a galvanometer. As the disk turned, the galvanometer showed that an induced current was produced. This was the first magneto-electric machine or dynamo. This discovery of electro-magnetic induction was thus the starting-point for the utilization of electricity on a large scale, and for the application of such power for lighting and traction.

A dynamo consists essentially of a suitable conductor, built up of many coils, which rotates in a magnetic field. The rotating conductor cuts through the lines of force of the magnetic field and an induced current is thereby set up in the coils of the rotating conductor. In each coil the induced current changes its direction during each revolution. Such a current is said to *alternate*. By means of a well-known device the alternating current may be made *direct* by reversing the current in each coil of the armature each time it passes a pair of conductors.

With Faraday's ring with two coils of wire (Fig. 150) it is possible to obtain a high electromotive force from a current given by a very few cells. Many experimenters after him sought to construct apparatus which should give a high electromotive force by inductive action of one circuit on another. It was H. D. RUHMKORFF (1803–77), a Parisian instrument maker, who in 1851 produced the type of coil still known by his name and so rendered practical the development of the electric motor.

About this time, when Faraday's researches were thus assuming practical significance, scientific men began to appreciate the exactness and preciseness behind much of his simple language. It is astonishing how many general theorems, the methodical deduction of which require the highest mathematical powers, Faraday attained by some sort of intuition without the help of mathematical formulae. Thus the first important scientific contribution of JAMES CLERK

MAXWELL (1831–79) was *On Faraday's Lines of Force* (1856). In it he sought 'to show how by a strict application of the ideas and methods of Faraday, the connexion of the very different orders of phenomena which he discovered, may be placed before the mathematical mind'. He followed the suggestion of William Thomson, later LORD KELVIN (1824–1907), who had been working at the subject since 1849. The analogies that Clerk Maxwell worked out were those of heat and of hydrodynamics. These gave rise to his conception of electric and magnetic effects as due to changes in the ether (1862) and to his great contribution *On a Dynamical Theory of the Electro-magnetic Field* (1864). In the latter he showed that electromagnetic action travels through space at a definite rate, in waves, and that these waves are, like those of light (pp. 371 f.), transverse to the direction in which the waves are propagated. Since he was able also to prove that the velocity of these waves is the same as that of light (1867), an electromagnetic theory of light thereby became possible.

(ix) *Undulatory Theory*

At the end of the eighteenth century there were in the field two rival conceptions of the nature of light, the emission theory and the undulatory or wave theory.

The emission theory is of great antiquity but was given modern scientific form by Newton. He treated a luminous body as emitting streams of minute corpuscles moving progressively in a straight line corresponding to the direction of the ray. Vision was supposed to be produced by the impact of these streams on the eye. The bending of the ray as it passes from air into a denser medium—as, for example, into glass or into water—is explained by assuming that as each corpuscle approaches the denser surface of the medium at any given angle it begins to be attracted towards it.

The undulatory theory of Christian Huygens (p. 228), advanced in 1678 and especially in his famous *Treatise on Light* (1690), treated all space as pervaded by a subtle and elastic medium, the *ether*, through which waves are propagated in all directions from a light-source. These undulations spread in a regular spherical form from the point of origin, just as waves produced by a stone dropped into water spread in circles (Fig. 155).

Huygens applied this theory to explain the phenomena of refraction. A source of light may be regarded as emitting a series of spherical waves in the ether. Any point A on the surface of such a wave (Fig. 155) may in its turn be regarded as a source of light. Every other point on the surface of the same wave, as for example B, C, or D, emits similarly its own spherical wave. At any distance from the

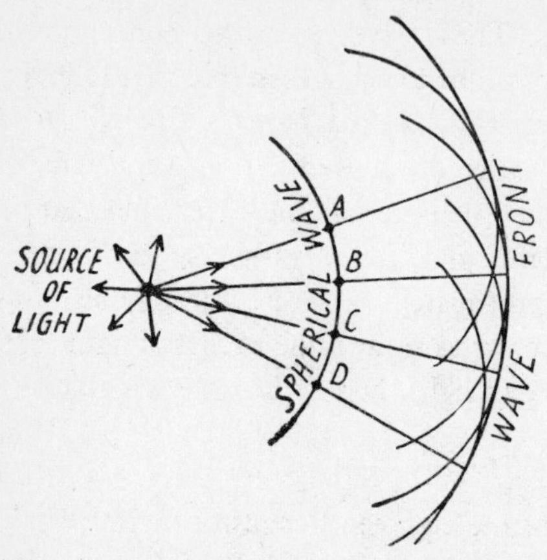

FIG. 155. Huygens's conception of 'wave-fronts'.

original source the surface of all these waves can be regarded as combining together to form what is called a 'wave-front'. If the source of light be sufficiently distant the wave-front is on so large a sphere that a small part of it may be treated as flat (or, in section, linear), while the lines radiating to it from the source of light may be treated as parallel.

We have now to consider, as did Huygens, the application of this wave theory to the known facts of refraction and notably to Snell's law (p. 229). Those facts require (as we shall presently see) that the velocity of propagation of light should be less in a denser than in a rarer medium. The change in the rate of propagation will produce a change in the direction of the wave-front.

In the diagram (Fig. 156) A, B, C are parallel rays derived from a distant source of light with wave-front a plane surface at right angles to their line of advance. They strike the surface of a denser medium obliquely, A reaching it at A_1, along wave-front $A_1 M$, before C reaches it at C_1. Suppose the velocity in the denser to be $\frac{2}{3}$ of that in the rarer medium. While C advances from M to C_1, A will reach a point A_2 which is $\frac{2}{3}$ as far from A_1 as M is from C_1. For another ray B, that strikes the surface at B_1 midway between A and C, the wave-front $B_1 N$ may be considered. Now C gets to C_1 at the moment when B having struck the surface at B_1 reaches a point B_2

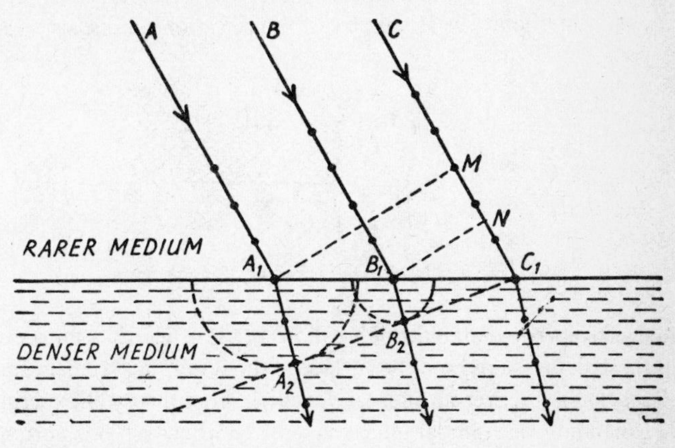

FIG. 156. Refraction in terms of Wave Theory. Beads on the lines mark equal intervals. Velocity of light in denser is represented as two-thirds of that in the rarer medium.

at $\frac{2}{3}$ the distance from B_1 that N is from C_1, or $\frac{1}{2}$ of $\frac{2}{3}$ the distance that M is from C_1. Thus $A_1 A_2$ is $\frac{2}{3}MC_1$ and $B_1 B_2$ is $\frac{1}{3}MC_1$. A_2 and B_2 are located on wave-front circles with centres A_1 and B_1 respectively and with radii equal respectively to $\frac{2}{3}$ and $\frac{1}{3}$ of MC_1. Thus the wave-front when C arrives at C_1 will be a straight line $A_2 B_2 C_1$, which only touches but cannot cut either of the two circles. That straight line is, in fact, a common tangent to all circles formed on the proportionate construction here considered. The angular change of direction of wave-front from $A_1 M$ to $A_2 C_1$ corresponds to the necessities of Snell's Law (p. 229).

The wave theory of light that prevailed in the nineteenth century was propounded at its dawn by THOMAS YOUNG (1773–1829). In two communications (1801), which place him in the forefront of scientific investigators, he set out his wave theory and its essential principle of *interference*.

Suppose [he said] a number of equal waves of water to move upon the surface of a stagnant lake with a certain constant velocity, and to enter a narrow channel leading out of the lake; suppose then, another similar cause to have excited another equal series of waves, which arrive at the same channel with the same velocity and at the same time as the first. One series of waves will not destroy the other, but their effects will be combined. If they enter the channel in such a manner that the elevations of the one series coincide with those of the other, they must together produce a series of greater joint elevations; but if the elevations of one series are so

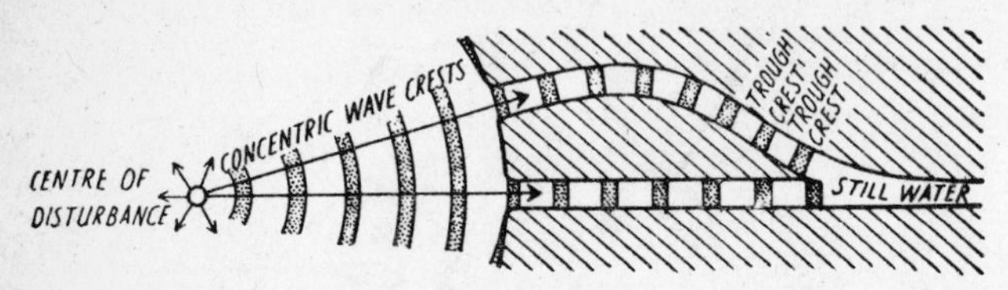

FIG. 157. To illustrate the principle of interference.

situated as to correspond to the depressions of the other, they must exactly fill up those depressions, and the surface of the water must remain smooth —at least, I can discover no alternative, either from theory or experiment. Now, I maintain that similar effects take place whenever two portions of light are thus mixed, and this I call *the general law of the interference of light.*

This view of interference is perhaps most simply presented if we picture waves from one centre of disturbance entering two channels of unequal length which subsequently meet. If at the meeting-point the waves are in opposite phases they will evidently neutralize each other (Fig. 157).

Newton had himself discussed the wave theory, and had dismissed it, saying:

If light consisted in motion, *it would bend into the shadow*, for motion cannot be propagated in a fluid in right lines beyond an obstacle which stops part of the motion, but will bend and spread every way into the quiescent medium beyond the obstacle. . . . A bell may be heard beyond a hill which intercepts the sounding body . . . but light is never known to bend into the shadow.

Young proved, however, that light does bend. The bend is extremely small, owing to the minuteness and immense speed of the waves, but it is greater in some mediums than in others, in water for example than in air (Fig. 158).

Young demonstrated this bending of light-rays by a simple experiment. Light reflected from the sun was admitted through a pin-hole in the side of a dark chamber, making a cone of light. In the pathway of this cone was interposed a narrow strip of card. Faint fringes of colour were seen on either side of the shadow thus cast on the opposite wall, while in the shadow itself was a sequence of faint dark and light upright bands, finishing off in a faint light band in the middle of the shadow. Since light normally travels equally in all directions, a part of it, passing on each side of the strip of card, must spread out behind it. But why should the light arrange itself in strips, and not fall equally all over the shadow? When an opaque object was placed so as to prevent the light from passing one of the edges of the card, the fringes disappeared. Therefore, so long as the light passes in one direction behind the card it spreads itself out equally, and only when two sets of rays from the two sides of the card meet do 'interference' bands appear. This is a close analogy to what happens in the case of water-waves.

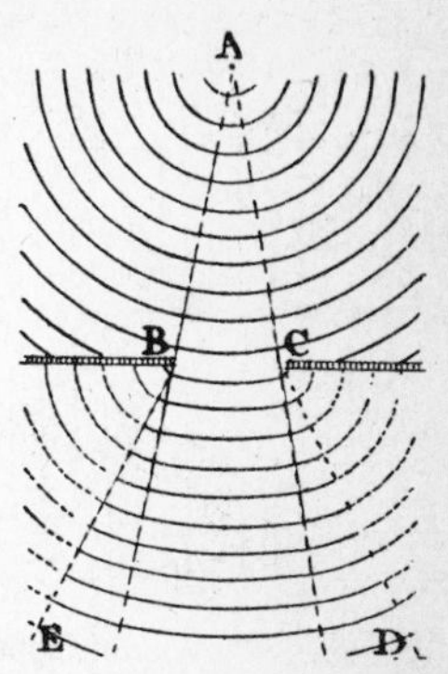

FIG. 158. Waves diverging from centre *A*, pass through aperture *BC*. They extend themselves on each side—that is, they 'bend into the shade'—so as to fill the space *BCDE* while affecting the parts outside this area much less or not at all. (Young's diagram.)

Light does not, however, always travel through a transparent medium equally in all directions. Thus it had long been known that light traversing two crystals of Iceland spar gives in general two streams of (usually) unequal brightness. The relative intensity of the two streams was known to depend on the relative positions of the crystals. In certain positions one stream disappears entirely. The French mathematician ÉTIENNE LOUIS MALUS (1775–1812) found that he could elicit results comparable to those of Iceland spar by light *reflected* from transparent surfaces. Misunderstanding the nature of the process, he called it *polarization* (1805), a misleading title which it still bears. Such phenomena as those investigated by Malus were inexplicable on the wave theory until it was given its modern form by the French experimenter AUGUSTE JEAN FRESNEL (1788–1827), in correspondence with Young.

In sound-waves, from which Young had drawn his picture of light-waves, the vibrating particles move in a direction parallel to the propagation of the wave. This is 'longitudinal' vibration. In water-waves the water-particles move up and down at *right angles* to the forward direction of the wave (Fig. 161). This is 'transverse' vibration. The 'ether vibrations' of a light-wave are transverse. They have, however, this complication, that the plane of vibration is only partly restricted, so that a ray of light may consist of waves vibrating in any plane passing through the direction of the ray.

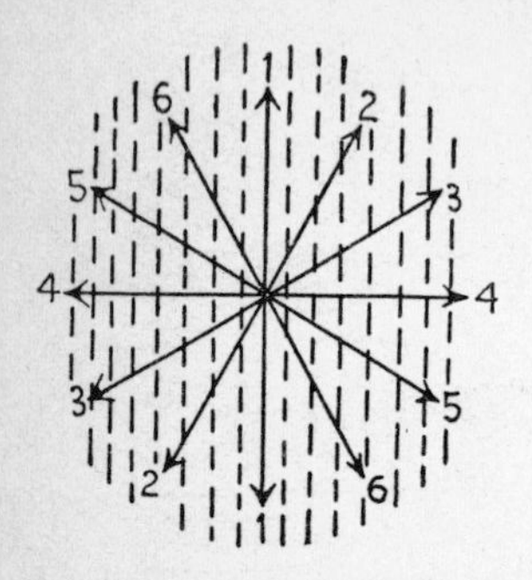

FIG. 159. Polarization of Light. The observer is supposed to view a light ray end-on as it advances toward him. The line of advance is along an axis represented by the central point. Vibrations in the ether take place in all planes through which the axis passes. These planes, from the observer's point of view, are seen as straight lines, of which six are represented. If the ray encounters a medium which acts as a grating (such as is represented by the dotted lines), permitting the passage of vibrations in only one plane (1 in diagram), the light is 'polarized'.

Graphically represented (Fig. 159), looking 'end-on' at a wave, we can visualize a series of short straight lines signifying the extremes between which the 'ether particles' vibrate. Although vibrating in any plane passing through the line of the advancing light, yet all vibrations are at right angles to the direction in which the wave advances, that is, for the purposes of our diagram, they vibrate in the plane of the paper. The action of Iceland spar upon the light-waves impinging on it may be compared to a set of railings with vertical chinks. Vibrations parallel to the rails will pass on between its chinks, but the remainder will be stopped. The light that passes on is said to be 'polarized'.

Fresnel also used the conception of interference to bring the undulatory view of light within mathematical range by making possible a quantitative estimate of the length of light-waves (1821). Two metal mirrors in almost the same plane (Fig. 160) reflect light from a pin-hole in the wall of a dark chamber on to a white screen. Looking into the mirrors from the screen the observer would see two 'virtual images' of the whole, as at *A* and *B*, and the optical effects are as though the light really proceeds from those points. By straightening the mirrors, *A* and *B* can be made to approach each other until, when

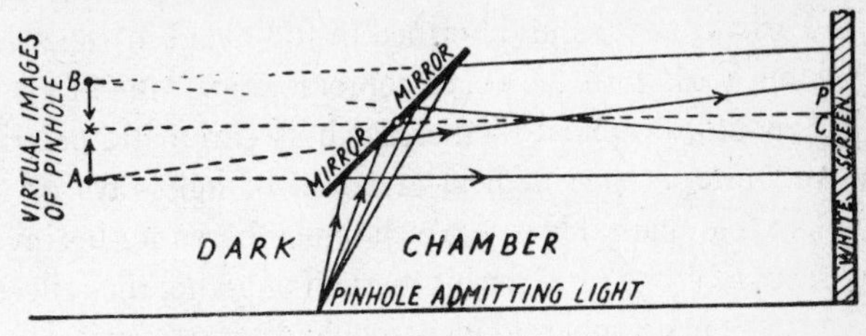

FIG. 160. Fresnel's Interference Experiment.

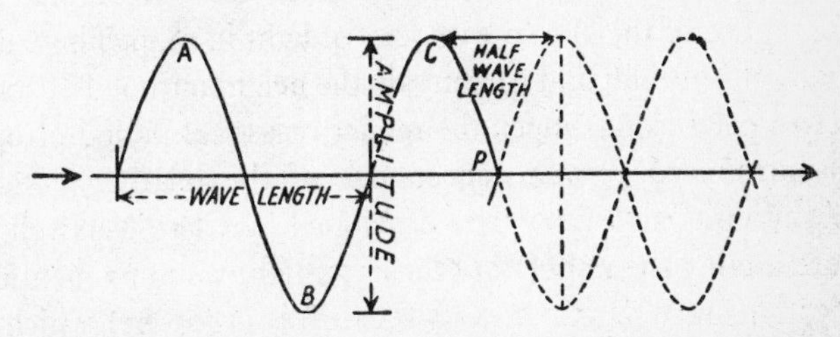

FIG. 161. *ABCD* represents a system of transverse waves propagated toward the right. At *P* it is joined by a second system, of equal amplitude and wave-length, but with oscillations half a wave-length later. There results a state of oscillatory rest or 'interference'. Should the waves of the second system have the same wave-length but unequal amplitude, the amplitude of the first would be reduced. If the second has a different wave-length, a more complex system will arise.

the mirrors are in the same plane, the points coincide into a single virtual image. A line drawn from this, vertical to the screen, meets it at *C*. Consider any point *P* on the screen in the area that receives light from both *A* and *B*. *PA* is longer than *PB*, but the difference becomes less the nearer *P* is to *C*. This difference, *PA* minus *PB*, can be calculated from the known conditions of the experiment. Now *P* sometimes shows a dark, sometimes a light, band. This will be according as the difference between *PA* and *PB* approximates to an odd or even multiple of a half wave-length; whether, in fact, the waves of light as from *A* and *B* strike the screen in the same or in opposite phases.

We have seen (pp. 368 f.) that the wave theory requires that the velocity of light in rarer media should be greater than in denser, the opposite being demanded by the emission theory. Thus a direct proof that light passes more rapidly through air than water would be a further confirmation of the wave theory. This was achieved by JEAN LÉON FOUCAULT (1819–68) in a very well-known series of experiments begun in Paris in 1850, and described in full detail in 1862. He had already done work with his exact contemporary, HIPPOLYTE LOUIS FIZEAU (1819–96), on allied themes, such as chromatic polarization of light and interference of heat-rays and of light-rays of greatly differing length of path. He is also well remembered for his invention of the gyroscope (1852) and for his method of giving the reflectors of optical instruments a spheroid or paraboloid form (1857). His name is, moreover, attached to several electrical devices. Fizeau had made determinations of the absolute velocity of light in 1849. These determinations of Foucault and Fizeau—in the neighbourhood of 300,000 kilometres per second—open the modern classical period of optics. Fizeau introduced certain conceptions of the relative motion of matter and ether that were later developed by Clerk Maxwell.

Before leaving the undulatory theory of light we must mention an investigation by WILLIAM ROWAN HAMILTON (1805–65) which well illustrates a noteworthy characteristic of the history of science, namely, that the importance of a discovery is often not evident until long after it has been made. In 1834 Hamilton had generalized the equations of Newtonian dynamics into a form in which the path of a particle could be represented as a *minimal* path; that is to say, the particle would move in a field of force along the course in which a certain mathematically defined quantity was smaller than in any other. Now this is characteristic also of light, as Hero (pp. 86 f.) and PIERRE FERMAT (1601–65) had shown centuries before. An analogy thus existed between the action of forces on a particle and that of the medium of transmission on a wave, and Hamilton worked out the details of the analogy in such a way that if one were presented merely with a curved path in space, one could represent the departure from straightness in terms either of deflecting forces or of varying properties in a surrounding medium. The matter remained a mere curiosity until this century when it provided the basis of wave mechanics, in which light and fundamental particles are shown to possess both wave and particle properties. This, however, is beyond our range.

(x) *Doctrine of Energy*

The History of Science submits, no more easily than the history of other subjects, to arbitrary time divisions. Nevertheless there are certain seminal scientific ideas, the appearance of which makes it possible for the historian to establish time boundaries sufficient for the division of his narrative. Such a one is the doctrine that any form of measurable physical activity is convertible into any other form, and that the total amount of such activity in the world is limited and remains the same. This Doctrine of Energy became accepted about the middle of the nineteenth century and opened a new era in the history of scientific ideas.

An important advance in this direction was made by the brilliant young Frenchman, SADI CARNOT (1796–1831). In his only publication, *Reflexions on the Motive Power of Fire* (1824), Carnot measured and defined work as 'weight lifted through a certain height'. He established quite clearly the principle that heat and work are reversible conditions and that the efficiency of a reversible engine depends on the temperatures between which it works. The pamphlet of Carnot attracted little attention during his lifetime. The principles involved were grasped some twenty years later by the Englishman J. P. JOULE (1818–89), a pupil of Dalton, who developed the subject with great experimental skill.

Joule's work began to assume significance in 1840, when he was emphasizing the idea of the importance of physical units. Those which he then adopted involved the conception of the transference of chemical into electrical activity in a measurable way. His unit of static electricity was the quantity needed to decompose 9 grammes of water, and his degree of current electricity the same amount propagated in an hour. He regarded the consumption of the metal in the electric battery as a source of energy analogous to that of the coal that drives the steam-engine.

In considering the dynamo as invented by Faraday, Joule was able to demonstrate a numerical relation between the chemical effect in the battery, the mechanical effect in the motor, and the electrical effect in the circuit. Thus, if a given weight of zinc be dissolved in acid, a certain measurable amount of heat is given off. Make the zinc an element in a battery and a measurably *less* amount of heat is produced in the course of its solution. If the current passes through a wire, it heats the wire. This amount of heat corresponds, he showed, to the difference between the heat produced by the simple solution of zinc in acid and that produced when it is dissolved as an element in a battery. Moreover, if the current drives a motor yet more heat is missing. The amount missing is proportional to the work done by the motor.

Joule's historic paper of 1843, *On the calorific effects of Magneto-Electricity and on the mechanical value of Heat*, brings out very clearly the relation between work and heat. It sets forth 'Joule's Equivalent', as it is now called, that is, the amount of work which must be transformed in order to give one unit of heat. This unit of heat was the amount needed to raise one pound of water one degree Fahrenheit. His unit of work was the amount required to raise one pound weight a height of one foot. His equivalent, as he then determined it, was 838 foot-pounds.

In the years that followed Joule pursued his idea with many refinements. Thus he measured the work required and the heat produced when water is driven through fine tubes, when air is compressed or allowed to expand, when a paddle-wheel is driven through water or through more viscous fluids, and so on. But not until 1847 did he give the first full and clear exposition of that principle now called *energy*, a term first applied in that capacity by William Thomson (Lord Kelvin) in his great paper, *Dissipation of Mechanical Energy* (1852).[1] Joule's superb exposition of 1847 had been given in the form of a popular lecture in a church reading-room! This great scientific pronouncement, after rejection by several journals, appeared in a Manchester weekly paper with the title: *Matter, Living Force and Heat*. His *living force* is, of course, what we call 'Energy'. He said:

Living force (*vis viva*) is one of the most important qualities with which matter can be endowed, and as such it would be absurd to suppose that it can be destroyed. . . . Experiment has shown that wherever living force is *apparently* destroyed, whether by percussion, friction, or any similar means,

[1] Thomas Young had used the word in an analogous sense in 1807.

an exact equivalent of heat is restored. The converse is also true, namely, that heat cannot be lessened or absorbed without the production of *living force* or its equivalent attraction through space. . . . Heat, living force and attraction through space (to which I might also add *light*, were it consistent with the scope of the present lecture) are mutually convertible. In these conversions nothing is ever lost.

In the same year appeared the little book of HERMANN HELMHOLTZ (1821–94), *Erhaltung der Kraft* ('Conservation of Energy'). In the same year, too, Joule came in contact with William Thomson, afterwards LORD KELVIN (1824–1907), who had long been interested in the transformation of heat. Helmholtz, in his famous pamphlet, in rejecting the possibility of perpetual motion, sought to establish the doctrine that through all transformations of energy the sum total of all energies in the universe remains constant. Thomson accepted the conclusions of Joule and Helmholtz and applied himself from 1848 onward to the mathematical implications of these doctrines. 'The first step toward numerical reckoning of the properties of matter', he wrote, 'is the discovery of a continuously varying action of some kind, and the means of measuring it in terms of some arbitrary unit or scale division. But more is necessary to complete the science of measurement in my department, and that is the fixing on something absolutely definite, as the unit of reckoning.'

Thomson reached his conception of a fixed point. He was familiar with Carnot's view of a reversible cycle and was one of the first to draw attention to it (1848), in illustration of the fact that the melting-point of ice is lowered by pressure. He saw that the amount of work performed by an engine does not depend directly on the thermometric scale value of the temperatures between which it is working. Thus, to take a simple example, the work done per unit of heat between 100° and 150° is not the same as that done between 150° and 200°. Thus before it is possible to reach a clear conception of the interchange of forces it is necessary to find some absolute scale which is not arbitrarily determined by the changes of state of a single substance as is that of the ordinary thermometer by the freezing-point and boiling-point of water. Now for an engine to be theoretically perfect, that is, for all its heat to be converted to work, it is necessary that the lower of the temperatures between which it works be the minimum possible. This minimum point Thomson called 'the absolute zero of temperature'.

Working between the temperature 0° and 100° Thomson found that for every 373 parts of heat put in at 100° the engine will return 273 parts into the receiver, converting 100 parts into mechanical work. In other words, if boiling-point under the stated conditions be taken as one fixed point and freezing-point be taken as another, then—treating the working range between these two points as 100—the lowest conceivable temperature, the zero of this absolute scale, would be −273°. This is the zero of an 'absolute thermometric scale'. That scale is concerned solely with the work done by the substance employed and has nothing to do with its physical properties.

The recognition of an absolute scale and of its implications with the doctrines of energy, of the transformation of forces, of the ether, and of atoms provides the foundations on which was built the impressive structure of classical physics during the second half of the nineteenth century.

6. BIOLOGY BECOMES SYSTEMATIZED

(i) *Early Classificatory Systems*

As the exploration of the globe proceeded, the number of kinds of living organisms known to science rapidly increased and became very large. Some system of codification and standardized description became an urgent need. Many attempts were made in this direction, but the successful and accepted scheme was that of the Swede KARL LINNAEUS (1707–78). Its pre-eminent convenience led to its rapid adoption to the exclusion of all other systems.

Linnaeus took the parts of a plant or animal in regular sequence and described them according to a recognized rule. This introduced what was almost a new international language, very condensed, very clear, and very easily learned. As 'botanical Latin' it has survived and maintained its usefulness. The method was a great improvement on the verbose and confused accounts usual till that time. It is best expounded in his *Philosophia Botanica* (1751).

Linnaeus also constructed a system of arrangement in which every known species of animal and plant had a position assigned to it. This involved grouping the *Species* into *Genera*, the Genera into *Orders*, and the Orders into *Classes*.

For plants the Classes and Orders were based on the number and arrangement of the parts in the flower. Linnaeus had a clear though not very accurate or searching conception of the sexual character of the floral elements. The number of 'stamens' or free male parts was his first consideration. Thus Linnaeus grouped plants with one stamen in the Class *Monandria*, plants with two in the Class *Diandria*, plants with three in the Class *Triandria*, and so on. Each Class was then divided into Orders, according to the number of 'styles', or free female parts, in the flower. Thus, the Class *Monandria* was divided into the Orders *Monandria Monogynia* with one style, *Monandria Digynia* with two, and *Monandria Trigynia* with three, and so on.

For animals, Linnaeus distinguished the Classes of Mammals, Birds, Reptiles, Fishes, Insects, and Vermes. The first four had already been grouped together by Aristotle as 'Animals with red blood' or, as we now call them, 'Vertebrata' or backboned animals. The remaining Classes, Insects and Vermes, contain, bundled together, all the Orders of animals without vertebrae or backbones. Here Linnaeus was behind Aristotle, who had broken up these groups more effectively (p. 47).

The contribution through which the name of Linnaeus will, however, always be remembered and is daily recalled by naturalists is his 'binomial nomenclature', the system of defining every known living thing by two Latin names, the first being that of its *genus* and the second that of its *species*. It will naturally be asked what is meant by these words. To this no one can give yet a clear or even an intelligible answer, though it is evident that an answer is slowly emerging from certain current work. Naturalists have been occupied for over two centuries with the more exact application of these terms without reaching any general definition of them. It is unparalleled that undefined terms should remain indispensable for so long. To find an answer is a main task of Genetics (pp. 483 f.), a twentieth-century science.

But although no one could or can define species or genus in general terms, Linnaeus had ideas concerning them which are of great historical importance. He held that species are constant and invariable, a view in which he differed from John Ray. 'There are just as many species as there were created in the beginning', wrote Linnaeus, and again, 'There is no such thing as a new species'. In this matter we have departed completely from his standpoint. Genera were definitively grouped species.

The *Systema Naturae* of Linnaeus is nevertheless a permanent contribution. It was first drafted in 1735, and he modified it and amplified it in its many editions. Of these, zoologists have agreed on the tenth, which appeared in 1758, as the permanent basis for the scientific names of living things. If a species is given its 'Linnaean name' by a modern naturalist, it means that adopted in this tenth edition. For plants the rules are rather more complex.

Linnaeus was an extremely stimulating teacher. He had a great number of enthusiastic pupils, many of whom went on expeditions to distant lands and discovered and described multitudes of species. He and his disciples, by concentrating their interest on external parts, which are specially valuable for classification of species, withdrew attention from the intimate structure and working of the living organism. The search for new species thus remained for long the chief aim of most naturalists, to the neglect both of anatomical and of physiological studies.

Much of the immense appeal of Linnaeus to his generation and those which followed was due to his appreciation of wild life. There have been few greater nature-lovers. His tradition can be traced especially in Britain. It is commemorated in the 'Linnean Society' (established 1790), and its impact happened to coincide in time and was ancillary to the literary movement known as the 'Romantic Revolt'. Natural History in Britain had long interested the country gentry and clergy. There came a time when these were reinforced by the scientific tastes of the rising and wealthy industrial class. Thus the study of Nature became 'fashionable'. Societies for it were founded in every major centre of population in Britain, from Kirkwall in the Orkneys to Penzance in Cornwall. Darwin, who approached his great task in the dual capacity of systematist and observer of wild life, was a typical product of this dual Linnaean tradition. Among its gifted literary exponents were the Rev. Gilbert White (1720–93) of *The Natural History of Selborne* (1789) and Charles Waterton (1782–1865) of the *Wanderings* (1823). Others associated with the movement were Banks (p. 395), Lyell (pp. 329f.), Murchison (p. 332), T. A. Knight (1759–1838), and the entomologist the Rev. W. Kirby (1759–1850) of Kirby and Spencer's *Introduction to Entomology* (1815–56).

Since the time of Linnaeus almost every important biological movement has left its mark on the system of classification current in its day. The classification of living things adopted by a biological writer may often be treated as an epitome of his views on many important biological problems, and especially on 'comparative' studies. This was notably the case with the system of Cuvier.

The French naturalist GEORGES CUVIER (1769–1832) wielded great authority and determined the general direction of biological, and especially of zoological, activity in the first half of the nineteenth century. His general approach, unlike that of Linnaeus, was analytic. He laid stress on the structure and relations of the inner parts rather than on their external characters.

Cuvier divided the animal kingdom into four great divisions, each of which, in his view (1817), was built on its own definite plan.

I. VERTEBRATA, with a backbone.
II. MOLLUSCA, slugs, oysters, snails, &c.
III. ARTICULATA, jointed animals, insects, spiders, lobsters, &c.
IV. RADIATA, all remaining animals.

In drawing up this scheme Cuvier was guided by his analysis of two main sets of functions. The heart and circulation provide, he considered, a centre for the 'vegetative functions' of growth, reproduction, &c., to which the breathing apparatus is accessory. The brain and spinal cord he regarded as presiding over the 'animal functions' which are associated with active movement and are served by the muscular system. We are here reminded of the 'vegetative' and 'animal soul' of Aristotle. The thought of Cuvier is, in fact, infused with that of his great predecessor. Though the conception of vegetative and animal functions have, since Cuvier's time, changed beyond recognition, much of our modern classificatory system is based on him and through him on Aristotle.

The Genevan botanist, AUGUSTIN PYRAMUS DE CANDOLLE (1778–1841), did for plants similar service to that of Cuvier for animals. He was a searching and patient investigator. Much of his classification of the higher plants (1824) survives in the systems developed by modern botanists.

(ii) *Main Subdivisions of Biological Study*

A feature in the biological outlook of the early nineteenth century was the slowness with which microscopical research took an important place. The great microscopists of the seventeenth century had singularly few successors in the eighteenth. Thus, when Humphry Davy needed descriptions of the microscopic structure of stems and leaves for his *Agricultural Chemistry* (1813) he had to rely on Grew's *Anatomy of Plants* (1682) of 130 years earlier. In the third decade of the nineteenth century technical improvements in the microscope began to be available, and awoke the interest of biologists. A number of observers were soon devoting themselves to the intensive study of microscopic organisms, and to the microscopic analysis of larger forms. From now on the microscope became the essential instrument of the biological sciences. Microscopic observations have since provided the building material for a dozen separate departments of biology.

In the first half of the nineteenth century philosophical naturalists were largely occupied in establishing 'affinities' between different types of organisms. These workers may be divided into fairly definite groups according to the character of the problems that they set themselves to solve. Of these groups five may be distinguished as of particular historical importance.

(*a*) Those concerned with comparing external characters of living forms nearly allied to each other; that is, in the work of establishing the nature and limits of species, genera, and families, and of their degrees of affinity. These are the systematists or 'taxonomists' (Greek *taxis*, 'arrangement'; *nomia*, distribution. The word was introduced by de Candolle 1813). A very great exponent of this study was Darwin. His concentration on its problems led him to his historic consideration of the *origin* of species. An investigator who worked on comparable lines but came to different conclusions was Louis Agassiz. Important botanical taxonomists were A. P. de Candolle and J. D. Hooker (p. 396).

(*b*) Those occupied in investigating the inner structure of contrasted forms, that is of forms belonging to widely separated groups as Orders and Classes. These are the *comparative anatomists* or *morphologists*. (The term 'Comparative Anatomy' was introduced in 1672 by Grew. Morphology, Greek *morphé*, 'form', was introduced by Goethe about 1817.) Typical exponents of this method were the versatile JOHANNES MÜLLER (1801–58) in Germany, RICHARD OWEN (1804–92), opponent of Darwin, and first director of the British Museum of Natural History, ROBERT BROWN (1773–1858) 'botanicorum facile princeps', CUVIER, and ÉTIENNE GEOFFROY ST. HILAIRE (1772–1844), opponent of Cuvier.

(*c*) Workers engaged in the comparative anatomy of fossil forms are known as *palaeontologists*. Among the greatest of these were RICHARD OWEN, and the palaeobotanist W. C. WILLIAMSON (1816–95). The word 'Palaeontology' was introduced into English by Sir Charles Lyell (1838).

(*d*) It was early realized that the structure of embryos revealed affinities that are less apparent in adults. Moreover, in certain respects, the knowledge of the formation of the parts in the embryo was found to make the structure of adult forms more intelligible. The beginnings of life had always excited wonder and curiosity. The investigation of embryos required, however, unusual kinds of skill, and 'embryologists' were early differentiated. (The term *embryologie* was admitted into the French language by the *Académie* in 1762. It did not enter English till the nineteenth century.) Important early embryologists were the Germans KARL ERNST VON BAER (1792–1876) and ROBERT REMAK (1815–65), the Swiss ALBRECHT KÖLLIKER (1817–1905), and the Swiss-American LOUIS AGASSIZ (1807–73).

(*e*) Quite apart from the schools of 'naturalists' and 'biologists', the first half of the nineteenth century saw a great extension of scientific interest in the analytical study of animal function by means of physical and chemical experiment. The exponents of this science of 'physiology' were mainly preoccupied with its medical applications. Such physiologists were not usually concerned to compare different forms. Choosing for preference those likest to man—the 'higher' animals—they devoted themselves rather to the examination of the parts and functions in their developed state. The results have been portentous in bulk, complexity, and interest, and have given rise to a picture of the animal machine which has deeply influenced the current conception of Man, and of his place in Nature. Among the greatest exponents of this department of science were Sir CHARLES BELL (1774–1842), JOHANNES MULLER (1801–58), and CLAUDE BERNARD (1813–78).

(iii) *Naturphilosophie*

The startling revelations of the microscopists and the 'mechanist' physiologists of the seventeenth century induced, especially in German thought, an era of speculative activity. The conception of the 'ladder of Nature' assumed a new importance. Aristotle had been content with its formal projection (p. 47). During the eighteenth century it became a rigid framework into which observations were fitted.

Certain microscopic observations had given rise to the false idea that the organism is already fully formed in the germinal original, that is in the ovum, of the female or, alternatively, in the spermatozoon of the male. The Genevan CHARLES BONNET (1720–93) raised this idea of 'preformation' to the rank of a scientific and philosophic doctrine (1762). Both this conception and the process of reproduction without fertilization (parthenogenesis, Greek, = virgin birth), which he rediscovered (1745), he made to serve theological ends.

In this peculiar intellectual atmosphere Bonnet and his followers developed a rigid interpretation of the conception of a 'ladder of nature'. Passing from the most subtle of the elements, fire, through air, water, and the densest, earth, this 'scala naturae' ascended through the finer minerals, such as crystals, to living things, proceeding through what were then regarded as the lowest of these, namely the moulds, via plants, insects, and worms, upward to fish, birds, mammals, and finally to man. The medieval and Christian view of 'man as the measure of all things'[1] was thus given a new significance by Bonnet and his school. 'All beings', he wrote, 'have been conceived and formed on one single plan, of which they are the endlessly graded variants. This prototype is man, whose stages of development are so many steps toward the highest form of being.' Each being was believed to be 'preformed' in the male or female 'primordium' or germ, the spermatozoon or ovum (pp. 283–4, Fig. 119).

[1] The phrase itself is Plato's and is quoted as from Protagoras (*c.* 481–411 B.C.) in Plato's *Theaetetus*.

Such views pass insensibly into the attitude, known later as *Naturphilosophie*, which became especially popular in Germany. Some of its developments in that country became fantastic to the verge of insanity. Yet there were several effective thinkers for whom this attitude became a useful approach to natural knowledge. Among these were two of the loftiest intellects that Germany has produced, Kant and Goethe.

The thought of the age was given a new direction by the Königsberg philosopher, IMMANUEL KANT (1724–1804) in his famous *Critique of Pure Reason* (1771). He had begun as a man of science, and it was from his treatment of scientific problems that his philosophical interest emerged. Beginning with a world of phenomena, of nature, of experience—the determinate world of the man of science—he gradually passed into the world of the intelligible, of ends, of the philosopher.

To most men then, and to most men still, these two worlds seem to confront one another. Men of science affirm this when they say that 'the study of purpose in Nature is inconsistent with the scientific aim, which is the adequate description of phenomena'. It was Kant's thought that the two attitudes are neither opposite nor irreconcilable. He reduces the problem to the discussion of the relation between our perception of things and their real nature. Our perceptions, Kant held, come into relation with the real nature of things through the character of our processes of thought. In other words, our thoughts work along Nature's own lines. Kant pointed out that, if we consider living organisms, we perceive that they are composed of parts which are comprehensible only as conditions for the existence of the whole. The very existence of the whole implies an end. True, says Kant, Nature exhibits to us nothing in the way of purpose. Nevertheless we can only understand an organism if we regard it as though produced under the guidance of thought for the end. The naturalist tacitly admits this when he considers the different organs or parts in relation to their function in the whole living organism.

The opposition, so familiar to the biologist, between the mechanist and teleological (pp. 489 ff *et passim*) or vitalist view, is, Kant held, due to the nature of our knowledge, or experience. But our thoughts must be distinguished from our experience. In thought we pass constantly from the view of the *part as mechanism* to a view of the *whole as purpose*, and back again. Nor do we separate these two views unless deflected by some specific doctrine that the parts are really separate. There is, Kant believes, a hidden basic principle of Nature which unites the mechanical and teleological. That principle is none the less real because our reason fails to grasp it or our powers to formulate it. So far as actual practice and use of language go, such a principle is, in fact, accepted by every biologist, the most convinced 'mechanist' no less than the most extreme 'teleologist'.

Kant's scientific influence is to be traced especially in JOHANN WOLFGANG VON GOETHE (1749–1832), whose pre-eminence as a poet and writer must not obscure his importance for science. Goethe did great service in emphasizing the fact that all organisms accord in structure to a certain quite limited number of patterns or plans. These represented for him 'ideas' in the mind of God. By searching for these ideas—'plans' or 'types' as they came later to be called—Goethe and his followers did much to stimulate the systematic comparison of diverse living things. Not the least of their services was that they thus persuaded biologists to abandon the point of view, derived from medical applications, which regards the structure of man as the type to which that of all other creatures must be referred.

Goethe expounded several doctrines of great importance, some of which are still of value. His most valuable scientific conceptions were the following:

(*a*) The genera of a larger group (Family, Order, Class, or Phylum) present something in the nature of variants on a common plan. These are all expressions of the same 'idea' or 'type'.

(*b*) The various parts of the flower are but modifications of leaves, a point he well brought out by studying monstrous form, as of the rose. The 'cotyledons' of germinating seeds (cf. the terms 'Monocotyledons' and 'Dicotyledons') are but the first leaves on the infant shoot and a bud-scale but a form of leaf (Fig. 162).

(*c*) Similarly all the parts of living beings are referable to one original model or 'primordium'. Thus not only is there a primordial species of animal and a primordial species of plant, corresponding to the animal 'idea' and the plant 'idea', but also there is a primordial part of each animal and plant. The bones of the spine provide a good

FIG. 162. Drawings to illustrate Goethe's views on the morphology of leaves. *To the left*, a monstrous rose. The stamens are replaced by petal-like expansions. The pistil is represented by a continuation of the stem on which are petal-like and leaf-like structures together with others of a character intermediate between the two. *In the middle*, a seedling bean to illustrate the homology of cotyledons with other leaves. *To the right*, a young chestnut branch just emerged from bud. It shows transition from scales, through leaf-like scales, to complete leaves.

illustration of this conception. These 'vertebrae', fundamentally of the same origin and structure, have different forms, and perform different functions in different parts of the backbone. All are variants on the 'primordial' vertebra (Figs. 163–6 and 168). Goethe believed that a like story might be told of other organs. This position is now untenable but was very stimulating in its day.

Goethe did sound work on animal morphology in his study on a wide range of mammals of the premaxillary bone and incisor teeth (1830). This bone is sometimes seen as separate from the rest of the

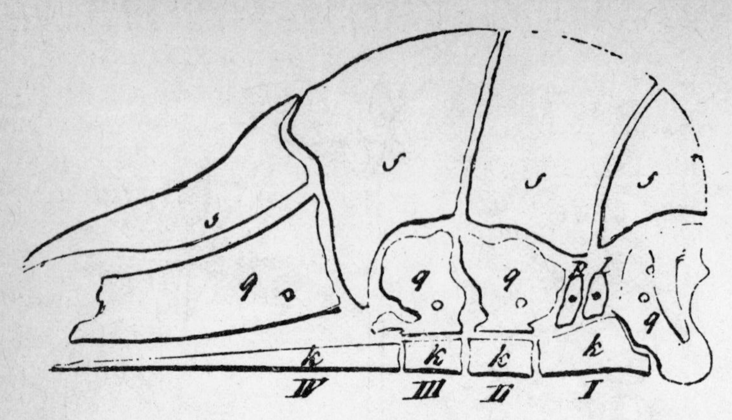

Fig. 163. Oken's diagram of the disarticulated skull of a sheep represented as equivalent to four vertebrae, 1819.

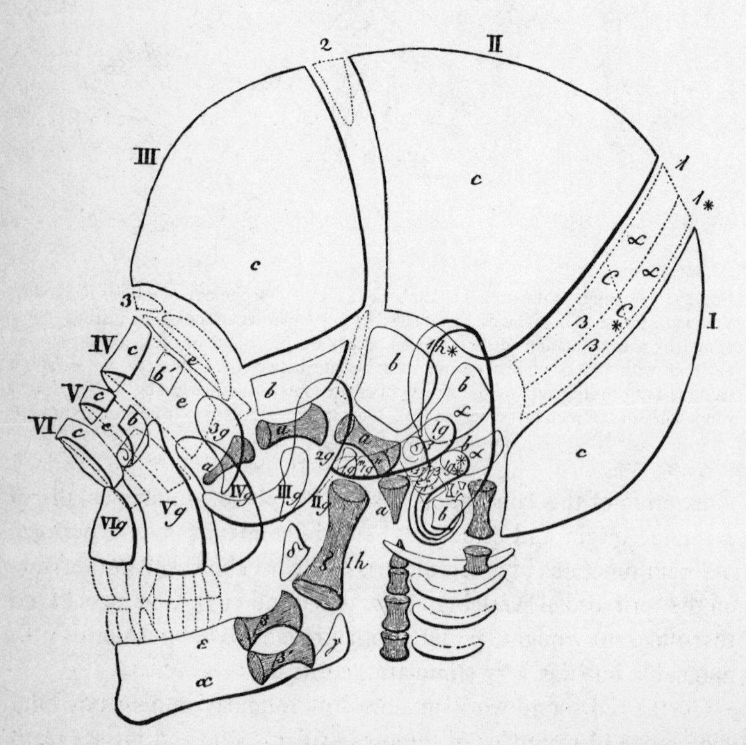

Fig. 164. Carus's diagram of human skull as equivalent to six vertebrae, 1828.

upper jaw, especially in young skulls. In this connexion his name is linked with the famous controversy on 'the vertebral theory of the skull'. It happens that the dried skull of young mammals, especially ungulates, tends to break into four sections. These were interpreted (wrongly) as corresponding to four vertebrae by the *Naturphilosoph*, LORENZ OKEN (1779–1851) in 1819 (Fig. 163). The idea was further developed by his follower, CARL GUSTAV CARUS (1789–1869), who described (1828) six vertebrae in the human skull (Fig. 164).

Owen accepted much of this (Figs. 165–6). Goethe claimed priority for this theory, dating his discovery to 1790 while musing on a sheep skull lying in sand. Though there is no trace of this in his published writings till 1824, there is evidence that he grasped it almost at the same time as Oken and for identical reasons. The theory was disproved by T. H. Huxley in 1858 but he did not quite empty it of meaning. Later in the century it became clear that in the Vertebrates—as in certain other segmented groups such as Crustaceans, Insects, and Annelids—there takes place an evolutionary process of *cephalization*, that is a concentration, specialization, and absorption of segments of the anterior appendages and nervous system in the head region.

Bonnet, Kant, Goethe, Oken, and their followers, the 'Nature Philosophers', were, in this mood, thinkers rather than observers, though their observational activities certainly cannot be despised. Some of their views, strange and strained as they now seem, and repellent as they are to many men of science, recur again and again during the centuries. For these Nature Philosophers the repetition of organs, 'segmentation', in many plant and animal types was a plan or 'idea' (in the Platonic sense, pp. 38–39) in the mind of God. A belated exponent of this was the anti-Darwinian, Richard Owen (Figs. 165, 166, 168. See also p. 394).

Whence comes the continued fascination of thoughts so little related to the daily task of the scientific observer? The essence of the thought of such men is that the processes of the mind reflect the processes of Nature. In this there is surely a truth, though it is presumptuous to suppose that we have any deep understanding of this parallelism. To say that 'the burnt child dreads the fire' is but to give a special instance of the wider statement that reason is generalized experience. Our minds, as much the product of evolution as our bodies, have in the ages developed as mirrors of the world in which we dwell; they are attuned to Nature. The mathematical thought of ages on the nature of certain curves elaborated a knowledge which Kepler

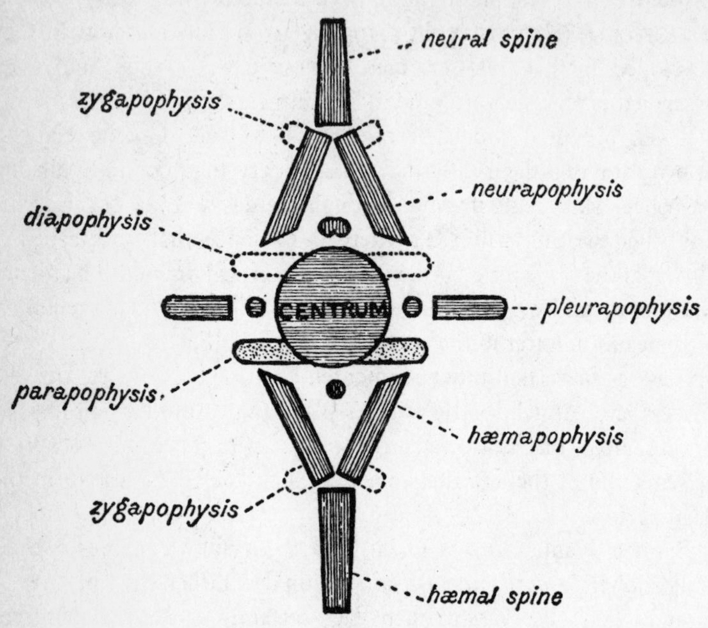

FIG. 165. Ideal typical vertebra after Owen.

and Newton fitted into the phenomena of planetary movements. The minds of the pre-Keplerian mathematicians were attuned to Nature. They were working on Nature's lines, though they knew it not. To say that we live in a rational world is but to say that by reasoning aright on accumulated knowledge, we may learn something about that world. This is as true for biology as for astronomy. By those whose minds were specially attuned to biological studies, truths have often been discerned which were verified later by experience. Aristotle and Harvey provide examples. This in itself is sufficient justification for that speculative attitude which is more productive during some scientific episodes than during others. It is often excessive but science cannot do without it.

(iv) *Correlation of Parts*

GEORGES CUVIER (1769–1832), 'the dictator of biology', was much more interested in structure than in function. He was essentially a

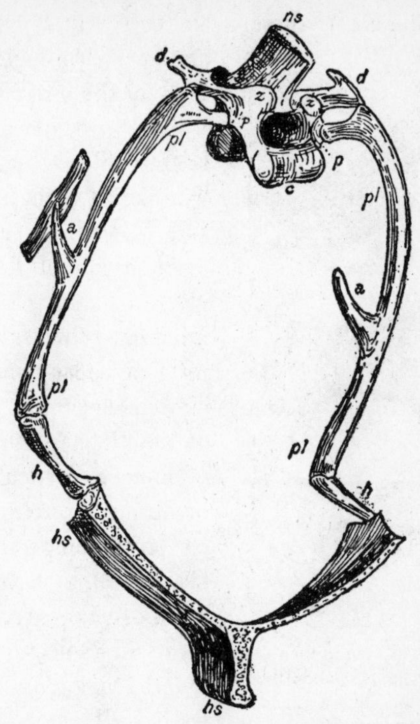

FIG. 166. Section of the bony thorax of a bird treated as a modification of an ideal vertebra, after Owen. Compare Fig. 165; most of the words there used are here represented by their initial letters.

'morphologist'. The main conception that guided him was that of the 'correlation of parts', the nature of which must be discussed.

Organs do not exist or function separately in nature, but only as parts of complete living things. In these living things certain relations are observed which are fundamental to their mode of life. Thus feathers are always found in birds, and never in other creatures. The presence of feathers is related to a certain formation of the fore-limb, with reference to its action as wing. Feathers are never found without wings, and in winged animals other than birds the wing structure is very different from that of birds, and it never has feathers (Fig. 167). But the wing structure peculiar to the bird is in turn related to certain formations of the limb-bones and breast-bone, with reference to the function of flight; these, again, to the form and movement of the chest; these, again, to the function of breathing, and so on throughout the entire body of the bird (Fig. 167).

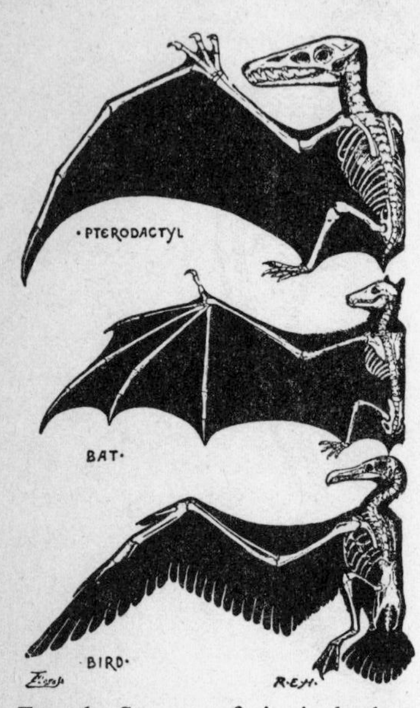

FIG. 167. Structure of wing in the three classes Reptiles, Mammals, and Birds. In each case the fore-limb forms the main part of the apparatus of flight. The same general plan of the bones of the limb is found in all. In this sense we trace *homology* in various parts. The other details are, however, so different in the three cases that it is evident that they have arisen quite independently and that the wing as a whole in the three classes is an *analogous* and not an *homologous* structure.

This principle of 'correlation' is traceable in the structure and working of each and every organ and indeed in every part of every organ of the bird. Thus, given a feather, it is possible to infer that its owner had a particular form of collar-bone, a particular kind of skeleton, a particular type of mouth, a particular structure of lung, a particular method of breathing, excretion, digestion, a particular temperature and heart-beat, even a particular kind of mind. Again, given a particular form of collar-bone, skeleton, mouth, lung, &c., we can infer a feather. If enough be known of the comparative morphology of the bird group, it is possible by the use of this principle to make astonishingly sweeping and accurate inferences.

Cuvier was far from being the first to apply his principle. In a sense it is obvious. If anyone were to find a severed hand, he would know that it had once been attached to the body of a human being, and not to that of an animal. He could make a very likely guess at sex, occupation, age, state of health, and the social position of the owner of the hand. This is nothing but the 'principle of correlation' which is the theme of most detective stories. Aristotle had, to some extent, been able to act upon this principle, but Cuvier, out of the great stores of his knowledge of organic forms, refined and extended the application of it far beyond any of his predecessors. In the hands of Cuvier's disciple, Owen (p. 394), the principle of correlation could often be brought to bear upon the merest fragment. From a little bit of leg bone, for example, even the 'leggy' nature of which none but an expert could guess, he succeeded in reconstructing an entire giant bird of a very aberrant type. His reconstruction was proved to be accurate by subsequent discoveries.

The principle of correlation has been of special value in the study of fossils, since these are usually fragmentary. Cuvier was therefore in a good position to elucidate the relationship between the living and the extinct forms. Thus arose the modern science of 'palaeontology' which owes to none so great a debt as to him. In his time large numbers of very strange fossil forms were being discovered.

The effect on the mind of Cuvier of these strange discoveries may itself seem strange. He realized that the evidence of geology showed that there had been a succession of different types of animal population, and he recognized that vast numbers of species, many no longer existing, had appeared upon the earth at different periods. Following Linnaeus, he was a firm believer in the fixity and unalterability of species. He had, however, to account for the extinction of many forms of life, and the new appearance of many other forms. His explanation was that the earth had been the scene of a series of great catastrophes, of which the last was the Flood recorded in Genesis. He expressly denied the existence of fossil man.

Cuvier did not commit himself to the doctrine of a special creation following each catastrophe. He suggested that on each occasion the earth was repeopled from the remnant that survived. This did not explain the regular succession of new species in geological time. He believed that these came from parts of the world still inadequately explored by geologists. His followers carried the matter farther, and elevated his teaching into a doctrine of successive creations. This came to assume fantastic forms even in the hands of serious scientific exponents, one of whom, as late as 1849, expounded the science of palaeontology on the basis of twenty-seven successive creations.

Cuvier's great work *Le Règne Animal* appeared in 1819. With various enlargements, modifications, and improvements by his pupils it remained standard for many years. To him and through his disciples Comparative Anatomy owes so much that the work may be said to be still standard. His personality lit up a zeal for comparative anatomy and palaeontology which lasted throughout the nineteenth century. Of the many inspired by this movement a typical representative was RICHARD OWEN (1804–92). He was also influenced by

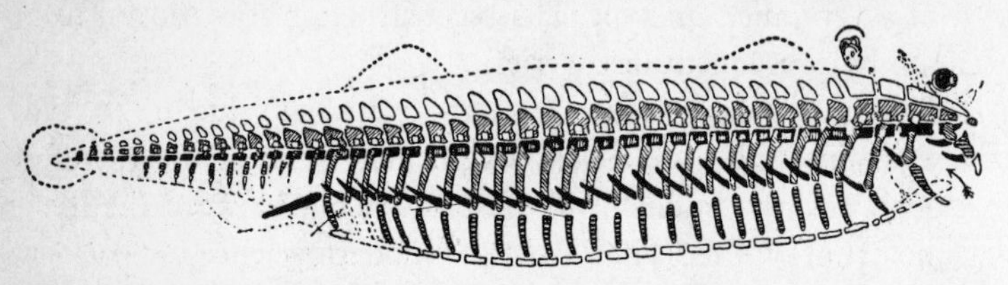

FIG. 168. The 'archetype' of the vertebrate skeleton, Owen 1846. In the view of Oken, Owen, and others the vertebrate skeleton could be analysed into a series of completely homologous and very similar segments. Of these, several of the anterior elements were supposed to be fused to form the skull. Each segment is equipped with a series of rib-like structures with appendages. From the fourth and fifth of these segments the jaws were thought to be formed. Other segments developed as limbs.

Naturphilosophie and was an obstinate opponent of Darwinian evolution (Figs. 165, 166, 168).

Owen embarked on an immense investigation of the teeth of vertebrates (1840–5). Teeth, being the hardest parts of the body, are found fossilized more often than any others. Thus his investigations led him into palaeontology, of which he became an admitted master. Among his best-known works in that department are those on the giant bird, the recent but extinct Dinornis of New Zealand (1846), and the much more ancient giant walking sloth, the fossil Mylodon of South America (1842).

In 1856 Owen became Director of the Natural History Department of the British Museum, and his activity and industry rose to the occasion. His great *Anatomy and Physiology of the Vertebrates* (1866–8) was based entirely on personal observation, and was the most important of its kind since Cuvier. The system of classification he adopted has not won favour, but as a record of facts the book was and is of very great value.

The activity of the comparative anatomists during the nineteenth century was immense. Many new Classes were described on the basis of fossil material. The teaching of Darwin, providing a framework into which comparative studies could be fitted, gave the effective stimulus to such work. The alliance of comparative studies with evolutionary doctrine focused attention on structure as distinct from function. Comparative anatomy in its turn became largely a study of developmental stages, and embryology became the comparative study *par excellence* (pp. 470 f.).

(v) *Distribution of Living Things*

In the exploratory voyages of the eighteenth century the practice was begun of carrying naturalists with equipment for observing and collecting. One of the earliest and most important of the expeditions thus provided sailed the Pacific between 1768 and 1776 under Captain JAMES COOK (1728–79). JOSEPH BANKS (1745–1820), a young amateur of great wealth and scientific competence, accompanied him and provided equipment. The staff included several artists, and a pupil of Linnaeus as botanist. The voyage yielded many plants and animals new to science. Cook's two other voyages were also very productive. Hardly less important biologically was the voyage of Captain Matthew Flinders (1771–1814) in H.M.S. *Investigator*. She left for the Pacific in 1801, never to return, but carried as naturalist ROBERT BROWN (1773–1858), one of the greatest botanists, with his admirable artist Ferdinand Bauer. Brown and Bauer, after many adventures, returned in 1805. They brought a large number of plants though others that they collected were lost. Those that survived included some 4,000 species new to science. The collection was added to the great herbarium of Sir Joseph Banks. On his death they passed to the British Museum of which Brown became the first botanical curator. An explorer and collector whose liberality and literary powers made him peculiarly influential was Alexander von Humboldt (1769–1859, p. 331). His scientific journeys in South America greatly advanced geography, geology, and biology.

Among such expeditions a most important place is taken by the voyage of the *Beagle* in 1831–4, which carried as naturalist the youthful CHARLES DARWIN (1809–82). His name is so associated with the evolutionary idea through which he profoundly influenced scientific, philosophical, political, religious, and ethical thought, that certain of his other claims are often forgotten. To appreciate his distinction, it is necessary to recall that, had he never written on evolution, he would still stand in the front rank among naturalists, and would have to be included in any history of science. Thus even during the voyage in the *Beagle* he reached conclusions that modified and extended the fundamental working principles of geology and geophysics.

In Darwin's record of experience in the *Beagle* in the famous *Journal of Researches* (1839) a special interest attaches to his observations on the highly peculiar animals and plants connected with oceanic islands. The Galapagos and St. Helena are good examples. Their extraordinary wealth of peculiar forms and the difference of these from those of the nearest neighbouring land—either continental or insular—are among the most striking phenomena in the distribution of living things. They, more perhaps than any other, suggested to Darwin his solution of the problem of the origin of species.

Sir James Ross (1800–62) in the *Erebus* and *Terror* explored Antarctica (1839–43). As naturalist there accompanied him JOSEPH DALTON HOOKER (1817–1911, p. 401), later in charge of the Botanic Gardens at Kew. Hooker was an industrious collector and skilled systematist. None of his numerous writings is of more weight than those (published 1844–60) on the flora brought back from the voyage with Ross. They include plants of the Antarctic area, as well as those of Tasmania and New Zealand, and laid the foundation of the systematic study of plant geography. Further Hooker showed the vast importance in the economy of Nature of the minute marine plants known as 'diatoms'.

The expedition of the *Erebus* was important for its revelation of a very varied fauna in a region hitherto unexplored, namely the depths of the sea. Four hundred fathoms were sounded by Ross, and life was proved to be abundant there. We now know that there is life in the open sea at every depth with a great concentration near the surface and at the bottom. Until about 1869, however, with the laying of the first Atlantic cable, it was not realized how vast and varied are oceanic fauna and flora, and how different are the conditions of life at the two levels. The effective knowledge of the ocean fauna dates from the work of the *Challenger* naturalists. They showed that most of the living matter in the world is contained in the microscopic plant forms that float at and near the surface.

The greatest of all biological explorations was that of this British Admiralty vessel in 1872–6. She carried full equipment for six naturalists under CHARLES WYVILLE THOMSON (1830–82). She travelled 69,000 nautical miles in the course of which every ocean and the least frequented parts of the world were visited, and hundreds of deep-sea soundings taken. The vast collections of the *Challenger* were investigated by a whole army of naturalists under JOHN MURRAY (1841–1914). The results were issued by the British Government in fifty large volumes. These provide the best-worked-out account of any biological expedition. The records of this voyage provided a solid basis for an effectively new science, Oceanography. They made it evident that, for any understanding of the life of our planet as a whole, an exact knowledge of the physical conditions of the sea is essential. Oceanography has since developed in a manner which demonstrates the interdependence of the biological and the physical sciences. A study which involves more than two-thirds of the earth's surface, and implicates the whole past and future history of the other third, is of primary importance to our conception of life as a whole.

The voyage of the *Challenger* was succeeded by that of the United States Government steamer *Tuscarora*, whose scientific staff investigated the floor of the Pacific. Other American and Norwegian expeditions followed in rapid succession. ALEXANDER AGASSIZ (1835–1910) was especially prominent in this work. Trained as an engineer, he was able greatly to improve the apparatus of oceanic investigation. Among his most remarkable results was his demonstration that the deep-water animals of the Caribbean Sea are more nearly related to those of the Pacific depths than they are to those of the Atlantic. He concluded that the Caribbean was once a bay of the Pacific, and that it had been cut off from the Pacific by the uprise of the Isthmus of Panama.

Many facts significant of the past or present configuration of the earth's surface have been revealed by the study of oceanography, and the subject has greatly modified our general conception of the world of life. Nor is this remarkable, since not only is the major part of the earth's surface covered by sea, but the general level of depression of the sea is greater than the general level of elevation of the land.

Oceanic plants dwelling near the surface were studied on the *Challenger* in conjunction with the floating fauna with which they dwell. The name *plankton* (Greek 'drifting') was invented for this whole community by VICTOR HENSEN (1835–1924) of Kiel (1888). The study of Plankton has become of great importance. Hensen, primarily a physiologist, began it while considering the production of nutritive substances under different meteorological conditions. He thus laid the foundations of the systematic study of the economics of the life of the ocean—*oceanic bionomics*, as we may call it. The subject is fundamental for our conception of the course of life as a whole upon this planet (pp. 406 f.).

The circumstances of life on the ocean floor, as revealed by the *Challenger*, and by later expeditions, are entirely different from those at the surface. The pressure at 5,000 fathoms is about 5 tons to the square inch as against 15 lb. at the surface. No sunlight penetrates there; below 200 fathoms all is dark. The temperature in the depths is uniform, and not much above freezing. There are few currents, and no seasons. Conditions are substantially uniform the world over, on the equator and at the poles. There is no vegetable life to build up the bodies of the animals that dwell there, and thus the animals prey only on one another, drawing their ultimate supplies from the dead matter that rains down from above. This last point may need modification in the light of recent knowledge.

The results of the deep-sea dredging have been in certain respects disappointing. Specimens of numerous new genera, and species of known families have been brought up. Many are interestingly specialized, but few are widely different in essential structure from more familiar forms. No 'missing links' have been discovered, no new Classes or Orders found. Evidently the depths have been colonized in comparatively recent geologic time.

The *Challenger* found, and further exploration has confirmed, that the species of plants and animals of the open ocean, whether on the surface or at the bottom, are mostly very widespread. An exception, on this point of distribution, must be made for the inhabitants of the extremest depths. The distribution of oceanic forms is determined by such factors as temperature, degrees of saltness, intensity of light, pressure, &c. (We now know that the ocean floor exhibits, in places, contours well nigh as marked as those of mountainous land.)

The extension of the knowledge of the conditions that prevail in the ocean and in its superincumbent atmosphere is leading to a new range of scientific ideas. As the laws of oceanic life were seen to come into relation with those of physical conditions, a most impressive physico-biological parallelism was distinguished which may one day provide a real 'physiology' of the ocean. The word *physiologia* was, in fact, originally applied to the material working of the world as a whole.

For a philosophical view of our planet as a whole a knowledge of the distribution of the life on land as well as in the sea is necessary. That different countries had different kinds of living forms had long been obvious. In the eighteenth century Buffon (pp. 503 f.) drew attention to 'natural barriers' delimiting flora and fauna. Lyell (1834, p. 503) convinced his readers that the present distribution of life is determined by past changes involving the major land-masses. The materials obtained by Darwin in the *Beagle* (published 1839–63) brought out striking facts in the geographical distribution of animals, both living and extinct. The peculiar way in which existing species were placed on the earth's surface was, however, a special object of interest to the traveller and collector ALFRED RUSSEL WALLACE (1823–1913), known for writings on South America and the Eastern Archipelago and for his connexion with Darwin's ideas (p. 507).

Wallace produced in 1876 his *Geographical Distribution of Animals*, still the most important work on the subject. He based his discussion on mammals, dividing the land-surface of the earth into six zoogeographical regions. These he named *Palaearctic*, *Nearctic*, *Ethiopian*, *Oriental*, *Australian*, and *Neotropical*. These have been retained in great part by more modern workers. The most important changes since his time are (*a*) the separation of Madagascar (Malagasy) from the Ethiopian region; (*b*) the general recognition that the Palaearctic and Nearctic regions are more nearly allied to each other than to any other region, and their union into a Holarctic region; and (*c*) the subdivision of the 'Australian' or Pacific region (Fig. 169).

Wallace demonstrated many remarkable faunal contrasts. None is more striking than that between the islands of Bali and Lombok, near Java. These islands are separated by a deep strait which at its narrowest is but fifteen miles. Yet, as Wallace remarked, they 'differ far more in their birds and quadrupeds than do England and Japan'. This strait, known as the 'Wallace Line', has been generally regarded

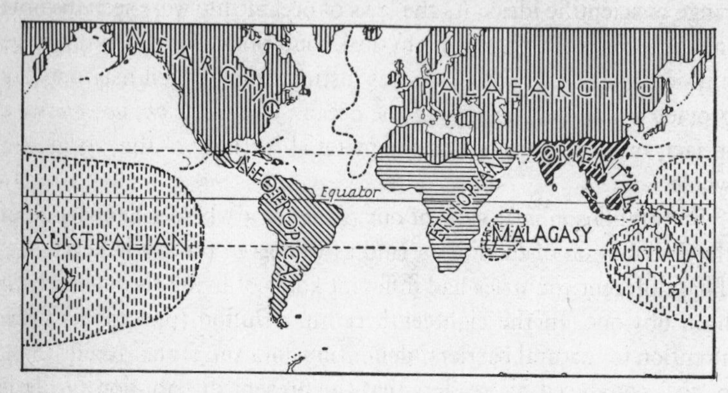

FIG. 169. The main zoogeographical regions. The 'Australian' region includes an immense number of islands, too small to appear on the map.

as delimiting the Oriental from the highly peculiar Australian zoogeographical region.

The zoogeographical regions into which the earth's surface can be divided must obviously depend upon the particular group of animals chosen, since different groups are of different geological age and have different modes of dispersal. It happens, however, that the division of geographical regions based on mammals accords closely with that based on perching birds, and is not vastly different from that based on certain invertebrate groups, e.g. the spiders, earthworms, &c. Very different from these, on the other hand, is the division based on such very ancient groups as reptiles or molluscs or insects.

The general principles that determine plant regions are similar to those of animals, but their application is considerably different. The subject has been broached mainly in connexion with the flowering plants. These are geologically younger than the groups on which zoogeographical regions are based. Moreover, temperature and moisture are of overwhelming importance in the life of plants. Further, the means of dispersal of flowering plants are more effective than those of most animal groups. The effects of this are sufficiently evident on oceanic islands.

A pioneer plant geographer was the German philosopher and traveller, Alexander von Humboldt (p. 331). Von Humboldt began his *Kosmos* (1845–7) when he was seventy-six, and he completed it in what he called the 'improbable years' which followed. This great book, now seldom read, did good service in emphasizing the relations between the forms and habits of plants and the character and soil of their habitat.

Certain resemblances between the flora of Africa, South America, and Australia had impressed Humboldt and other naturalists. In 1847 J. D. Hooker (p. 396) suggested in explanation a land connexion between South America and Australia as late as Jurassic times. Various names, forms, and areas have been ascribed to this imaginary continent.

Attempts to delimit definite plant regions have been less successful than those of the zoogeographers. A simple scheme is to divide the earth's flora into three primary areas: (*a*) the North Temperate Zone (*b*) the Tropical Zone, and (*c*) the South Temperate Zone. The northern tropic cuts off (*a*) from (*b*) with considerable accuracy. The southern tropic separates (*b*) from (*c*) with less precision.

(*a*) The North Temperate Zone contains most land. It is continuous save for the geologically recent break at the Behring Straits. It is characterized (i) by needle-leaved cone-bearing trees; (ii) by catkin-bearing and other trees that lose their leaves in winter; and (iii) by a great number of herbaceous plants that die down annually.

(*b*) The Tropical Zone occupies areas widely separated by intervening ocean. It is characterized (i) by giant Monocotyledons, notably the palms, by the Banana family, and by the enormous grasses known as 'bamboos'; (ii) by evergreen polypetalous trees, and by figs; (iii) by the relative rarity of herbaceous plants which, in this region, are mostly parasitic on other plants.

(*c*) The South Temperate Zone occupies very widely separate areas of South Africa, South America, Australia, and New Zealand. It is characterized by a number of peculiar Natural Orders, mostly of shrub-like habit. Many are intolerant of moisture. Individual species are very numerous and often very restricted in area of distribution.

Geographical regions are biologically interesting less in themselves, than as revealing or summarizing the history of the various biological groups which may flourish in them. Thus the distribution in space of living forms is ultimately referable to their distribution in time. The discussion of the one is of little profit without the other. The first systematic efforts to correlate the two sets of facts for plants were made by WILLIAM CRAWFORD WILLIAMSON (1816–95) of Manchester, who came early under the influence of William Smith (p. 329) and began his work on plants in 1858. Williamson demonstrated that in coal are to be found gigantic woody forms similar to the higher existing flowerless plants, such as horse-tails, ferns, and club-mosses.

Knowledge of the geological succession of plant forms became astonishingly detailed, and the floristic landscape at various periods and in various parts of the world was confidently restored. Moreover, owing to the fact that plant cells have definite and thick walls which may be preserved in fossils, it is sometimes possible to examine the minute structure of fossil forms. In many cases even the reproductive processes are susceptible of close examination. Such studies have produced theories of the lines of descent of plants.

7. PHYSICAL INTERPRETATION OF THE LIVING ORGANISM

(i) *Beginnings of Modern Physiology*

Throughout all history there has been an opposition, alike in philosophy and in science, between the interpretation of the nature of life in terms of mechanism and that in terms of some other entity. In the first quarter of the eighteenth century this conflict came into very clear view.

GEORGE ERNST STAHL (1660–1734, p. 281), set himself especially against the mechanism of Descartes. His vitalism passed into mysticism. Yet he deeply influenced science until Lavoisier by his *phlogiston*—the 'matter of fire' supposedly lost in combustion. This involved paradoxes since some substances, notably metals, become heavier on combustion. The rejoinder that phlogiston had 'negative weight' found analogy in the then current view that the 'elements', fire and air, rise as against earth and water. It was held too (though easily and anciently refuted) that a living body gains weight at death when breathing ceases, for some still thought that air has 'negative weight'—perhaps with the current phrase 'dead-weight' (*poids mort*) in mind.

Almost exactly contemporary with Stahl was his rival at Halle FRIEDRICH HOFFMANN (1660–1742), who was no less skilled a chemist. In Hoffmann's view the body is like a machine. Nevertheless he separated himself, on the one hand from the pure mechanists of the school of Descartes and Boerhaave by claiming that bodily movements are the exhibition of properties peculiar to organic matter and, on the other, from the Stahlian vitalists by denying the need to invoke a sensitive soul. 'Life', he wrote, 'consists in the movements of the blood. This circular movement maintains the integrity of that complex which makes up the body. The vital spirits which come from the blood are prepared in the brain and released therefrom to the nerves. Through them come the acts of organic life which can be reduced to the mechanical effects of contraction and expansion' (1718).

An important participator in the controversy was HERMANN BOERHAAVE (1668–1738), professor of medicine at Leyden, and one of the greatest physicians of all time. He, too, was skilled in chemistry. His admirable *Institutiones medicae* (1708) remained the standard account of physiology for half a century. In this work Boerhaave goes systematically through the functions and actions of the body, seeking to ascribe chemical and physical laws to each. He does lip-service to the influence of mind on body, but in practice is as completely mechanist as Descartes. Thus he still believed that something material passes down the nerves to cause movement by distending the muscles. He set the tone to physiological thought for at least a century.

Throughout the eighteenth and nineteenth centuries nearly all important physiological investigators were medical men. An exception was the exemplary parish priest, the Rev. STEPHEN HALES (1677–1761), who made many important advances in both animal and vegetable physiology. His work on the functional activity of plants was the most important until the nineteenth century. His *Vegetable*

Staticks (1727) records many experiments, devised to interpret plant activity in terms of physical forces. Thus, measuring the amounts of water taken in by the roots and that given off by the leaves, he estimated what botanists now call 'transpiration'. He measured the force

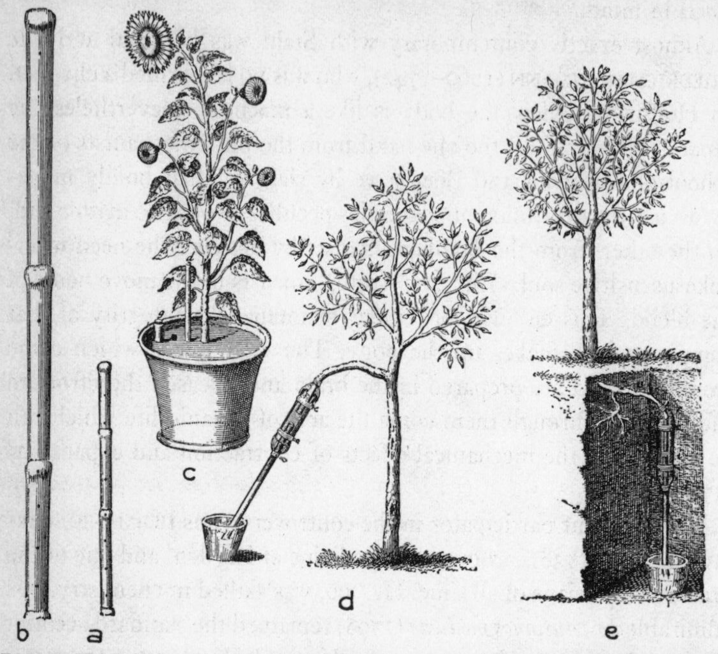

FIG. 170. From Hales, *Vegetable Staticks* (1727). *a* and *b* exhibit his method of showing the region of growth. A young stem is pricked with holes at known equal distances apart. The distance between the holes is measured in the stem when older. Growth takes place mainly in the middle part of the internodes. *c* is a sunflower planted when younger in a known weight of earth in a pot on the top of which a metal plate is fixed. Air and water are admitted through a tube in this. From the dry weight of such a plant when grown, it can be proved that something material is absorbed into its substance from the air. *d* is an apparatus for measuring the negative pressure resulting from transpiration. The apparatus fixed to the branches is, in effect, a 'manometer'. *e* is a similar apparatus for measuring root pressure.

of the upward sap-current in the stems. He sought to show that these activities of living plants might be explained in mechanical terms with reference to their structure. An interesting contribution by Hales was his demonstration that the air supplies something material to the substance of plants. This we now know to be carbon dioxide (Fig. 170).

Hales tried to give a quantitative expression to the conception of the circulation of the blood in animals. He showed that just as there is a 'sap-pressure' that can be measured, so there is a 'blood-pressure' that can be measured and that it varies according to circumstances. It is different in the arteries and the veins; different during contraction of the heart from what it is during its dilation; different with a failing and with an active heart; different in large and in small animals. He measured, too, the rate of flow in the capillaries of the frog. These experiments and conclusions of Hales initiated the quantitative phase of the science of animal physiology.

In contrast to the secluded career of Hales is that of ALBRECHT VON HALLER (1708–77), a Swiss of noble birth and ample means. He exhibited literary and scientific activity almost unparalleled in range and volume. His great *Elementa Physiologiae* (1759–66) set forth his conceptions of the nature of living substance and of the action of the nervous system. These formed the main background of physiological thinking for a hundred years.

Associating life with movement and muscular contraction, Haller concentrated on the muscle-fibres. A muscle-fibre, he pointed out, has in itself a tendency to shorten with any stimulus, and afterward to expand again to its normal length. This capacity for contraction Haller called 'irritability' which he recognized as an element in the movement of various organs, and notably of the heart and intestines. The salient features of irritability are (*a*) that a very slight stimulus produces a movement altogether out of proportion to the original disturbance, and (*b*) that it will continue to do this repeatedly, so long as the fibre remains alive. We now recognize irritability as a property of all living matter.

Besides its inherent force of irritability, Haller recognized that a muscle-fibre can develop another force which (*a*) comes to it from without, (*b*) is carried from the central nervous system by a nerve, and (*c*) is that by which muscles are normally called into action after the death of the organism as a whole. This is the 'nerve force'. It provides one way of arousing irritability.

Having dealt with movement, Haller turned to feeling. He showed that the tissues are not themselves capable of sensation, but that the nerves are the channels or instruments of this process, and that all the nerves are gathered together into the brain or spinal cord. These views he supported by experiments involving lesions or stimulation of the nerves and of different parts of the brain. He ascribed special importance to the outer part or cortex, but the central parts of the brain he regarded as the essential seat of the living principle, the soul. Although his view on the nature of the soul lacks clarity, he separates such conceptions sharply from those which he is able to deduce from actual experience.

(ii) *Foundations of Bionomics*

Light was thrown on the vital activities of plants by the chemist JOSEPH PRIESTLEY (p. 337). In his *Experiments and Observations on Different Kinds of Air* (1774) he demonstrated that plants immersed in water give off the gas which we term 'oxygen'. He observed, too, that this gas is necessary for animal life. His contemporary, Lavoisier (pp. 339 f.), made quantitative examinations of the changes during breathing (1774). These displayed the true nature of animal respiration, and proved that carbon dioxide and water are the normal products of the act of breathing.

In the meantime JAN INGENHOUSZ (1730–99) was introducing the highly important concept of the balance of animal and vegetable life. He was a Dutch engineer who worked in London with Hunter, and in 1779 published his *Experiments upon Vegetables, discovering their great power of purifying the common air in the sunshine and of injuring it in the shade and at night*. It contains a demonstration that the green parts of plants, when exposed to light, fix the free carbon dioxide of the atmosphere. He showed that plants have no such power in darkness, but that they give off, on the contrary, a little carbon dioxide. This most significant discovery is the foundation of our whole conception of the economy of the world of living things. Animal life is ultimately dependent on plant life. Plants build up their substance from the carbon dioxide of the atmosphere together with the proproducts of decomposition of dead animals and plants. Thus a balance is kept between the animal and the plant world. The balance can be observed in the isolated world of an aquarium.

The biological contribution of JOHN HUNTER (1728–93), is peculiarly elusive and difficult to present. His older contemporary Linnaeus and his young contemporary Cuvier were both occupied in classifying organisms. To do this they sought always differences. It was similarities, however, that attracted Hunter. He experimented on and anatomized over 500 species. He set out to trace systematically through all these the different phases of life, as exhibited by their organs, their structure, and their activities. But his main work was his museum. A spirit informs it which is as different as possible from the 'magpie instinct', the motive of many older collections. Here every object has its place and its reason for being included. Hunter created the modern idea of a museum by his conception of a collection to illustrate the varieties of structure and function, right through the organic series.

Hunter was ever seeking the general principles that underlie the dissimilarities in organic forms. The most general of all is that mysterious thing called life. Life is never exhibited by itself, but is seen in the various activities of living things. As a surgeon Hunter naturally stressed, among these, the power of healing and repair. This power is peculiar to living things, and cannot be paralleled in the non-living world. He considered that, whatever life may be, it is something held most tenaciously by the 'less organized' beings. It must therefore be independent of structure and must be somehow an attribute of a substance which all organic forms contain. These ideas lead to the conception of *protoplasm*, the substance, simple in appearance, yet inconceivably complex in ultimate structure and composition, without which life is never found. Hunter did not use the word 'protoplasm', which was invented (in 1846) fifty years after his death. But he was reaching out toward the conception of a common material basis of life (pp. 414 f.).

The orderly observations of vital phenomena by naturalists such as Hunter, Linnaeus, and Cuvier were given an entirely new direction by the chemical workers of the next generation. Respiration had already been made chemically intelligible by Priestley, Lavoisier, and Ingenhousz. Many other processes of the living organism were now chemically interpreted by Liebig and his school.

JUSTUS VON LIEBIG (1802–73), professor of chemistry at Giessen, was an exceedingly stimulating teacher who had an immense following and did much to introduce laboratory teaching. He greatly improved the methods of organic analysis and, notably, he introduced a method for determining the amount of *urea* in a solution. This substance is found in blood and urine of mammals, and was the first organic compound to be prepared from what were then regarded as inorganic materials. It is of very great physiological importance, for it is regularly formed in the animal body in the process of breaking down the nitrogenous substances, known as 'proteins' (p. 415), characteristically found in association with all living substances.

The name of FRIEDRICH WÖHLER (1800–82) is associated with the reconstitution of urea (1828). This substance, the chief final product of nitrogen metabolism in mammals, was discovered in 1773 and investigated by William Prout (1785–1850) in 1815. Wöhler formed urea from ammonium cyanate which then could not be formed from its elements. This, though an important advance, therefore, cannot be called, as it often is, 'the first synthesis of a substance produced in a living organism'. With his colleague, Liebig, Wöhler showed that a complex organic group of atoms—a 'radicle' as it is now called—is capable of forming an unchanging constituent which can be traced through a long series of compounds. A radicle may behave throughout as though it were an element (1832). The discovery is of primary importance for our conception of the chemical changes in the living body.

From 1838 onwards Liebig devoted himself to attempting a chemical elucidation of living processes. In the course of his investigations he did pioneer work along many lines that have since become well recognized. Thus he classified articles of food with reference to the functions that they fulfilled in the animal economy (fats, carbohydrates, proteins), and he taught the modern doctrine, then little recognized, that all animal heat is the result of combustion, and is not 'innate'.

Very important was Liebig's teaching that plants derive the constituents of their substance, their carbon and nitrogen, from the carbon dioxide and ammonia in the atmosphere, and that these are returned by the plants to the atmosphere in the process of putrefaction. This development of the work of Ingenhousz (p. 406) made possible a conception of a 'circulation' in Nature. That which is broken down is constantly built up, to be later broken down again. Thus the cycle of life turns, the motor power being derived ultimately from the heat of the sun.

By far the major part of existing living matter is contained in green plants. These also provide the ultimate source of aliment for the entire animal kingdom. The economic significance of the sources from which the substance of plants is replenished cannot, therefore, be exaggerated. A most important source is carbohydrate, especially in the form of starch, the formation of which is associated with the green matter itself.

We now know that starch is built up in the plant from the carbon dioxide absorbed from the atmosphere; that starch formation is a function of the living plant-cell, intimately connected with the green substance; and that the process is active only in the presence of light. ('Chlorophyll', Greek = 'leaf green', was coined in 1817.) Steps toward the modern view of photosynthesis were made by the French experimenter HENRI DUTROCHET (1776–1847). A key to the working of the living organism is the process by which the gases of the atmosphere come into contact with the tissues. In animals the general character of this is fairly evident, especially in such as breathe actively. Plants, however, were long in giving up their secret. Dutrochet showed (1832) that little openings on the surface of leaves—'stomata' (Greek, plural of *stoma*, 'mouth') as he called them—communicate with spaces in the substance of the leaf, but it was sixty years before the stomata were generally recognized as the normal channel of gaseous interchange. Dutrochet also knew from Ingenhousz that the plant as a whole gave off oxygen and absorbed carbon dioxide, and he showed that only those cells that contain green matter are capable of absorbing the carbon dioxide (1837).

We turn to consider the origin and fate of nitrogenous substances in living things. Davy, Liebig, and others were well aware of the importance of nitrogen in the substance of plants. Liebig showed that nitrogen is taken into the plant by the roots in the form of ammonium compounds and nitrates. He made the general process of nutrition intelligible by a wide generalization of the utmost importance. Rejecting the old idea that plants grow by the absorption of humus, he claimed that carbon dioxide, ammonia, and water contain in themselves all the necessary elements for the production of vegetable matter and that these substances are also the ultimate products of their processes of putrefaction and decay (1840).

JULIUS SACHS (1832–97) of Würzburg was immersed from 1857 onward in problems of plant nutrition. He demonstrated that the green matter of plants, chlorophyll, is not diffused in tissues but contained in certain special bodies—'chloroplasts' as they were later (1883) named. He showed also that sunlight plays the decisive part in determining the activity of chloroplasts in absorption of carbon dioxide. Further, chlorophyll is formed in them only in the light. Moreover, in different kinds of light the process of carbon dioxide assimilation goes on with different degrees of activity. These views and discoveries of Sachs were brought together in his treatise on botanical physiology (1865).

The French mining engineer, JEAN BAPTISTE BOUSSINGAULT (1802–87), applied himself persistently, and, in the end, successfully to the nitrogen problem. During the fifties he succeeded in proving that plants absorb most, at any rate, of their nitrogen not from the atmosphere but from the nitrates of the soil. He showed further that plants can grow in soil devoid of organic or carbon-containing matter, provided that nitrate be present, and that therefore the carbon in plants may be derived only from the carbon dioxide of the atmosphere.

(iii) *Cell Theory*

While chemists were interpreting in their own terms the processes by which living things build up the substance of their bodies, microscopists were investigating the details of those bodies that were invisible to the naked eye. The mystery of the unexplored lay still over the world of microscopic beings with their bizarre forms and entrancing strangeness. By some they were fancifully endowed with complex organs that they do not possess, but the 'minima naturae' were more generally regarded as the 'simplest' and 'most primitive' of beings wherein the secrets of life might most hopefully be sought.

Such inquiries were prosecuted especially with those minute creatures—'animalcula' was the old name for them—that appeared, seemingly 'spontaneously', in infusions of various kinds. The term *Infusoria* soon, however, came to include certain other minute organisms that present superficial resemblances to the animalcula of infusions, notably Rotifers (1764). The limits and definition of the *Infusoria* were long disputed.

As so often, the discussion was barren until directed along lines which corresponded to a concrete and intelligible theory. It came gradually to be realized that all non-microscopic and certain microscopic organisms are aggregates, each unit (cell) of which has some degree of individual life. Not until this position was reached could the Infusoria be properly defined and the term restricted to unicellular forms to the exclusion of cell aggregates (1841). Again, as so often in scientific history, this position was repeatedly approached and even temporarily occupied before it was actually won. Such a pioneer attempt was that of the wildly speculative *Naturphilosoph*, LORENZ OKEN (1779–1851), who in 1805 compared Infusoria to the 'mucous vesicles [cells] of which all larger organisms are composed', spoke of 'the infusorial mass or *Urschleim* [protoplasm] of which larger organisms fashion themselves', and claimed that such organisms are equivalent to 'agglomerations of Infusoria'.

The conception that 'cells' of various forms and functions, but variants of a common plan with a greater or less degree of independent life, form the basis of larger organisms came slowly to be accepted doctrine. The progress occupied the first half of the nineteenth century. The nomenclature of the earlier part of this period is naturally confused. The term 'cellula' dates back to Hooke (1665), who, however, applied it only to the walls of plant-cells. The word 'cell' is frequently used by late eighteenth- and early nineteenth-century writers to describe the microscopic divisions perceptible in most tissues with suitable treatment. The central body of the 'cell' substance, its controller, is the *nucleus*. This term was first applied to the important structure now known by that name in 1823. Robert Brown (p. 395) in 1831 realized that the nucleus was a regular feature of plant-cells and he normalized the use of the word.

The great Czech naturalist JOHANNES EVANGELISTA PURKINJE (1787–1869) in 1835 drew attention to the close analogy of the packed masses of cells in certain parts of animals with those in plants. FELIX DUJARDIN (1801–62) of Toulouse, a most penetrating observer, entered in that year upon a critical examination of microscopic forms. Two conceptions of primary importance emerged from his researches. First, he clearly distinguished unicellular organisms as such, and adequately delimited the 'Infusoria'. Secondly, he discerned that life is always associated with a substance of mucilaginous consistence with very definite optical, chemical, and physical characteristics. Purkinje, who worked on comparable lines, gave to it the name *protoplasm* (1839; Greek = first formed, originally a theological term). It became recognized that the living parts of all cells were composed of it.

The first adequate presentation of the knowledge of the cell as a body of doctrine (1839) was made by THEODOR SCHWANN (1810–82), a pupil of Johannes Müller. He extended the discussion to the ovum or egg which is the beginning of the animal or plant body. In some animals, as the hen, the egg is very large, being distended with food substance—the yolk—and surrounded by a larger and protective substance—the white or albumen. In other eggs, as the frog's, the amount of yolk and albumen is much less. In yet others yolk and albumen are reduced to a minimum, as in the microscopic eggs of mammals then recently discovered (1828) by von Baer (p. 471). Schwann discerned that all these are essentially cells and exhibit the characteristic elements of cells—nucleus, protoplasm, cell membrane, &c.

The development of the egg into the young animal (or plant) proceeds by division of the egg cell. This process of 'segmentation' is particularly evident in the earliest stages of development, and had been casually noted in a variety of organisms by several early naturalists. Schwann treated the process as a normal part of embryonic development. He showed that the continued division of the egg or 'germ-cell' gives rise to the organs and tissues, and he distinguished on a cellular basis five classes of tissues:

(*a*) Tissues in which the cells are independent, isolated, and separate. Such is the blood.

(*b*) Tissues in which the cells are independent but pressed together. Such is the skin.

(*c*) Tissues in which the cells have well-developed walls that have coalesced to a greater or less degree. Such are cartilage, teeth, and bones.

(*d*) Tissues in which the cells are elongated into fibres. Such are tendons, ligaments, and fibrous tissue.

(*e*) Tissues 'generated by the coalescence of the walls and cavities of cells'. Here he included muscles and nerves.

Schwann now passed to a general statement of his belief as to the cellular origin and structure of animals and plants. His conclusion may be expressed in terms still valid. Thus:

(*a*) The entire animal or plant is composed either of cells or of substance thrown off by cells.
(*b*) The cells have a life that is to some extent their own.
(*c*) This individual life of all the cells is subject to that of the organism as a whole.

The synthesis of the ideas of protoplasm, unicellular organisms or 'protozoa', and egg or germ-cell was made by MAX SCHULTZE (1825–74). He devoted himself to a study of tissues—'histology'—in a wide range of animals. In 1861 he gave the definition of a cell as 'a lump of nucleated protoplasm', and in 1863 defined protoplasm as 'the physical basis of life'. He showed that protoplasm presents essential physiological and structural similarities in plants and animals, in lower and higher forms, in all tissues wherever encountered.

Great influence on the biology of the second half of the nineteenth century was exercised by the liberal Berlin professor RUDOLF VIRCHOW (1821–1902). His main contributions are set forth in his *Cellular Pathology* (1858), in which he analyses diseased tissue from the point of view of cell-formation and cell-structure, and enunciates the now familiar idea that the body may be regarded 'as a state in which every cell is a citizen. Disease is a civil war, a conflict of citizens brought about by external forces.' Further: 'Where a cell arises, there a cell must have been before, even as an animal can come from nothing but an animal, a plant from nothing but a plant. Thus in the whole series of living things there rules an eternal law of continuous development, nor can any developed tissue be traced back to anything but a cell.'

Virchow crystallized the matter in his famous aphorism, *Omnis cellula e cellula* ('Every cell from a cell'). Place this beside *Omne vivum ex ovo* ('Every living thing from an egg') of Harvey,[1] and *Omne vivum e vivo* ('Every living thing from a living thing') of Pasteur.

[1] Harvey's actual phrase is *Ex ovo omnia* ('All things from an egg') (p. 286).

These are three of the widest generalizations to which biology has attained. They were all confirmed within about ten years around the middle of the nineteenth century, for though Harvey's was stated much earlier, he had not the evidence on which to base it.

(iv) *Protoplasm*

From when it was recognized that a similar substance—now called 'protoplasm'—underlies all vital phenomena, there has been much interest in its composition. At that stage the subject could not be greatly advanced. Something was learned of what protoplasm took in or extruded and something of the local reactions of protoplasm to ingested or applied substances. But living protoplasm was beyond the chemist's reach for, when treated chemically, it ceased to be living nor was there any effective theory of its activities. Thus for most of the nineteenth century attention was directed rather to the behaviour of living cells, as observed under the microscope, to reactions of their integral parts to fixatives and dyes, or again to cell products and inclusions.

Dead protoplasm consists of a very complex mixture of numerous substances. Of these the bulkiest is water. The others are largely made up of the complex nitrogenous groups known as *proteins* and their derivatives, of the *lipoids* or fats, and of the *carbohydrates* or starchy substances. The general significance of these three types was first made definite by Justus von Liebig about 1840 (p. 407).

Living protoplasm is liquid. Nevertheless, an elementary acquaintance with its behaviour shows that it exhibits a considerable degree of 'viscosity', that is it has some of the properties of a sticky or of a jelly-like substance. Modern views of the intimate structure or composition of living protoplasm have become closely linked with a comparison of its behaviour with that of other substances in the *colloid* ('glue-like') state. The study of the colloid state, one of the many areas by which the old sciences of chemistry and physics have become merged, was initiated by THOMAS GRAHAM (1805–69) in 1850 while Master of the Mint in London. The term was already in use, but he applied it to a particular state of matter. He divided soluble substances in general into the two great classes, *colloids* and *crystalloids*. He observed that certain substances (*a*) pass very slowly into

solution, (*b*) do not crystallize, and (*c*) cannot diffuse, or diffuse very slowly, through organic membranes. Of these substances glue is the type, hence the name *colloid*. In this class are starch (compare starch paste), white of egg, gelatin (the basis of most table jellies). Opposed to these in all three respects are the crystalloids.

Graham was aware that certain substances—silica for instance—could exist as either colloid or crystalloid. He recognized, too, that instability was a characteristic of colloids. Moreover, he perceived that most colloids are of organic origin. He foresaw certain later views of the nature of vital activity in his conception that the surface energy of colloids 'may be looked upon as the probable primary source of the force appearing in the phenomena of vitality'.

The knowledge of the essential nature of colloids was but little extended until the twentieth century. Investigators of our own generation have given a physical interpretation to the differences between the colloid and crystalloid states.

Among the colloids, biologically the most important is the vast class known as proteins. They are necessary to the building up of protoplasm. Dead protoplasm largely consists of them. They are not only essential for growth and repair of living substance, but can be used by the living organism as a source of energy and of heat, though the carbohydrates and fats share this function with them. Chemically the proteins are all built up of very large molecules.

The modern chemistry of the proteins is based on the work of EMIL FISCHER (1852–1919) from 1882 onwards. He demonstrated that they are built up of linkages of numbers of molecules of the substances known as amino-acids. These are characterized by the presence in each molecule of one or more NH_2 ('amino') groups and one or more COOH ('carboxyl') groups. The former gives them basic, the latter acid qualities. According as one or the other predominates, the amino-acid acts as base or acid. They can become immeasurably complex by associating with each other in varying intimate ways. A modern mechanist view of life pictures all vital activity as a continuous change and interchange of the conditions and relations of amino-acids. These, it was held, act through local changes in the degree of viscosity. Many other phenomena of the living cell were interpreted as due to changes in degree of viscosity.

Another aspect of protoplasmic activity is that of 'enzyme' action. The word (Greek 'in yeast') was introduced in 1878 by WILLY KÜHNE (1821–1901) to distinguish a class of organic substance which activates chemical change. Such an enzyme can act on an indefinite amount of material without losing its activating power. The living body produces a large number of enzymes. These are remarkably specific in their action.

Within the protoplasm, though not of it, are numerous materials, the so-called 'food substances', which are often of relatively simple composition. Under this heading are to be included sugars and their derivatives, fats, and the 'reserve' proteins. The problem of the nature of protoplasm thus resolves itself into that of the nature of the matrix in which a vast variety of controlled reactions are taking place, and the ways in which the matrix can influence these reactions. The chemical processes at any moment within a single cell are many and varied. They must somehow be spatially separated from one another within that minute laboratory that we call a cell.

IX

CULMINATION OF THE MECHANICAL VIEW OF THE WORLD

(*c. 1850–c. 1900*)

1. NOTE ON DELIMITATION OF THE PERIOD

THE history of no form of human activity can be separated sharply into periods marked by exact dates. In conventional history the record of reigns of sovereigns, or dynasties, or great constitutional changes, or revolutions, is often chosen, or again the narrative is divided into centuries. These devices serve but poorly for social history, or for the history of art, literature, religion, or science. Nevertheless, it happens that for science the dates 1850–1900 are more definite than most, though it must be remembered that the relative rate of advance of the various sciences varies greatly even within the best chosen period. Like the other divisions in this volume the dates 1850–1900 are chosen merely for convenience in description.

The latter end of our period is well defined—so far as these things can be—by certain changes and emphases which profoundly affected both the physical and the biological sciences. Shortly before 1900 it was demonstrated that there were particles less massive than atoms. The old 'billiards ball' view was becoming untenable. With this went the ether, to which, in the nineteenth century, mutually contradictory properties had been ascribed. The discovery of the electron gave to the atom a nature and action comparable to those of a solar system, while the electron itself came to be regarded as a natural unit of electricity. Thus the classical data of physics and chemistry had to be restated. The quantum theory (1901) made restatement urgent, and acceptance of the general theory of relativity (1916) was soon to change the very conception of the universe. The beginning of the period is less sharply defined than its end. The fundamental doctrine of the nineteenth century, that of energy, was long in gestation but was developed in the decade before 1850. It was first clearly enunciated under that name in 1852.

For the biological sciences 1900 is a convenient ending, because in that year the neglected papers of Mendel on discontinuity in inheritance (1864–6) were rediscovered (pp. 485 f.). From 1900 they redirected the main course of evolutionary inquiry. Soon after, two other main biological branches became redirected. Physiology, previously chiefly concerned with the separate action of the organs, began to treat of their interrelation, especially through hormones. Again, microbiology, which had received an enormous impetus from the doctrine of bacterial and protozoal infection, began now to consider 'ultra-microscopic' or filter-passing 'viruses', the nature of which is still under discussion. For biology the beginning of the period is only tolerably defined by Darwin's *Origin of Species* (1859), for organic evolution had a long prehistory as had almost every other important biological doctrine of which we shall speak.

2. NATURE OF THE GALILEAN-NEWTONIAN REVOLUTION

The history of science strikingly exemplifies the curious interaction of continuity and revolution that characterizes all major historical development. Its record would be comparatively simple if we had merely to set down a steady growth of knowledge, each decade adding something to the established discoveries of previous times, the whole process being one in which progress

> slowly broadens down
> From precedent to precedent.

It would be comparatively simple also if change came by definite steps, a new idea abruptly displacing an old, developing in various directions, until it more or less suddenly yields to a successor which repeats the process in a truer and ampler form.

But what actually occurs is very far from being either of these things. Revolutions occur, but for the most part they are not immediately recognized as such. It seems as though a new method of approach can gain admittance to science or to philosophy only by transforming itself into something like the shape of one already existing. It must be attuned to the times. Its success depends on its maintaining itself, sometimes even in disguise, until its achievements are so great and so incongruous with the form in which they have been expressed that recognition of its true shape becomes, at last, inescapable. It is then apparent that there has been a revolution, but the real revolution was far earlier. What has been going on for perhaps centuries is suddenly understood.

All this is nowhere more evident than in the history of the physical sciences in the twentieth century. The revolution that occurred in the seventeenth century has become at last more fully understood. Here it is not our task to chronicle these final events. It is necessary, however, to indicate something of their character, for it is the privilege of the historian to survey his material from a viewpoint inaccessible to those whose work formed part of it, and it is his function to interpret as well as to describe.

Two aspects of the work of Galileo and Newton call for special mention. In the first place, the immediate object of their study was not the material world as such, but our *experience* of the material world, the phenomena, by which we say that it makes itself known to us. The laws set forth by Galileo and Newton were laws of *motion*, not of moving bodies. Secondly, these motions were taken as data from which general principles were, if possible, to be inferred; they were not to be regarded as necessarily exemplifying previously accepted general principles.

Both these novel characteristics of seventeenth-century physical science were forced into the framework of existing ideas. Thus, the law of gravitation, which prescribes the motion of a mass-point in the neighbourhood of another—i.e. the motion of a point which might, as the centre of gravity of a number of bodies, be in empty space—was automatically interpreted as prescribing the motion of a *planet*, a material body, round the sun, another material body. And because the solar system happened to consist of a very few conspicuous planets in a large otherwise empty space, the discrepancy between the two essentially different conceptions remained hidden. Astronomers had always been concerned with planets, so the new venture, to be understood, had also to be seen as concerned with them.

Similarly, the course of the new science, as a progression from the observation of a few phenomena to a world-outlook, was transmuted into a world-outlook imposed on those phenomena, at least for purposes of ordinary thinking about them. It had been the habit of philosophers to see all detailed things *sub specie aeternitatis.* Hence, no sooner was the conception of inert bodies passively following the dictates of blind forces seen to be applicable to the motion of mass-points, than it was immediately generalized into a world-philosophy. Instead of being accepted as what it was—a generalized statement of certain particularly simple motions—it became a principle of universal determinism or materialism and to that principle phenomena and experiences quite unrelated to such motions were held to be subject.

The progress of the physical sciences from Newton to the end of the nineteenth century was thus anomalous. The revolutionary method was practised in its purity, but interpreted as though there had been no revolution. It is to this cause that we owe the great conflicts between science and religion, and science and humanistic feeling, that reached their culmination in the second half of the nineteenth century. Science was justified by its success, and it was incompatible with religion and with art; therefore religion and art were illusions or fancies. What was not realized was that the success of science was due to the faithfulness of its practice, while its destructiveness arose from the error of its philosophy which saw that practice as though it were the outcome of a world-view with which it was in fact fundamentally incompatible.

An illuminating confirmation of this incompatibility is the fact that, 'the whole of physics can be derived from certain postulates of impotence' (E. T. Whittaker, 1947), i.e. statements of things which it is not possible to do. For instance, the laws of thermodynamics may be expressed in the form that it is impossible to construct a perpetual motion machine or to obtain work from the heat of the coldest body in a system without a more than compensating loss. Again, the laws of electro-magnetism can be founded on the axiom that it is impossible to set up an electric field in an enclosed space by charging the outside of its bounding surface. These great generalizations are now universally acknowledged to be the high-water marks of our period, and soon after the twentieth century had begun the restricted theory of relativity (that it is impossible to detect motion through the ether) summarized the efforts of the previous half-century in this direction. Now these conclusions are not admissions of failure; they are triumphant successes. And when success has to be expressed in the form of failure there is only one deduction to be drawn: it is being expressed in terms of a false outlook. That is the fundamental truth about nineteenth-century classical physics.

A more superficial aspect of the same truth has great importance for our task of recording the history of the period. General impotence is not discoverable by single experiments. You can arrive at it only by repeated efforts, using all conceivable means to achieve the end, until you have exhausted all the possibilities. Only then are you justified in treating your generalization as a postulate.

The second half of the nineteenth century saw an enormous broadening of the field in which science laboured. Where previously one man had made a great discovery, a hundred now failed to make one, but nevertheless each added his mite to the sum-total of experiences on which discoveries were later to be based. It was in 1869 that for the first time a journal (*Nature*) was successfully founded for those engaged on the frontiers of science, and it went on to become one of the most important channels for scientific intercommunication. In such circumstances it is impossible to maintain the practice, entirely appropriate to the story of earlier times, of recording the history of science chiefly on biographical lines. The importance of the period lies not so much in great discoveries as in a multitude of smaller achievements, the perfecting and extended application of instruments, and the exemplification of general ideas in particular cases.

No attempt will be made, therefore, to follow each step in the onward march. If the greater names are to be mentioned, the smaller must perforce be omitted, notwithstanding the indispensability of their contributions to the process as a whole. It is still possible, however, to treat the greater divisions of the subject—physics, astronomy, chemistry, biology—separately, for though they are now so often merged as to make their frontiers at least difficult to distinguish, it is also true that they were still largely autonomous throughout the nineteenth century. Even the branches of physics itself—sound, light, heat, magnetism and electricity, and the traditional mechanics—

could still be profitably studied in partial isolation. There is little to record in mechanics proper, since the eighteenth century had well nigh exhausted the possibilities of the Newtonian scheme, and the great aim of the nineteenth-century physicists was to establish the other departments of physics on a mechanical basis.

3. PHYSICS

(i) *Sound*

In the study of sound, this effort was entirely successful, and it suffices to mention the great work of JOHN WILLIAM STRUTT, 3rd Baron RAYLEIGH (1842–1919), *The Theory of Sound*, which appeared in 1877–8. In it, the phenomena of sound, in practical completeness, are interpreted in terms of longitudinal mechanical waves in matter, proceeding from the source of sound to the receiver with a velocity equal to the square root of the ratio of the elasticity to the density of the transmitting medium. The greater the frequency of vibration (i.e. the smaller the length of the wave), the higher is the pitch of the note that is experienced. On the physiological side, knowledge of the subject was comprehended in the work of Helmholtz (1821–94), *The Sensations of Tone* (1862).

(ii) *Heat*

Almost as complete success was achieved in the explanation of heat in mechanical terms. It has already been recorded (p. 352) that, through the development of the doctrine of energy, the eighteenth-century view of heat as a material fluid had gradually yielded to the idea that heat consisted of the motion of small parts of bodies. But the earlier views of this kind were necessarily rather imprecise because there was no very definite idea of the discontinuous structure of matter. This atomic notion had indeed been introduced into chemistry early in the nineteenth century by Dalton (p. 343), but it took many years to establish itself, and chemical phenomena were then regarded as so distinct from those of physics that ideas originating in chemistry stood little chance of consideration in the treatment of physical problems, though the material bodies involved in the two sciences were precisely the same.

The earliest successes in the mechanical interpretation of heat—or *thermodynamics*, as it has come to be called—were accordingly obtained not by applying the laws of motion to the particles of bodies, but by measuring the amount of work obtainable from a given amount of heat or vice versa, irrespective of what the nature of heat might be. The two outstanding aspects of this investigation, resulting in the establishment of the equivalence in amount of the heat and work concerned in the transformation (the first law of thermodynamics) and the asymmetry between the one transformation and its opposite, by which energy tended to be degraded into heat unavailable for transformation (the second law of thermodynamics), have already been described (pp. 375–8).

The great development of the succeeding years in this field was the expression of these laws, and of the detailed phenomena concerned in them, in terms of the motions of *molecules*, of which all bodies (especially gases) were conceived to be composed. The term 'molecule' in this connexion is not to be regarded as denoting something identical with the chemical molecule conceived by Avogadro (p. 345). The whole significance of Avogadro's conception lay in the relation between the molecule and the atoms of which it was composed, but the molecule of heat theory was essentially indivisible, an ultimate, tiny, spherical, elastic mass descended from the atoms of Democritus and the hard, massy particles of Newton rather than the composite systems of the chemists.

The earliest considerable development of heat theory in terms of the molecular constitution of matter was premature. It was made by JOHN JAMES WATERSTON (1811–83). His work was submitted to the Royal Society in 1845 but rejected as its importance was not realized. It was rediscovered in 1892 by Lord Rayleigh and published. Much the same ground had therefore to be covered and extended independently, and the most prominent figures in the process were JAMES CLERK MAXWELL (1831–79), LUDWIG BOLTZMANN (1844–1906), and JOSIAH WILLARD GIBBS (1839–1903). The fundamental conception was that any particular gas[1] was composed of an enormous number of absolutely identical particles in violent motion. Their sizes were

[1] And solids and liquids as well, but since the properties of gases were much simpler and better known, the theory was applied at first almost exclusively to them, and became known as the *kinetic theory of gases*.

negligible compared with the volume of the spaces between them, and when they collided, as they were constantly doing, they rebounded from one another without loss of energy. When the gas was heated the molecules moved faster; when it was cooled they moved more slowly.

In these terms all the thermal properties of a gas could be represented. Its heat was proportional to the kinetic energy of the molecules; its temperature to the average kinetic energy of each molecule. At any moment, of course, the velocities (and therefore the kinetic energies) of the various molecules covered a considerable range. The motion, in magnitude and direction, of each molecule was continually changing by collisions with other molecules and with the sides of the containing vessel, yet the numbers were so large that the average kinetic energy remained uniform unless heat was gained from the outside (by impact of more energetic molecules on the containing vessel or by the receipt of radiant energy) or lost to the surroundings (by transfer of energy to slower-moving external molecules through the medium of the molecules in the containing vessel or by radiation). The momentum of impact of the molecules on the walls of the containing vessel constituted the pressure exerted by the gas on its surroundings. The ordinary gas laws—those of Boyle and Charles—as well as some curious and unexpected relations afterwards verified by observation (such as the fact that the viscosity of a gas is independent of its pressure) were readily deduced from the kinetic theory. No one doubted that the picture presented was a *reality*, i.e. that if we could magnify the gas enormously we should see the molecules behaving exactly as they were conceived to do. Nevertheless, what the molecules themselves were like was completely unknown. Thus, despite the impression of reality, all the phenomena were, in fact, explained in terms of their *relations* with one another, not with their essential nature.

An important feature of this conception was the essentially statistical character of the explanations. For instance, if a vessel contained two gases, one hot and one cold, with a partition between them, and the partition were removed, the motions of the molecules would at once cause them to mix. The collisions would result in the faster molecules hastening the slower, and the slower retarding the faster, until eventually the average energy of the molecules would be uniform throughout the vessel.

This corresponded to the observed fact that the temperature would become the same at all parts of the vessel. But, as Maxwell pointed out (1871), there is no compulsory law that this must happen. If a minute, invisible being (which has become known as 'Maxwell's demon') were inside the vessel and by moving a frictionless valve allowed only the faster molecules to pass it in one direction and only the slower in the other, we should have one side of the vessel again containing hot gas and the other cold gas. Now it *might* happen, in the endless chances of endless time, that the actual motions of the molecules would bring about this situation without the demon, though our experience is that this does not happen. Hence, the kinetic theory is essentially bound up with the conception that the apparently inevitable laws of nature are actually no more than laws of probability. Things happen as they do since the probability that they will happen otherwise is extremely small. (The philosophical implications of this conclusion are outside our consideration.)

The kinetic theory of gases agrees with the known laws only so long as it is assumed that the sizes of the molecules may be neglected in comparison with their 'mean free path', i.e. with the average distance which a molecule travels between successive collisions. If, now, we compress a gas, we reduce the mean free path without altering the molecules themselves, so a point will be reached at which this assumption breaks down. This agrees with the observed fact that Boyle's law (p. 272) ceases to hold good at high pressures. We must then suppose that we are approaching the liquid or solid state in which the molecules are packed close together. The mathematical difficulties of calculating what then would happen are too great for an exact solution, particularly as the molecules might exert forces on one another which become important when they are close together. Yet an experimental investigation was within the resources available, and the most effective work of this kind was by THOMAS ANDREWS (1813–85), who during the 1860's carried out some beautiful experiments on the relation between the volume and pressure of carbon dioxide at various temperatures (Fig. 171).

Consider first the curve (that farthest to the right) for the temperature 48·1° C. At this temperature the conditions of the kinetic theory are satisfied, and the curve is close to that required by Boyle's law, namely, a rectangular hyperbola, the volume varying inversely as the pressure. Now consider the curve (that farthest to the left) for the temperature 13·1° C. At very low pressures the conditions of the kinetic theory are again satisfied, and as the pressure is gradually in-

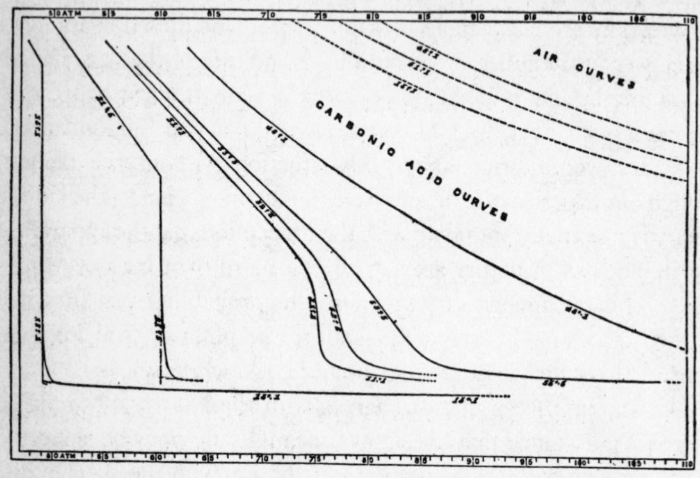

FIG. 171. Andrews's curves (1869) showing how the volume of a given mass of air and carbon dioxide changes with pressure at various constant temperatures. The pressure, in atmospheres, is shown horizontally and the volume vertically.

creased the volume at first decreases as Boyle's law demands. But at the 'knee' a sudden change occurs. Attempts to increase the pressure are frustrated by the gas changing into liquid. This continues, as the vertical line shows, until all the gas is liquefied. The pressure can then rise again, but, the liquid being almost incompressible, the curve follows a line of almost constant volume.

The intermediate curves show how the transition occurs between these extremes. The vertical part of the curve diminishes with increasing temperature until it becomes a mere kink in the hyperbolic curve, and the point at which it just disappears marks what is known as the *critical temperature*. It has a specific value for each gas. If a gas is below its critical temperature, then it is possible to liquefy it by increasing the pressure; but if not, increase of pressure fails to reduce it to a liquid since the effect of the closer proximity of the molecules is overcome by their greater speed of movement.

The discovery of a critical temperature was of the greatest importance in the practical problem of liquefying gases. It showed that before this could be done the gas had to be cooled below its critical temperature, and the great difficulty of liquefying the so-called 'permanent' gases, such as hydrogen, nitrogen, oxygen, lay in the fact that their critical temperatures were very low. Faraday in 1823 had liquefied certain gases, including chlorine, by pressure alone, but it was not completely clear before the work of Andrews why the same methods failed for other gases, notwithstanding that as early as 1822, CHARLES CAGNIARD DE LA TOUR (1777–1859) had partly anticipated Andrews's work.

The first success with the permanent gases came in 1877, when LOUIS PAUL CAILLETET (1832–1913) and RAOUL PIERRE PICTET (1846–1929) independently succeeded in liquefying oxygen. Progress in the nineteenth century culminated in the liquefaction of hydrogen (whose critical temperature is −238° C.) by JAMES DEWAR (1842–1923) in 1898. He also invented the now well-known vacuum flask in which the liquefied gases could be preserved for considerable periods without evaporation.

It should be added that the experiments, interpreted in terms of the kinetic theory, indicated that there are indeed forces between the molecules which, though small, are an important factor in some methods of liquefying gases. In the solid and liquid states, in which the molecules must be supposed to be nearly in contact, these forces are sufficient to hold them together. Attempts to represent the experimentally established continuity between the liquid and gaseous states, which Andrews's experiments revealed, have produced only approximate formulae, of which that of JOHANNES DIDERIK VAN DER WAALS (1837–1923) is the best known.

(iii) *Optics*

In all these investigations, though mathematical difficulties arose, there was nothing to cast doubt on the belief that the phenomena could be interpreted completely in mechanical terms. In the fields of optics and electricity, however, things were otherwise. No one appears to have doubted that such an interpretation was possible—indeed necessary. Thus, Lord Kelvin, in a well-known passage, stigmatized all physical knowledge which could not be embodied in a mechanical model as 'meagre and unsatisfactory'. But the phenomena obstinately refused to admit of any consistent expression in the language of mechanics.

The establishment of the undulatory theory of light (p. 374) demanded an all-pervading ether through which the waves could travel with the well-determined velocity now universally represented by the symbol *c*. The mechanism of wave propagation in a material medium was known from the theory of sound, and although sound waves were longitudinal while light waves had to be regarded as transverse (p. 372), this merely implied that the values of the relevant mechanical constants were different. But the values required were most mystifying. The absence of longitudinal waves, for instance, meant that the compressibility of the ether must be zero, while the great velocity of light required that its elasticity should be vastly greater than its density. Moreover, it would be expected that bodies, such as planets, moving at high speeds through the ether, would experience some resistance and so be slowed down, but no such effect was observable.

Again, phenomena which appeared to involve action at a distance, such as gravitational attractions and electrical and magnetic attractions and repulsions, must, if everything had a mechanical explanation, be regarded as actions through a medium connecting the bodies concerned. The principle of economy of hypotheses indicated that this could be no other than the luminiferous ether. The ether must accordingly be strong enough to bear the enormous stresses which gravitational phenomena imply. Thus, for example, to sustain the pull of the sun on the earth it could be calculated that the equivalent of a billion steel pillars, each 30 feet in diameter, would be required, and this and still greater stresses the ether must be supposed able to bear. To combine such a variety of apparently incompatible properties in a single mechanical medium was indeed a formidable task.

The earlier models of the ether were in purely mechanical terms; the medium was assumed to be a kind of elastic solid, albeit a remarkably aberrant one, and its properties were assigned so as to yield the phenomena as closely as possible. After the rise of Maxwell's electromagnetic theory, however (p. 434), electrical concepts became available, which provided still greater scope for ingenuity and correspondingly greater complexity in the models devised. But the ultimate aim of a mechanical interpretation of nature remained unchanged. Electricity and magnetism also were believed to be mechanical at bottom—though their nature had so far eluded detection—so there was no loss of faith in the true scientific ideal, as it was conceived, in this development.

It was clear that the ether—though not material in the ordinary sense and though material bodies passed through it without resistance—could yet be modified by matter, since different bodies had different refractive indices (i.e. the velocity of light through them had various values (p. 374)) and some were completely opaque. The models proposed are conveniently distinguished from one another by the views which they implied of the origin of this modification. They fell into two broad classes—those which located it in the inertia, or density, of the ether, and those which favoured the elastic properties. A third device was to suppose that both the inertia and the elasticity of the ether itself were unaffected by matter, but that the interaction between ether and matter produced the phenomena observed.

Looking back from an entirely different point of view on the variety of proposals, it is hard to decide which is the more remarkable, the ingenuity displayed by the investigators or the complete absence of suspicion that they were attempting an impossible task. Both the greatness and the limitations of Lord Kelvin are displayed in full measure in his lifelong efforts to solve this problem, culminating in a model 'composed of spheres, bars, caps and flywheels' to quote Whittaker's description, which, even if it had fully satisfied the mathematical requirements, would have strained to the utmost the credulity of a simple follower of Newton, who wrote that 'Nature is pleased with simplicity'. But, as is usual in such cases, it was the letter rather than the spirit of Newton that was followed, and apart from an occasional dissentient voice, such as that of Kirchhoff, HEINRICH RUDOLF HERTZ (1857–94), or ERNST MACH (1838–1916), the unanimity of the belief that the ether was completely describable in ordinary mechanical terms was unbroken. Newton had rejected the suggestion that atoms cohered because they were hooked as a begging of the question, but no one hesitated in the name of Newton to invoke without limit caps and flywheels in the ether.

One problem, however, was independent of the actual constitution of the ether, and seemed to present no theoretical difficulty but to be merely a matter of experimental ingenuity. This was the question of the velocity of the earth—or, for that matter, of any other body—through the ether. Before Copernicus, everyone supposed the earth to be at rest, and the idea of 'rest' was not analysed. Copernicus transferred the state of rest from earth to sun, and when Newton introduced his idea of immovable absolute space, and distinguished it from the relative space 'which our senses determine by its position to bodies', it was natural to regard a state of rest as being that of immobility in absolute space. When, further, that absolute space was filled with a luminiferous ether, rest became definable still more specifically as immobility with respect to the ether. The Copernican idea would now have been expressible as the doctrine that the sun was at rest in the ether, but since Halley had discovered that the stars had proper motions, and W. Herschel had ascribed such a motion to the particular star which is our sun (p. 308), this simple view could not be accepted without question. It became a matter for experiment, and experiment only, to determine how some one body moved in the ether, and then the motion of any other body could be determined from its motion relatively to that.

For practical reasons the earth was the only body on which the appropriate measurements were possible. Many experiments were devised and performed to this end. The most notable was that carried out by ALBERT ABRAHAM MICHELSON (1852–1931) and EDWARD WILLIAMS MORLEY (1838–1923) in 1887 following a preliminary performance by Michelson alone in 1881, and repeated afterwards by various observers. In these experiments a beam of light was divided into two parts which travelled to and fro along two equal paths at right angles to one another and then reunited. It was easy to calculate that if the earth were travelling through the ether in the direction of one of the beams, the motion of the apparatus during the passage of the light would make the effective paths of the beams unequal, and this would be deducible from the interference pattern which they would form on reuniting. To everyone's astonishment, this experiment, like all others directed to the same end, showed that the earth was at rest in the ether.

If this experiment could have been performed in the sixteenth century, this would of course have been interpreted simply as a proof that Copernicus was wrong and Ptolemy right, but by 1887 such an interpretation had become incredible on other grounds. There was, however, no acceptable alternative until in 1892 GEORGE FRANCIS FITZGERALD (1851–1901) and shortly afterwards HENDRIK ANTOON LORENTZ (1853–1928) independently pointed out that, on the electrical theory of matter, then becoming fashionable, a material body moving in the ether should be contracted in the direction of its motion by just the amount necessary to cancel out the effect of the difference in the paths of the two beams of light. This contraction, which has become known as the *FitzGerald contraction*, also explained the failure of other experiments to show the motion of the earth, and it was readily accepted.

It was left for ALBERT EINSTEIN (1879–1955) in 1905 to give an explanation. Its nature is beyond our scope, but it has since been universally accepted under the name of *the restricted principle of relativity*. It implies that the ether, if we suppose it to exist, cannot possess the property of having identifiable parts, so that it is meaningless to ask whether the earth is here or there in it, and so meaningless to ask whether the earth is moving in it or not. The theory of relativity has the same mathematical requirements as the FitzGerald contraction, but the ideas underlying it are profoundly different.

This last fact is one of many examples of a most striking characteristic of the history of science, namely, that—to use a rough but graphic expression—the right results are often accepted for the wrong reasons. The assumption of an ether conveying waves with a definite velocity and possessing many of the ordinary properties of material bodies yields results that are often verified by observation although the picture thus yielded must be essentially false. Notable among these results was the theoretical demonstration in 1884 by Ludwig Boltzmann (1844–1906), confirming and making more precise an earlier experimental result by JOSEF STEFAN (1835–93), that the full ethereal radiation from a body is proportional to the fourth power of its absolute temperature. This law is of profound importance in the theory of temperature radiation, which shares with gravitation the distinction of being an absolutely universal phenomenon. Two bodies anywhere in space will affect one another's motions and temperatures, and the Stefan–Boltzmann law prescribes the course of the latter effect. The exact experimental confirmation of this law came in 1897 from the work of OTTO R. LUMMER (1860–1925) and ERNST PRINGSHEIM (1859–1917).

The ether may be regarded as the supreme catalytic agent in nineteenth-century physics. It directed the course of development in both optics and electromagnetic theory along the right lines, only to disappear in the twentieth century as a figment of the imagination or, as some prefer to express it, to remain as a name for something essentially different from the medium which the Victorian physicists contemplated.

(iv) *Electromagnetism*

We turn to developments in electricity and magnetism, initiated by the work of Faraday. Superficially, electric and magnetic actions are similar to those of gravitation. Two electric charges, for instance, exert a force on one another which is directly proportional to the charges, inversely proportional to the square of the distance between them, and acts along the line joining the charges. The same thing is true of magnetic poles. But when we come to the interaction between electricity and magnetism the analogy breaks down. To take but one of the puzzling phenomena which Ampère and Faraday brought to light, a moving electric charge (i.e. an electric current) exerts a force on a magnetic pole in its neighbourhood which is inversely proportional to the distance, not to its square, and is perpendicular both to the direction of the current and to the shortest line from the pole to that direction (pp. 359 f.).

Phenomena such as these called for a new attitude to electromagnetism, and this was provided by Faraday in his conception of lines of force (p. 364). As we have said, this was not a new conception, but to a generation intent on explaining the whole of nature in terms of Newtonian forces acting at a distance it was unwelcome and therefore not appreciated until Maxwell expressed it in mathematical form and showed its potentialities.

Now that the mathematical interpretation of Faraday's conceptions regarding the nature of electric and magnetic forces has been given by Clerk Maxwell [wrote von Helmholtz in 1881] we see how great a degree of exactness and precision was really hidden behind the words, which to Faraday's contemporaries appeared either vague or obscure. . . . I have no intention of blaming his contemporaries, for I confess that many times I have myself sat hopelessly looking upon some paragraph of Faraday's descriptions of lines of force, or of the galvanic current being an axis of power.

The essential point of Faraday's innovation was that he transferred attention from the charges or current or poles to the spaces between them, i.e. to the *field*, as it is now universally called. The choice was expressed by Maxwell: 'We may conceive the physical relation between electrified bodies, either as the result of the state of the intervening medium or as the result of a direct action between electrified bodies at a distance.' Faraday had chosen the former; Maxwell undertook the investigation of the nature of a medium in 'a state of mechanical stress', as he expressed it, which would produce the phenomena observed. There was to be no departure from the Newtonian ideal of interpreting all physical phenomena in terms of mechanical forces. What was new was the concentration on the mechanical properties of the continuum between the charges rather than on the charges themselves. Field physics, in contrast to particle physics, began its sway.

Up to a point all went well. The appropriate stresses and strains could be formulated so as to account for a number of observations. It became clear that the medium concerned, like the luminiferous ether, was modified by the presence of matter, since, for example, the capacity of a condenser is changed by changing the dielectric separating the plates. But the crucial point of Maxwell's theory lay in a conception which he introduced, not from observation but apparently from one of those inexplicable intuitions of genius which, though without foundation in fact, lead to the right result. Electricity flows freely along a conductor, but apparently does not flow at all along a non-conductor, or dielectric. Maxwell conceived that it *begins* to flow through a dielectric in which a difference of potential is set up, but cannot continue, and brings the dielectric (or the ether within the dielectric) into a state of strain. This incipient current he called a *displacement current*. By treating it as an ordinary current, with the appropriate magnetic effects, he was able to arrive at a set of differential equations ('Maxwell's equations') representing the whole science of electromagnetism. These, in their developed form, were published in 1865 in his *Dynamical Theory of the Electromagnetic Field*. The history of the subject since Maxwell is that of a progressive verification of the equations and working out of their implications, together with an abandonment of the mechanical ideal and the displacement current which gave rise to them.

Maxwell's theory was received with a somewhat sceptical admiration. Kelvin, whose greatest ambition was to achieve the mechanical explanation of electricity, admitted its ingenuity but found it unconvincing. But it achieved immediately one most remarkable result. It implied that the discharge of a condenser, for example, was not an instantaneous but an oscillatory phenomenon, in which waves of electric and magnetic force were emitted into the surrounding space. The surprising fact emerged that the velocity of these waves, calculated entirely from known electric and magnetic quantities, was equal to the measured value of the velocity of light. This suggested two things: first, that the medium in which electric and magnetic forces were seated was identical with the luminiferous ether, and second, that light was an electromagnetic phenomenon. Instead of being a train of actual movements of particles of a sort of elastic solid, light waves had to be pictured as rapid alternations of electric force in one direction and magnetic force in another, these directions being perpendicular to one another and to the direction of propagation of the wave.

The great nineteenth-century problem was thus at the same time made simpler and more difficult. It was simpler in that the reduction of light and of electromagnetism to mechanics was now reduced to a single problem: it was more difficult in that the solution of that problem was harder than ever. But whatever might be thought of its ultimate truth, Maxwell's theory profoundly influenced physics in two directions. Firstly, it stimulated the interpretation of optical phenomena, such as dispersion, reflection and refraction, &c., in the new terms, and led among other things to an important theorem by JOHN HENRY POYNTING (1852–1914) which may be described very broadly as a harmonization of the theory with the principle of conservation of energy. Secondly, it led to experimental investigations which culminated in the production by HEINRICH RUDOLF HERTZ (1857–94) in 1886–8 of waves from an electric discharge, travelling with the velocity of light and exhibiting reflection, refraction, and other phenomena of light waves, all in accordance with Maxwell's theory, of which Hertz gave an interpretation of his own.

Hertz was able to calculate the length of the waves, and found that they were considerably longer than those of light. These Hertzian waves were of the kind later used in wireless telegraphy and telephony and are now revolutionizing astronomy by their detection in the radiation from the most distant spaces.

A further consequence of Maxwell's theory—that light should exert a pressure on a surface on which it falls—was confirmed experimentally by PETR NIKOLAIEVICH LEBEDEW (1866–1912) in 1899.

The review of nineteenth-century physics thus far given shows two opposing tendencies. In the realm of thermal phenomena we have seen the introduction of the idea of particles achieving successes which threw into the background the older ideas of the continuous flow of heat and simple 'energetics' based on whole undivided quantities of heat and work. On the other hand, in electricity and optics the study of isolated electric particles and magnetic poles and the conception of light corpuscles yielded to the field conception which seemed to relegate them to the limbo of finally discarded illusions. The question thus posed itself to the more philosophically minded: 'Was Nature actually continuous or discontinuous?' The kinetic theory of gases and the atomic theory in chemistry spoke undeniably for discontinuity; the wave theory of light and electromagnetic field theory spoke equally undeniably for continuity.

This equivoque was perplexing in the extreme, but there was no actual contradiction. It was not logically impossible, though it offended the aesthetic sense, that Nature should be dualistic, and one could therefore treat matter as essentially corpuscular and electricity as essentially diffused throughout space, with a blameless if not an easy conscience. But even this qualified satisfaction was not to remain for long. In another field of electrical investigation—that of the discharge of electricity through gases—observation spoke more and more loudly in favour of electrical corpuscles, and confusion became worse confounded.

The phenomena of electrolysis—the discharge of electricity through liquids (p. 357)—had related electrical to chemical action in such a way that an atomic theory of matter seemed naturally to imply an atomic theory of electricity also. However, so long as it stood by itself this anomaly might reasonably be expected to clear itself up in the course of time. But the discharge of electricity through gases presented a problem which could not be so easily set aside. At atmospheric pressure gases do not conduct electricity except under very high tension, when a momentary very intense discharge takes place, as in the lightning flash. But when the pressure is reduced to values well within the range of the air pumps available early in the nineteenth century, continuous conduction occurs at fairly low voltages, though not in accordance with Ohm's law (p. 358). These discharges produced beautiful glows in the so-called 'vacuum tubes' used and exhibited colours varying with the gases concerned.

Such electric glows in vacuum tubes were for long a source of entertainment without understanding. Faraday made some observations that convinced him of the fundamental importance of the subject, but the first investigator to obtain results confirming this was JULIUS PLÜCKER (1801–68), through whose work and that of his pupil, JOHANN WILHELM HITTORF (1824–1914), and of EUGEN GOLDSTEIN (1850–1930), it became clear that the discharge was accompanied by the emission of rays of some kind from the negative pole, or *cathode*. These rays, which became known as *cathode rays*, cast shadows on the inside of the tube if an obstacle were placed before the cathode. They were found to proceed in straight lines normally from the cathode and to produce fluorescence on the walls of the tube.

The nature of these rays was for long a subject of much speculation. WILLIAM CROOKES (1832–1919) regarded them as a 'fourth state of matter'—i.e. something different from solid, liquid, or gas—and opinions on the matter eventually resolved themselves into two distinct classes: the rays were believed by most continental physicists to be ethereal radiations of some kind, while the English physicists preferred to regard them as particles.

The researches of many investigators culminated in the establishment by JOSEPH JOHN THOMSON (1856–1940) in 1897 of the fact that the rays must be regarded as particles. He called them *corpuscles*, but the name was soon changed to *electrons*, following a suggestion of GEORGE JOHNSTONE STONEY (1826–1911) for the unit of electric charge which seemed to be revealed by the facts of electrolysis. Thomson was able to determine both the velocity and the ratio of charge to mass of the particles. Their mass was about $\frac{1}{2000}$ of the smallest mass previously known, namely, that of the hydrogen atom. Their charge, which was of the kind that had arbitrarily been called negative, was equal to that of the hydrogen ion in electrolysis. The mass and charge of the electron were found to be the same whatever gas was used in the discharge tube, so that similar electrons were apparently constituents of all matter. Moreover, from an earlier investigation of Thomson's, in which he had shown that a body on being charged with electricity acquired a calculable increment of mass, or inertia, it appeared that the *whole* of the mass of one of these particles arose from its charge, so that they could be regarded actually as particles of electricity.

The new horizons opened up by this work together with the almost simultaneous discovery of X-rays and radioactivity, can only be mentioned here; their implications belong to the physics of the twentieth century. The immediate effect, however, was twofold. On the one hand, all ideas concerning the nature of electricity were thrown into complete confusion. Two sets of apparently irrefutable deductions from experiments led to opposite conclusions, namely, that (*a*) electricity was something diffused throughout space, and (*b*) electricity was at the same time something concentrated in tiny corpuscles. This anomaly was to be accentuated later and has reached a partial resolution only with the recent advent of wave mechanics. On the other hand, the suggestion was irresistible that matter was at bottom composed of electricity. A constituent of all atoms owed its mass entirely to its electric charge, and although the remainder of the atom—the positively charged part—was not so readily isolated (it was later found also to be corpuscular), it was hard to doubt that mass in general was an electrical phenomenon.

We may fittingly close this account of nineteenth-century physics with an oft-quoted remark made by Lord Kelvin in 1900. 'The beauty and clearness', he said, 'of the dynamical theory which asserts heat and light to be modes of motion, is at present obscured by two clouds.' One of these was concerned with the constitution of the ether—problems of continuity—the other with particle physics—problems of discontinuity. Kelvin believed that these clouds would eventually be dissipated and the dynamical theory vindicated. Subsequent progress has made this as nearly impossible as anything in science can be said to be. It is the dynamical theory (in the sense in which Kelvin understood the phrase, which is inadequately expressed by the mere assertion that heat and light are 'modes of motion') that has been dissipated.

(v) *Spectroscopy*

We have purposely omitted from this description of physical science one important subject, namely, that which, originally called *spectrum analysis*, is now universally known as *spectroscopy*. The reason is that, although essentially a physical investigation and a part of the study of light,[1] its applications have been predominantly in astronomy and chemistry. It thus fittingly serves as an introduction to the history of those branches of science.

The two essential preliminary steps towards the foundation of a science of spectroscopy were Newton's discovery of the spectrum in 1666 and Kirchhoff and Bunsen's discovery in 1859 that, in the gaseous state, each chemical substance emits its own characteristic spectrum (pp. 315 f.). Newton, allowing sunlight which had entered a shuttered room through a small hole, to fall on a prism, found that it emerged on the opposite side as a band of colour, changing from red at one end, through orange, yellow, green, blue, and indigo—to choose his own divisions of a continuously varying strip of colour—to violet at the other. He made experiments which convinced him that the original light consisted of a mixture of these colours which the prism

[1] It was through the spectrum that, in the twentieth century, the quantum theory became established.

served to separate (see Fig. 133). After his time it was gradually discovered that such a band of colour was emitted by any sufficiently

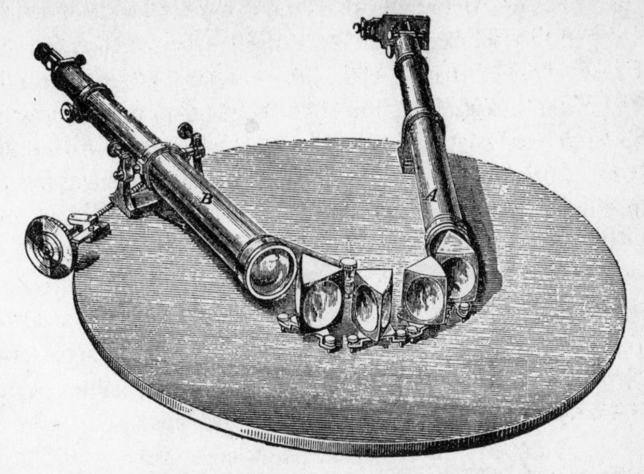

FIG. 172. Kirchhoff's spectroscope of 1858.

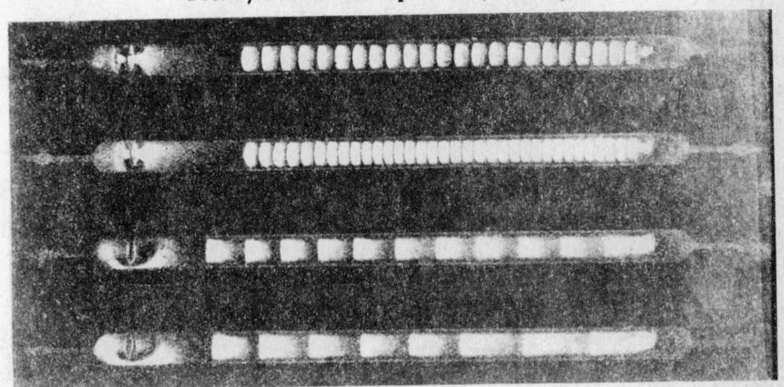

FIG. 173. Successive appearances of an electric discharge through hydrogen as the pressure is gradually reduced. The pressure in the top discharge is 0·11 mm. of mercury, and that in the bottom 0·037 mm.

hot substance so long as it remained solid or liquid. A vapour, however, gave a number of isolated colours, separated by dark spaces and Kirchhoff and Bunsen's great contribution lay in the establishment of the fact that the pattern of these colours depended in a perfectly characteristic way on the gas concerned.

The importance of this for chemical analysis and for the study of celestial sources of light is obvious. We consider first, however, the problems raised by spectroscopy in itself, as an optical phenomenon. It was soon evident that the 'visible' spectrum, as it is called, was only part of a far wider range of radiations, a part to which the human eye happened to be sensitive. William Herschel in 1800 found, by their heating effect, that rays (the *infra-red* rays) existed beyond the red end of the spectrum, and in 1852 GEORGE GABRIEL STOKES (1819–1903) discovered that radiations beyond the violet end (the *ultra-violet* rays) produced fluorescence in certain substances. The later application of photography enabled these additional rays to be studied with the same ease as those that are visible, and the development of the wave theory of light permitted their wave-lengths to be measured. The unit for this purpose, introduced by ANDERS JÖNS ÅNGSTRÖM (1814–74), was the ten-millionth of a millimetre, at first called the *tenth-metre* and now the *angstrom*. The visible spectrum was found to extend from about 7,600 angstroms (7,600 Å.) in the red to about 3,900 Å. in the violet. Knowledge of the extensions beyond these limits gradually increased as time went on.

Two problems presented themselves—first, the mechanism of origin of the radiations; second, the relations between the wave-lengths occurring in a given spectrum. On the first point, it was naturally assumed that each radiation (or *spectrum line*, as it came to be called, because the spectroscope, the instrument used for its production, included a narrow slit through which the light passed) arose from a mechanical vibration in the atom. This, however, encountered the difficulty that some spectra contained many hundreds of lines, and it was difficult to conceive of such a simple thing as an atom was thought to be, executing such a variety of vibrations. This difficulty was accentuated when the second problem was studied, for instead of the wave-lengths of the radiations being related according to the harmonic ratios familiar in the similar problem in acoustics, they showed at first no relation at all, and when in 1885 one was discovered by JOHANN JAKOB BALMER (1825–98) for the simplest of all spectra—that of hydrogen—it was definitely at variance with that indicated by the acoustical analogy. Moreover, largely through the work of JOHANNES ROBERT RYDBERG (1854–1919), many other spectra could be analysed into *series* of lines whose wave-lengths or frequencies could be represented fairly closely by a formula, but a formula of a type which was quite inexplicable in terms of ordinary mechanical vibrations.

A further complication was introduced by the discovery by JOSEPH NORMAN LOCKYER (1836–1920) that the spectrum of a single substance could be changed if the energy used in producing it were increased. Since the atom was believed to be unchangeable this was inexplicable, and for long it was regarded with inexcusable scepticism. Lockyer himself ascribed it to a dissociation of atoms at high temperatures into simpler constituents. As the electron had not then been isolated, this was summarily rejected by chemists and physicists alike.

The solution of this, as of so many nineteenth-century problems, awaited the abandonment of Newtonian mechanics as the fundamental truth concerning natural processes.

4. ASTRONOMY

We turn to the application of the spectroscope to astronomy. Here it virtually brought about a rebirth, for to a large extent it annihilates distance. Except that the source of light cannot be experimented with, this instrument is as effective with a star or a distant nebula as with a vacuum tube in the laboratory, provided that the body sends us enough light for the spectrum to be observed. It was therefore inevitable that spectroscopy should dominate astronomy as soon as the indications of the spectrum became intelligible.

But here at first lay a great difficulty. In the laboratory a glowing solid or liquid gives a continuous band of colour (*a continuous spectrum*, as it is called) which is useless for analysis, but a glowing gas gives a set of isolated bright lines, separated by dark spaces, which are characteristic of the chemical nature of their origin. The vast majority of stars, however, including the sun, give neither of these things, but a continuous spectrum crossed by a large number of dark lines—the Fraunhofer lines (pp. 314 f.). Many of these were found to coincide in position with bright lines obtained from known elements in the laboratory, so there was little doubt that they arose from those elements in the stars, but the reason why they were dark instead of bright remained mysterious until 1859. In that year Kirchhoff (following a very similar investigation, unknown to him.

by Foucault in 1849 which was interpreted by Stokes in 1852) showed that if light from a very hot source, which by itself would give a continuous spectrum, were passed through a cooler but still luminous vapour before entering the spectroscope, the continuous spectrum would be crossed by dark lines in positions corresponding to those of the bright lines which the vapour by itself would emit. He was accordingly led to the general law, which now bears his name, that a luminous vapour is able to absorb the same radiations as, under the same conditions, it emits. The door was then wide open for the entry of spectroscopy into astronomy.

The first-fruits were naturally a physical theory of the constitution of the sun—and, of course, of stars in general. Kirchhoff pictured our luminary as possessing a bright surface—the *photosphere*, as it is called—yielding a continuous spectrum, surrounded by an atmosphere of incandescent vapours which absorbed their characteristic radiations and so produced the dark lines seen in the spectrum. Examination of those lines showed that most of them at least emanated from known terrestrial elements in which the metals figured prominently. The same thing was true in general of the stars. It was just this knowledge of the composition of the heavenly bodies that the 'positivist' philosopher, AUGUSTE COMTE (1798–1857), had given as an example of eternally unattainable knowledge. The science of *astrophysics* had come to birth—first as an application of terrestrial knowledge to the determination of conditions in the stars, and later, when that had reached the necessary stage of development, as a means of investigating the behaviour of terrestrial elements under stellar conditions of temperature and pressure far exceeding what could then be produced by man.

The sun, however, was more complex than Kirchhoff's simple model indicated. During total eclipses, when its photosphere was covered by the moon, huge red flames, known as *prominences*, had been observed rising upwards from scattered regions on the sun's limb, too high to be eclipsed, and surrounding the whole orb was the pearly white *corona* which extended sometimes for some solar diameters beyond the limb. These features could be observed only during eclipses, but successive observations showed that the prominences changed both shape and location and the corona changed shape and extent. The course of these changes could only be guessed so long as the appendages could not be observed continuously. But in 1868 Lockyer and PIERRE JULES CESAR JANSSEN (1824–1907) independently showed that, by using a spectroscope of large dispersion, the spectrum of the atmospherically diffused sunlight could be sufficiently weakened to enable the spectrum lines of the prominences (which were thus shown to consist of glowing gases) to be observed at any time. Further investigation revealed that the prominences rose from a continuous envelope of glowing red gas (called the *chromosphere*) surrounding the sun. At the base of this was the so-called *reversing layer* containing the majority of the substances whose absorption caused the Fraunhofer lines.

The chromosphere itself consisted mainly of hydrogen, calcium, and something giving a yellow spectrum line not previously observed on the earth. Lockyer, in the light of continued observations of this line and of other unknown lines which were later found, concluded that it arose from a new element, which he named *helium*. The majority of astronomers, however, believed that 'helium' was some already known element under abnormal conditions. It was not until 1895 that helium was discovered by WILLIAM RAMSAY (1852–1916) on the earth and found to be actually a new element. The corona defied all attempts to observe it, except during a total solar eclipse, until this century.

Two obvious lines of research now lay open. First, the investigation of the spectra of individual heavenly bodies; second, the classification of stellar spectra (which showed considerable variations in the lines they contained and in their relative intensities) with the object of discovering relations between them. So far as the sun was concerned, the most immediate progress was made in the application of the spectroscope to sunspots. These objects, discovered by Galileo, had been a source mainly of speculation until HEINRICH SCHWABE (1789–1875), who had undertaken systematic counts of them with the aim of finding an infra-Mercurial planet if such existed, discovered after many years of observation beginning in 1826, that their number fluctuated systematically, reaching a maximum about every $11\frac{1}{3}$ years.

This 'sunspot period', as it has come to be called, was later found to be correlated (often, of course, very indirectly) with the most diverse terrestrial phenomena. Outstanding among these were terrestrial magnetic elements, and the relation indicated that the variations of terrestrial magnetism followed with extreme closeness the variations of the sunspot numbers, and particular magnetic storms coincided in time with the occurrence of active spots.

RICHARD CHRISTOPHER CARRINGTON (1826–75) and GUSTAV SPÖRER (1822–95) independently showed that the spots were confined to two belts of solar latitude extending roughly from about 5° to 40° in the respective hemispheres of the sun, and that during a cycle their places of outbreak, beginning in the higher latitudes, steadily approached the equator. All this indicated that sunspots were symptoms of some important periodic fluctuation in the sun's constitution, but theory had not advanced far enough to indicate what this might be. The spectroscope, however, as the wealth of knowledge which it could provide gradually became realized, indicated that the spots were at a lower temperature than the rest of the sun and were regions in which solar atmospheric material was dragged down to the photosphere.

As applied to other individual stars, perhaps the outstanding application of the spectroscope was to the determination of motions in the line of sight (*radial velocities*, as they are called) by an application of Doppler's principle (p. 316). The pioneer in this work was WILLIAM HUGGINS (1824–1910) who in 1868 measured the radial velocity of Sirius. His work was followed by that of HERMANN CARL VOGEL (1841–1907) and JULIUS SCHEINER (1858–1913) who, by using photographic methods, obtained more numerous and more precise results (1890). The motions of stars *across* the line of sight—their *proper motions*, as they are called—which had first been detected by Halley (p. 308), combined with the radial velocities, made possible the determination of the actual speeds and directions of movement in space of those stars whose distances were known. The gradual accumulation of trustworthy stellar parallaxes thus laid the foundation for a detailed study of the systematic movements of the stars among themselves. This, however, did not come to fruition in the nineteenth century.

A particularly useful application of the Doppler principle was the detection of double stars too close to each other for resolution by the telescope. If the components of such a star revolve round one another in or near a plane passing through the earth, they will be respectively approaching and receding from the observer during each revolution, and the spectrum lines seen will accordingly be double, since those of one star will be displaced towards the violet and those of the other towards the red. The first star to be detected as double in this way was Mizar in the Plough, by EDWARD CHARLES PICKERING (1846–1919) in 1889. Since then many 'spectroscopic binaries', as they are called, have been classified.

One of the most striking successes of the application of the spectroscope to the study of individual bodies was the elucidation of the nature of nebulae. These objects, of which by far the greater number then known had been discovered by Sir William Herschel, had been the subject of much discussion since his day. Some objects which appeared nebulous in a small telescope had been resolved into clusters of stars on higher magnification, so Herschel was at first inclined to believe that they were all stellar in character. Later he revised his opinion and declared his belief in the existence in the heavens of 'a shining fluid' which was 'not of a starry nature'. As telescopic power improved, opinions fluctuated between these possibilities until in 1864 Huggins (p. 444), observing a nebula in the constellation Draco, found that it showed a bright line spectrum, indicating a gaseous constitution. Other nebulae, but not all, showed a similar spectrum, and the lines appearing indicated that the gases present were mainly hydrogen, helium, and an unknown gas or gases, whose lines were found in no other sources of light. It was tentatively named *nebulium*. The apparent nebulae which did not show a gaseous spectrum could still be regarded as unresolved clusters of stars—their spectra were continuous, with an indication of the stronger of the absorption lines found in stellar spectra. Later research has shown that some of them act as reflectors of the light of stars in their neighbourhood and appear luminous on that account, while others are whole stellar systems, or 'island universes' as they have been called, which lie right outside our galaxy.

The other line of research—statistical investigation, as we may call it—lay in the accumulation and classification of as many stellar spectra as possible. This was well begun by ANGELO SECCHI (1818–78). He classified the spectra of some 4,000 stars into four types, indicated by the Roman numerals I to IV. Type I comprised the white or blue stars in which strong lines of hydrogen dominated the spectra; type II were yellow stars, such as the sun, containing lines of many elements; type III were red stars, like Betelgeuse, whose spectra showed 'bands' or 'flutings', soon to be interpreted as arising from chemical molecules, in contrast to the lines generated by atoms; and type IV, which were rare, were extremely red and showed different flutings. Secchi's classification was adopted with some modifications by Vogel and later transformed in the extensive Henry Draper Catalogue of the Harvard College Observatory, the first volume of which appeared in 1890. In this, the now universally adopted classification, the stars are grouped in a number of classes somewhat related to Secchi's. The great majority of spectra belong to a continuous sequence (now known as the 'main sequence'), in which the chief divisions are denoted by the letters B, A, F, G, K, M.

The chief characteristic of this classification is that the transition from one group to the next appears to be perfectly continuous. Astronomy, with its company of innumerable and infinitely various stars, is the only physical science that closely resembles biology in the statistical problems it presents. Historically the emergence of the idea of biological evolution almost automatically suggested the similar possibility of stellar evolution. The idea was therefore irresistible that the sequence of stellar spectra indicated successive stages in the life of a single star, and that the sky accordingly presented us with stars of all ages, from those only recently born to those dating presumably from the Creation. This idea was supported by the fact, which gradually came to light with the accumulation of knowledge of stellar motions, that the velocities of the stars increased steadily from the B- to the M-type, indicating that the development was a general characteristic and not merely a peculiarity of the spectra alone.

The question therefore arose: 'Which way does the development go? Does a star come to birth as a B- and die as an M-type object or the reverse?' The most obvious indications were in favour of the former. The colours of the photospheres indicated that the B were the hot and the M the cool stars, and a star would be expected to cool with age. This meant that it would gain speed with age, and although there was no obvious reason for this, it was not improbable that the continual action of the gravitational attraction of the rest of the stars would have this result. The general view was therefore that a star was born as a hot, sluggish B-type object which began and continued to cool and move faster until it reached senility in the M-type condition. Presumably it contracted in size during the process, but there was no evidence of this.

One momentous consequence seemed to be entailed by this view. If a star changed during its life from the B- to the M-type, and if, as seemed certain, the spectrum indicated the chemical composition of the star (or at least of its atmosphere), then apparently matter was created largely in the simple form of hydrogen, the lightest of all elements, and gradually got built up as the star cooled into the heavier elements and ultimately into the compounds shown by the M-type spectra. There were thus not 80 or 90 different types of atom created at the beginning of the world, but these apparently eternally unchanging things were capable of transformation under physical conditions transcending those then within human reach.

This 'transformation' of elements ran counter not only to the knowledge but also to the instincts of the physicists and chemists of the day, and, probably for this reason, most astronomers preferred to say nothing about the evolution of the stars. They concentrated on accumulating more observations in the almost inexhaustible field before them. We thus have the curious circumstance that, while the evidence of continuity among existing types was far more complete in astronomy than in zoology and botany, it is in the latter sciences that the explanation in terms of evolution dominated scientific thought at this period, so that even without the adjective 'organic', the word 'evolution' was automatically taken to refer to living forms alone (pp. 500 f.).

There was one exception, however, to the general attitude. Lockyer, who was quite ready to admit change in the atoms or molecules which others regarded as the foundation stones of the universe (p. 441), had no such inhibitions concerning the effect of physical conditions on atoms. He was indeed prepared, from the evidence of his experiments, to assert that atoms were broken up even in the laboratory, every time that a change of physical conditions—mainly of temperature—produced a change of spectrum. From other considerations he was led to assign to meteorites a far more important place in the scheme of things than that generally allotted to them, and in 1887–8 he propounded a most comprehensive scheme of celestial evolution which he summed up in the words: 'All self-luminous bodies in the celestial spaces are composed either of swarms of meteorites or of masses of meteoritic vapour produced by heat.'

Lockyer conceived that at the beginning of things the whole of space was uniformly filled with meteorites moving in all directions. A chance grouping would act as a centre which by gravitation would attract others and so would grow. At the same time the more and more frequent collisions among them would generate heat. A comparatively small loose swarm would become a comet, a much larger loose swarm a nebula, a denser swarm would grow ultimately into a star.

According to Lockyer the course of a star's life was thus as follows. It was born as a large swarm of meteorites, cool but getting hotter by collisions until it was hot enough to show an M-type spectrum (Lockyer made his own classification, but we speak in the now customary terms). Further condensation through gravitation, and production of heat through more frequent collisions, led the star upwards through the sequence of spectral types until at the B stage it was completely vaporized. Thereafter, with no more collisions to raise the temperature, it cooled by radiation, passing again in its decline through the same sequence of spectral types until it became again an M-type star just before passing into eternal death as a cold solid body, spectrumless as were the meteorites with which it began. On this view, therefore, an M-type star might be very young or very old, and Lockyer chose certain details of the spectra, which he regarded as indications of the very different physical states of the two kinds of star, as criteria for distinguishing between these possibilities. He supported his theory by laboratory observations of the changes in the spectra of meteorites enclosed in vacuum tubes and subjected to different conditions of temperature and electrical discharge.

This all-comprehending scheme was ahead of its time. Its most revolutionary aspect—the idea that a star began as a cool body, became hot, and then cooled again—was universally accepted early in the twentieth century, but on other grounds than Lockyer's. His idea of dissociation of atoms into simpler units also became accepted but again on other grounds and with a quite different type of dissociation. His arguments from laboratory observations, though sound in principle, were vitiated through an insufficient knowledge of the pitfalls facing the inexperienced spectroscopist. Though his vision surpassed that of his contemporaries, none of the details of his theory has survived.

Lockyer's interest in meteorites was aroused by the wonderful display of the Leonid meteors in 1866, which followed a similar display in 1833 and appeared from previous records to indicate a periodic occurrence every 33 or 34 years. His view that these bodies might be more important in celestial evolution than appeared on the surface was strongly encouraged by the discovery about the same time of an undoubted connexion between meteors and comets. The periodic recurrence of meteor showers—there were many such showers, occurring at specific dates during the year, each appearing to emanate from a particular point in the sky—gave rise to the idea that these bodies existed in swarms pursuing definite orbits round the sun, so that a shower would be seen when the earth was at a point in its orbit which intersected the orbit of a swarm. The Leonids, for example, which appeared on 12 August, were seen every year on that date, but with special brilliance on certain occasions about 33 years apart, from which it was concluded that the swarm, which was continuous all round its orbit, had a condensed region which the earth crossed at those times (Fig. 174).

Comets, on the other hand, tended to appear sporadically, but many of them moved in elliptical orbits which could be calculated and which enabled their reappearances to be predicted. The great discovery which was made in the 1860's, first of all through the work of GIOVANNI VIRGINIO SCHIAPARELLI (1835–1910), was that the orbits of certain comets were identical with those of certain meteor swarms. This indicated a close connexion between meteors and comets, and the conclusion soon became inescapable that a comet was, at least

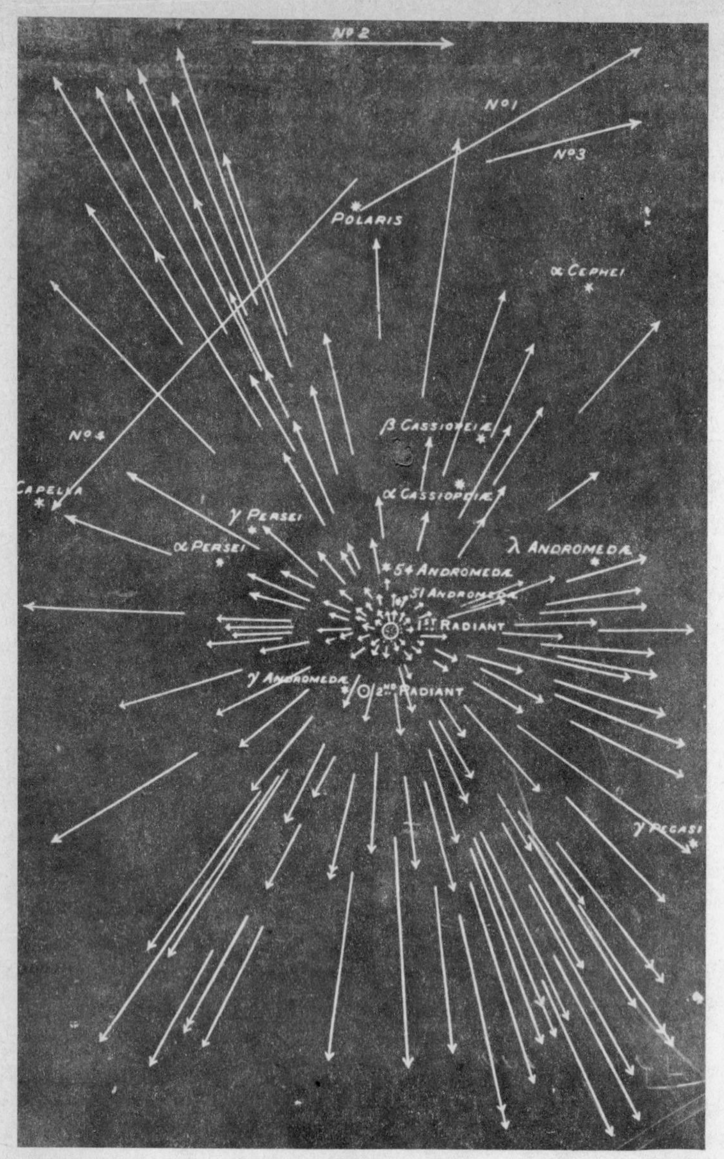

FIG. 174. Paths of meteors observed in the swarm of 27 November 1872, showing their 'radiant point'. Their divergence is an effect of perspective. Actually they all proceed parallel to one another from the direction of a point in the constellation Andromeda.

originally, a fairly dense swarm of meteors. Confirmation was afforded by Biela's comet which, after several appearances, vanished as a comet and reappeared as a swarm of meteors (Fig. 175). The

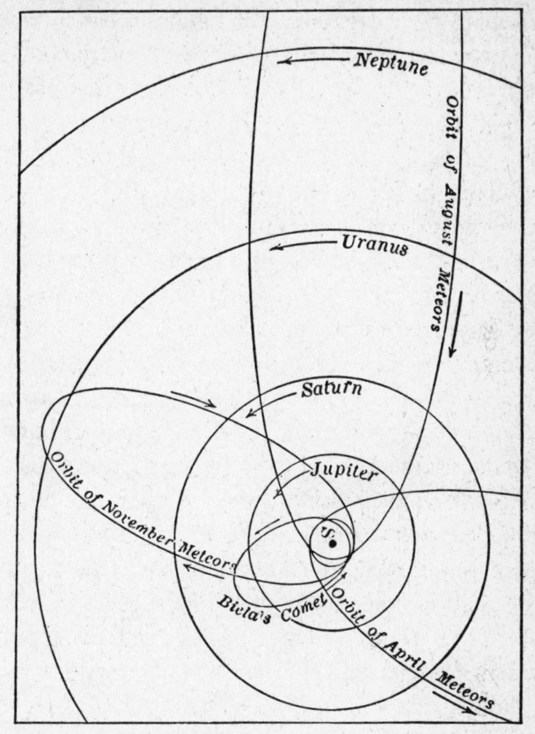

FIG. 175. Orbits of the more prominent meteoric swarms.

spectroscope was of less assistance in this matter than in most astronomical problems, since meteors, being raised to incandescence only through friction with the earth's atmosphere, show largely the spectrum of that atmosphere. Comets, on the other hand, show considerable variations in their spectra as they approach or recede from the neighbourhood of the intensely hot sun. They are, however, at all times low temperature bodies compared with stars. Their tails—which appear only when they are near the sun—were found to be always directed away from the sun, and their existence was attributed first to electrical repulsion and later, when Maxwell's electro-magnetic theory of light appeared, to the pressure of solar radiation.

While all this detailed work was proceeding, knowledge of the structure of the universe as a whole, which had been Sir William Herschel's main preoccupation, was naturally largely in abeyance. Just as the discovery of the telescope brought astronomy back from the celestial spheres to the sun and planets, so that of the spectroscope brought it back from the stellar system to individual luminous bodies. There was, however, one line of research which did afford some knowledge of the distribution of the stars in space. Stellar photometry was placed on a sound basis by the establishment in 1856 of a precise magnitude scale—i.e. a scale for indicating the brightnesses of the stars—by NORMAN ROBERT POGSON (1829–91), and by the exact measurements of brightness on this scale of many thousands of stars by E. C. Pickering and CHARLES PRITCHARD (1808–93). It was then possible, from counts of the numbers of stars of successive magnitudes, to infer how the density of distribution of stars varied, if at all, as one receded from the earth in various directions. It appeared that for the smaller distances there was no appreciable change, but that after a certain distance had been passed the stars began to thin out, though much more slowly in the direction of the Milky Way than in other directions. The suggestion was therefore that we were near the centre of a disk-shaped galaxy, of the kind already deduced by Herschel (Fig. 130), but no very precise measurements of the distance to which the system extended were possible. It was sufficiently evident, however, that the universe was of an awe-inspiring vastness; nevertheless the conception was minute compared with the dimensions we envisage today.

We must not leave astronomy without mention of two items of less spectacular but fundamental importance. Although the spectroscope (or spectrograph, as it became when photography was applied to it) directed the main line of progress, the much older telescope remained indispensable, and indeed the spectroscope was almost useless in astronomy without it. The second half of the nineteenth century saw a great expansion of telescopic astronomy in the United States, where the 36-inch Lick and 40-inch Yerkes telescopes—both visual instruments—were then erected and still remain the largest refractors in the world. The great photographic reflectors were still to come. The size of a telescope determines its light-grasping power and so the depth in space to which it can penetrate. Every increase in diameter of the object-glass therefore increases the fraction of the universe that we can survey.

In this period also the perennial work of compiling larger and better catalogues of stars was continued, but on a much broader basis. In the older catalogues the positions and eye estimates of the brightnesses of the stars were all that could be recorded. Now their spectra, proper motions, radial velocities, parallaxes, precise magnitudes, and so on, all called for measurement and compilation. The catalogues which were published give but a faint indication of all the work which was going on in preparation for those of the future. The great *Bonner Durchmusterung*, planned by Bessel and completed by his pupil, FRIEDRICH ARGELANDER (1799–1875), which appeared in 1859–62, is outstanding among catalogues of northern star places, and this was supplemented for the southern stars in 1900 by the *Cape Photographic Durchmusterung* originated by DAVID GILL (1843–1914).

5. CHEMISTRY

The other great field of application of the spectroscope was that of chemistry. Although its role here was far less prominent than in astronomy, it forms an appropriate beginning for our consideration of this subject, for the prospects opened up by Kirchhoff and Bunsen's discovery in 1859 seemed almost unlimited. Here was a simple, universal and apparently infallible method of analysis, for before the multitudinous ways in which physical, as distinct from chemical, conditions can influence spectra had been understood, every modification of a spectrum seemed to denote a chemical difference. Furthermore, a sovereign means was now available for the discovery of new elements, even when present in the most minute quantities. Spectrum lines not already found in the spectra of known substances seemed to denote the presence of something previously unknown, and it was in this way that the elements rubidium and caesium were discovered by Bunsen himself in 1860 and other workers later discovered thallium (1862), indium (1864), and gallium (1875). We have mentioned the spectroscopic discovery of helium in the sun in 1868, to be confirmed spectroscopically in 1895 (p. 443). The other inert gases found at the latter time also owe their discovery in large part to their unique spectra.

The first of the inert gases to be discovered on the earth was argon. In 1892 LORD RAYLEIGH (1842–1919) pointed out that atmospheric nitrogen was denser than that from other sources, by an amount too large for experimental errors. He planned with WILLIAM RAMSAY (1852–1916) a series of investigations which resulted in Ramsay's isolating in 1894 a new atmospheric gas which he named *argon*. Shortly afterwards he discovered Lockyer's helium in nitrogen from a chemical source, and then, by fractional distillation of argon, helium and two further gases, krypton and xenon, were found to be constituents of the air. A whole new family of gases with no detectable chemical properties had been brought to light.

The word 'family' in this connexion had by that time a precise meaning. In 1850 there was no reason to suppose that there was any definite limit to the number of elements or any relation between them. They were simply Nature's independent units of substance, as distinct from one another as the species of living beings were believed to be before the hypothesis of organic evolution was taken seriously. It is true that in 1815 Prout (p. 345) had suggested that all other elements were built up of hydrogen, but the fact that their atomic weights were not simple multiples of that of hydrogen seemed to discredit this idea. In 1864, however, JOHN ALEXANDER REINA NEWLANDS (1837–98) found that if the then known elements were arranged in the order of their atomic weights, their chemical properties showed a tendency to recur as though they fell into groups of eight. He called this a *law of octaves*. His views did not attract serious attention. But in 1869 the Russian chemist, DMITRI IVANOVICH MENDELÉEF (1834–1907), with better data, repeated the scheme, leaving gaps where needed to preserve the sequence of properties. He thus gave the idea a much more plausible appearance. As for the gaps, he predicted the properties of elements which might fill them. His view was justified by the discovery of gallium in 1875, scandium in 1879, and germanium in 1886. Thereafter there could be no doubt that the elements did in fact fall into definite groups or families, of which prominent examples are the alkalis (lithium, sodium, potassium, rubidium, caesium) and the halogens (fluorine, chlorine, bromine, iodine). Ramsay's discoveries indicated such a family, then new to the 'periodic table of elements', as it came to be called.

It was clear, however, that the arrangement was not simply that of a regular repetition of properties at every eighth element. In a few instances the order of atomic weights had to be violated to preserve the sequence of chemical properties, and there were certain places in the table where the intrusion of independent unrelated groups appeared to have occurred. The most striking of these was that of the *rare earth metals*, the members of which continued to be discovered at intervals. These elements were so alike in their chemical properties that separation was very difficult. Only the spectroscope afforded a ready means of distinguishing them one from another. A full understanding of these mysterious phenomena awaited knowledge of the structure of the atom in terms of fundamental electrical particles, which came with the next century.

In the later nineteenth century chemistry and physics were far more separate than now. The atomic theory of the chemist and the kinetic theory of gases of the physicist, though they both contemplated the same substances as composed of tiny discrete units, were not brought into relation with one another. What the physicist called a 'molecule' was often more akin to the chemist's atom. The two ideas were independent, and only later became reconciled (pp. 341 f.).

Notwithstanding the absence of any theoretical basis for the regularity revealed by the periodic table, it seemed pretty clear that the number of elements was in fact limited to something less than 100. It was otherwise with chemical molecules. The possible combinations of atoms to form compound molecules, though evidently not unrestricted, seemed indefinite in number and offered an apparently inexhaustible field for the chemist's exploration.

The most fundamental problem was that of determining why some elements united into compounds and others did not, and why the former united in definite proportions which varied with the atoms concerned. The first effective step towards the solution of this problem was taken by EDWARD FRANKLAND (1825–99) in 1852. He assigned to each kind of atom a property called *valency*, which represented its capacity for combining with other atoms. Taking hydrogen as the unit, for example, the four compounds, methane (CH_4), ammonia (NH_3), water (OH_2), and hydrofluoric acid (FH), indicate that the valencies of carbon, nitrogen, oxygen, and fluorine are respectively 4, 3, 2, and 1. These elements head four successive families in the periodic table, and when that table was established valency was seen to be among the properties in which members of a family resembled one another. Frankland pictured each atom as possessing an imaginary *valency bond* represented by a straight line (later, dots were used); thus the carbon atom could be denoted by =C=, and in the compound methane, the hydrogen atoms could be placed at the free ends of the bonds, forming the symbol

$$\begin{array}{ccc} H & & H \\ & \diagdown\diagup & \\ & C & \\ & \diagup\diagdown & \\ H & & H \end{array}.$$

Compounds in which each of the carbon bonds terminated in another atom, as in this case, were said to be *saturated*, but unsaturated compounds exist, such as ethylene (C_2H_4), which could be represented as

$$\begin{array}{ccccc} H & & & & H \\ & \diagdown & & \diagup & \\ & C & - & C & \\ & \diagup\ | & & |\ \diagdown & \\ H & & & & H \end{array},$$

or, more usually, by combining the unattached bonds into one, thus

$$\begin{array}{ccccc} H & & & & H \\ & \diagdown & & \diagup & \\ & C & = & C & \\ & \diagup & & \diagdown & \\ H & & & & H \end{array}.$$

Unsaturated compounds easily become saturated by taking up other atoms for the purpose; thus we have the stable compound $C_2H_4Cl_2$ (ethylene dichloride), represented by

$$\begin{array}{ccccccc} & & Cl & & Cl & & \\ & & \diagdown & & \diagup & & \\ H & - & C & - & C & - & H \\ & & \diagup & & \diagdown & & \\ & & H & & H & & \end{array}.$$

This idea, though of course it *explained* nothing, went a long way towards a systematic *description* of the nature of chemical combination, and created expectations, which were usually realized, as to whether a particular molecule was likely to exist or not.

In selecting carbon as an example we have chosen one of outstanding importance, since the study of its compounds forms what is known as *organic* chemistry. Of fundamental significance among the carbon compounds is *benzene*, discovered by Faraday in 1825. The chemical formula for benzene is C_6H_6, and when written in terms of valency bonds it becomes

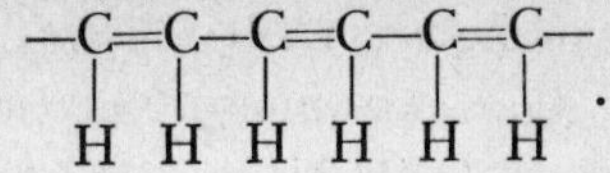

$$\begin{array}{c} -\mathrm{C}{=}\mathrm{C}{-}\mathrm{C}{=}\mathrm{C}{-}\mathrm{C}{=}\mathrm{C}- \\ \;|\quad\;\; |\quad\;\; |\quad\;\; |\quad\;\; |\quad\;\; | \\ \mathrm{H}\;\;\mathrm{H}\;\;\mathrm{H}\;\;\mathrm{H}\;\;\mathrm{H}\;\;\mathrm{H} \end{array}$$

. The three 'double bonds' indicate the unsaturated nature of the compound and its consequent ability to form an extraordinarily large number of derivatives. A step of vital importance was taken in 1865 by FRIEDRICH AUGUST KEKULÉ (1829–96) who, by uniting the two loose bonds at the ends of the chain, formed the famous 'benzene ring', thus:

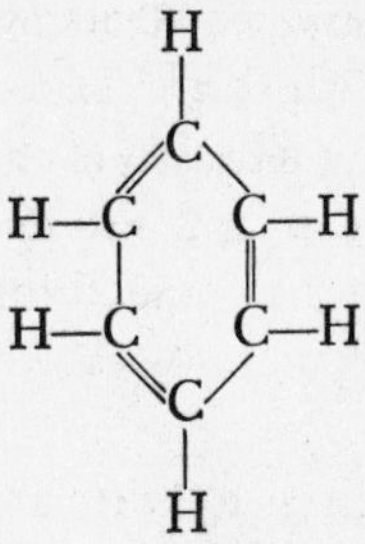

$$\begin{array}{ccccc} & & \mathrm{H} & & \\ & & | & & \\ & & \mathrm{C} & & \\ & /\!/ & & \backslash & \\ \mathrm{H}{-}\mathrm{C} & & & & \mathrm{C}{-}\mathrm{H} \\ | & & & & \| \\ \mathrm{H}{-}\mathrm{C} & & & & \mathrm{C}{-}\mathrm{H} \\ & \backslash\!\backslash & & / & \\ & & \mathrm{C} & & \\ & & | & & \\ & & \mathrm{H} & & \end{array}$$

This is the basis for an untold number of compounds, obtained by substituting for one or more of the hydrogen atoms a variety of other atoms or groups. Included among these are the 'aniline dyes'.

The viewpoint thus established prepared the way for a great advance in synthetic organic chemistry, for it gave a clear indication of the kinds of compounds which it might be advisable to try to construct. Of the hundreds of thousands of 'organic' compounds now known, the majority have been made artificially, and the stimulus for this great advance came from the clearly visualizable model provided by the structural formulae that Kekulé described.

A further development of that model laid the basis for another great department of the subject known as *stereochemistry*. There are many instances of substances known as *isomers*, which have the same chemical composition but nevertheless exhibit differences in chemical behaviour. An example elucidated by LOUIS PASTEUR (1822–95, p. 461) in 1848 was that of the tartaric acids. There are three of these, identical in composition and therefore in molecular weight. Pasteur found that in solution two of them rotated the plane of polarization of light passing through them, but in opposite directions, while the third had no effect on such light. Moreover, the crystals of the former two ('optically active' as they are called) are mirror images of one another. This was interpreted later in terms of molecular structure independently by JACOBUS HENDRICUS VAN T'HOFF (1852–1911) and JOSEPH ACHILLE LE BEL (1847–1930) in 1874. If the idea of valency bonds be extended from a plane to three-dimensional space, the four bonds of a carbon atom can be represented as pointing towards the corners of a tetrahedron of which the atom is at the centre. The other atoms in the molecules can then be arranged round the carbon atoms so as to produce structures (*right handed* and *left handed* as they are called) which are related to one another as are an object and its image. The application of this idea has thrown a great amount of light on isomerism and also on *isomorphism*—the occurrence of crystals having the same form but different chemical composition.

If the space we have given to chemistry seems to indicate that there was less, or less important, activity in this subject than in physics and astronomy, the appearance must be regarded as illusory. Chemistry was no whit behind the other subjects in prominence, but its progress at this period necessitates a more compact presentation. One of the largest branches of chemistry is physical chemistry, and while in a detailed history a great deal would have to be recorded of the progress in this field, there is room in this work for mention only of the underlying principles, which are physical in character. It thus comes about, for example, that while the principles of thermodynamics (pp. 375–8) were established by physicists, the bulk of the applications of the second law have been chemical.

This is but one aspect of a more general feature of the physical sciences in this half-century. By its very nature, physics is concerned with the root principles which must ultimately burgeon in chemical and astronomical phenomena as well as in those more specifically called physical. A stage must therefore be reached in the development of chemistry and astronomy at which each of those subjects begins to close in, so to speak, towards physics, and to submit its discoveries to interpretation in physical terms. But this stage must necessarily be reached far sooner in chemistry than in astronomy, for in every respect—the number of elements compared with the number of stars and nebulae, the ranges of temperature and pressure within which molecules are stable compared with the corresponding ranges for the heavenly bodies, the facilities available for the study of objects which can be handled and manipulated in the laboratory compared with those that can be applied to inaccessible points of faint light, and so on—in every respect the speed of progress of chemistry towards absorption in fundamental physics is bound to be far greater than that of astronomy. And so it comes about that the nineteenth century sees chemistry converging towards its consummation and astronomy still rapidly diverging.

From the moment the periodic table was constructed, the end of chemistry as a basic independent science came within sight. Chemical reactions had yielded all the basic knowledge that lay in their possession. The next step was the elucidation of atomic structure, for when once the atom is understood its possibilities of combination can be deduced and the properties of the products become potentially predictable. And atomic structure is a problem for the physicist, the beginning of the solution of which came with the discovery of the electron. Twentieth-century physics has revealed a periodic table in which the detailed electronic structure of each element is set out, and on this the valency of the atoms depends. The future contribution of the science of inanimate things towards our understanding of the secret processes of nature comes under the heading of generalized physics.

Two things, however, must be emphasized. First, not the whole of science is of this fundamental character, and there still remain, and will for long remain, many problems in which chemical methods will be used. While the theoretical possibility of predicting the existence and properties of all non-living substances is already envisaged, we are still very far from solving the practical problem of converting that possibility into an actuality. The chemistry of molecules, as distinct from that of atoms, is still a highly specialized field in which work of vital importance in medicine and other activities of deep human concern is being prosecuted. The history of physics for a long time to come will not be without a chemical chapter.

Secondly, it was impossible for physics to accommodate chemistry and astronomy so long as it was dominated by the nineteenth-century conviction that the ultimate explanation of things must be expressed in terms of Newtonian mechanics. No description of the atom in those terms could throw any light on the periodic table or, more immediately, on the physical phenomenon of radiation of energy. Chemistry and nineteenth-century physics are distinct subjects.

Our period ends with the beginning of a new physics in which mechanical forces recede into the background and quite other conceptions take their place. But this was not, as the nineteenth-century physicists would have thought had they been granted the privilege of seeing into the future, a revolution in the conception of scientific explanation: it was a belated acknowledgement that that conception, as it arose in the seventeenth century, had been misunderstood. Indeed, the new physics came directly from a deliberate return to the two features of seventeenth-century physics indicated on p. 419. The principle of relativity was made possible by the realization that the ultimate data of physics were not material objects but our measures of material objects, i.e. our experiences of the world and not the world itself. The quantum theory of the atom was made possible by the rejection of the idea that bodies consisted of inert particles moved by independently specifiable Newtonian forces, and the substitution of principles of behaviour directly dictated by the phenomena observed regardless of their apparent inconsistency with phenomena of a different kind.

It is likely that we are now well on the way to a period of advance in the physical sciences comparable with that which ended with the nineteenth century, but on a vastly greater front and with a vastly greater speed of movement, and the prospects are that the present principles of relativity and quantum physics, when they are properly understood and formulated in terms harmonious with one another, are adequate for the study of the whole of inanimate nature. What change will be necessary to unite this generalized physics with biology and psychology remains to be seen, but there is no reason to doubt that the process which began in the seventeenth century, which we call science and of which this is an outline of the history, will be able to continue to this achievement. There is nothing more fundamental than experience, and the study of that, uninhibited by any principles other than those of reason itself, would seem to provide a means for the attainment of all knowledge possible to man.

6. MECHANIZATION OF BIOLOGY

(i) *Microbiology*

To microscopists of the later seventeenth century, 'spontaneous generation' of forms whose eggs were visible to the naked eye was becoming incredible (pp. 282 f.). For microscopic creatures the possibility remained though in the later eighteenth and early nineteenth century many workers cast doubt on it. The weakness of their position was that they were seeking to prove a universal negative. 'Common experience' could still reasonably hold that minute creatures might arise *de novo* in certain media and notably in infusions. Hence the name *Infusoria* for these microscopic organisms (410–11).

As the nineteenth century advanced the position was transformed by improvements in the microscope itself. Instruments involving the achromatic principles of Dollond (pp. 300 f.) were produced by the Italian G. B. AMICI (1786–1863) who was at once mathematician, technical expert, and biologist, and J. J. LISTER (1786–1869), the father of Lord Lister. Microscopes of this type were placed on the market by the optician CARL ZEISS (1816–88) of Jena. His partner and heir ERNST ABBE (1840–1905) perfected the microscope so far as the nineteenth century was concerned. These advances attracted much attention and the microscopic world was extensively explored by many workers. Of these C. G. EHRENBERG (1795–1876) and the colleagues W. H. DALLINGER (1842–1909) and J. DRYSDALE (1817–92) demand special mention. Ehrenberg explored the world of microscopic life with unique industry and, despite errors in principle, illustrated it with unique skill and beauty. Dallinger and Drysdale studied a single organism with an intensity for which the term heroic is not inept (p. 464).

But the great triumph of microbiology was the work of LOUIS PASTEUR (1822–95). He was by training and profession a chemist and his first scientific successes were on the optical action of tartrates produced in fermentation of grapes (p. 462). This caused him to consider the ferments themselves, and in due course to consider other organisms.

Many before Pasteur had associated fermentation with organisms. Thus Schwann (pp. 412 f.), for example, in 1837 revealed that yeast consists essentially of a mass of plant-like beings the presence of which is a condition of the alcoholic fermentation of sugar. Twenty years later Pasteur set himself to the study of other types of fermentation taking the first important step with his classic *Memoir on the fermentation known as 'lactic'* (1857). He showed that in sour milk there is a substance that produces lactic acid. He isolated this substance, added it to fresh milk, and watched it act. It was a mass of bacteria of a species now familiar. This work led him to acquire skill with the now greatly improved microscope and he became expert in exploring the life-history of minute organisms.

At this time the general views of fermentation were those of Liebig (pp. 407 f.) who regarded it as a peculiarity of organic matter and 'of the nature of death'. Pasteur realized that fermentation demanded the presence of living organisms.

From the souring of milk Pasteur turned to other fermentations and notably that of sugars into alcohol in wine-making. He found that essential elements are yeasts or moulds which he found on the grape-skins. If these organisms were absent, deteriorated, or of the wrong kind, the fermentation was either arrested or on abnormal lines. In tracing the life-history of these organisms, he found that their mode of growth was closely controlled by their physical and chemical environment. Thus he passed to the diseases and defects of wine, most of which, he found, were due to abnormal ferments. He soon saw that a great variety of organisms is associated with fermentation. Not all are yeasts. Some are much smaller and grow often in chains. These were bacteria (*Studies on wine*, 1866).

Pasteur was by now convinced that fermentation, decomposition, and putrefaction are all vital processes, that innumerable minute organisms are responsible for all and that their germs are carried by the air. In this he had predecessors but none so skilled or so persistent. An obstacle to the acceptance of his views was the demand they made on the germ-carrying capacity of the air. Some critics suggested that, according to him, the air must be a solid mass of germs. Some held that the organisms found in fermenting and decomposing matter were the products of those processes and not their cause.

By 1859—the year of publication of the *Origin of Species*—Pasteur was engaged in controversy on the origin of life. Discussion turned on the bacteria. These were then considered the lowest form of life and were known to propagate by division but the question remained 'Were they *also* spontaneously generated?' If a sterilized flask of broth 'went bad', and bacteria were found in it, must they have had their first origin from without? Must all living things, even those so minute and lowly, be derived only from other living things? Life must begin somewhere; why not at this lowest stage? An answer was impossible in terms of formal logic, but Pasteur answered in a way that was to bring doubt to rest.

The crux, he recognized, is the germ-carrying capacity of air. It was well known that if a flask of broth be closed, sealed, and then adequately heated for a while, the broth will not decompose until the flask be unsealed. The criticism had been that, in heating, the contained air, necessary for the bacteria, was so altered as not to support bacterial growth. Now if, as Pasteur believed, the beings associated with fermentation come from the air, it should be possible to find them there. He therefore drew air through gun-cotton filters and then dissolved the cotton. The deposit that settled at the bottom revealed microscopic bodies like those seen in fermentation, while similar filters through which air already filtered had been drawn, showed no such bodies. Therefore the unfiltered air contained these bodies.

But it was still said: 'Yes, these structures that you demonstrate are in the air, but they are not alive. Not they, but something far more minute and subtle, generates the swarms of living things in fermenting substances.'

To this, Pasteur found a most satisfying answer. An infusion of a fermentable broth is introduced into a flask. The neck, very narrow and long, but left open, is drawn out in an S-shape. Flask and contents are boiled repeatedly and the flask then left undisturbed in still air. It can so remain for days, weeks, months—even, as we now know, for years. Its contents do not ferment. Now the neck is severed. The fluid within is thus exposed to the fall of atmospheric dust. In a few hours it is fermenting (Fig. 176). Organisms are demonstrable in it under the microscope (1861). During the century that has since elapsed, it has been shown repeatedly that, if living organisms be excluded, there is no fermentation, putrescence, or other production of minute life. It is all a question of adequate technique.

The germ-carrying power of the air was given more exact expression by the physicist JOHN TYNDALL (1820–93). He devised methods for determining the physical conditions of aerial purity and demonstrated the errors in technique that had led some to oppose Pasteur.

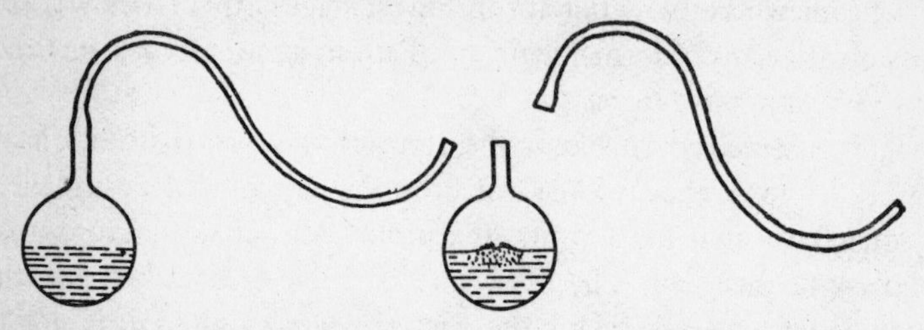

FIG. 176. Pasteur's crucial experiment to prove that fermentation or putrefaction results from the action of air-borne organisms. The flask, containing a putrescible fluid is subjected to prolonged heating to destroy all organism. It is then left in still air with the mouth open. Days, weeks, months, even years, pass without sign of putrefaction. No organisms reach the broth, since any that enter the open mouth fall on the floor of the S-shaped neck and remain there. Sever the neck so that organisms can fall from the air on to the surface of the fluid and they multiply. In a few hours putrefaction sets in. This is shown by the formation of a film or scum on the surface just below the severed neck. Microscopically the broth is seen to teem with organisms.

His work (1876–81) marks the final abandonment of the doctrine of spontaneous generation. Whether spontaneous generation has ever taken place with other forms of life or elsewhere than on this earth are questions on which science can, at present, hardly express an opinion. The word *biogenesis* (Greek, 'life-origin') was coined by T. H. Huxley in 1870 to express 'the hypothesis that living matter always arises by the agency of pre-existing living matter'. The opposite hypothesis is *abiogenesis*.

The work of Tyndall was supplemented by the conjoined researches of Dallinger and Drysdale (p. 461) on an individual organism (1873–8). They kept a single flagellated 'monad' under continuous observation for days and nights on end, using a magnification up to 5,000 diameters. This minute creature, associated with sepsis, had a length of 1/4000th of an inch and divided every few hours. After several divisions it encysted. They saw the cysts rupture and liberate spores so minute as to be just visible. In 5 hours a spore attained adult form and size. They found that the adults are destroyed by a temperature of 150° F. but that the spores can sustain a fluid heat of 220° F. and a dry heat of 250° to 300° F. Some have doubted the technical possibility of the partners' vigil but their work, like Tyndall's, influenced the development of antiseptic and later of aseptic surgical methods.

Long before Pasteur, it had been repeatedly suggested that infectious disease might be ascribed to minute organisms and to their passage from one host to another. A pioneer of this view was the amateur microscopist, AGOSTINO BASSI (1773–1856), of Lodi near Milan. In 1835 he gave the first demonstration that a vegetable micro-organism could be a cause of a certain disease of silkworms. He proved that it was transmissible from moth to worm, and that the transmitting material was a minute fungus. In 1840 the Berlin anatomist, JAKOB HENLE (p. 495), set forth in detail his theory that infectious diseases are caused and conveyed by invisible forms of life. During the next two decades the idea was taken up by several others.

In the sixties the French silk industry was suffering from an epidemic among silkworms. They were dying or forming poor cocoons. Such few moths as were hatched showed signs of disease. Microscopists found numerous oval corpuscles in the caterpillars and moths. Pasteur took up the investigation. Though he knew of Bassi's work Pasteur's task was complicated because he was dealing with two different diseases but ultimately he showed that each is always associated with its own special organism (1862).

The general nature of infection was still in debate but here were two infections causatively associated with microscopic organisms. The hint was very valuable. Primarily, it enabled Pasteur to recommend measures that saved the industry. Secondarily, it led to discoveries which have opened out new departments of science, initiated a new era in medicine, and given a new view of the world of life.

The next disease on which light was thrown was anthrax, which is very fatal in cattle and sometimes transmitted to man. As with silkworm-disease, rodlike bodies had been found in the tissues of cattle suffering from anthrax. A French observer, CASIMIR DAVAINE (1812–82), had shown that in the blood of animals with anthrax such bodies occur in numbers proportional to the gravity of the attack and that their blood conveys the disease, when inoculated in as little as a millionth of a drop (1863–8). A German worker, C. J. EBERTH (1835–1927), filtered the blood of stricken animals and injected the filtrate into healthy animals which failed to develop the disease (1872). He thus proved that the infective substance could be separated from the blood.

Bacteria at this time began to attract the attention of botanists, especially of FERDINAND COHN (1828–98) of Breslau. He recognized many species and sought to classify them according to their form and habitat. His work (1872–6) clarified the current attitude toward these organisms which are a group of lowly plants, perhaps allied to the fungi, but without any known sexual processes. Like fungi, they have no chlorophyll and therefore cannot build up carbohydrates from the carbon dioxide of the air (pp. 408 f.). Bacteria depend for their sustenance upon the organic media in which they live and in which they set up rapid and profound chemical changes. They fall into two great classes, the *aerobic* which flourish only in the presence of oxygen and the *anaerobic* which cannot so flourish. The anthrax organism which, for a bacterium, is large and conspicuous, is aerobic.

The life-history of the anthrax bacillus was unravelled by ROBERT KOCH (1843–1910), probably inspired by the theories of his teacher, Henle (p. 465). In 1872 anthrax broke out among the cattle near Breslau. Koch found that in infected animals the spleen was specially affected. In examining minute fragments of it microscopically he saw the anthrax bacilli as Davaine and Eberth had seen them (above). He found that the disease could be conveyed to the mouse, which made experimental investigation simple. After propagation for twenty generations, he recovered the characteristic organisms from the last cultivation.

A fragment of spleen of a mouse infected with anthrax was sown in a drop of blood serum. This was kept at body temperature for 18 hours, after which it presented a characteristic microscopic picture. In the centre, where air could not penetrate, the bacilli had grown very slowly and were mostly separate from each other. Farther out, where the oxygen supply was better, they had grown longer and many were end to end. The nearer to the edge, the longer were the threads thus formed. In some threads in contact with outer air, little round refringent bodies had formed. These 'spores' were very highly resistant to heat, chemical agents, drying, &c. The spores are of the utmost significance for the history of the disease (Fig. 177).

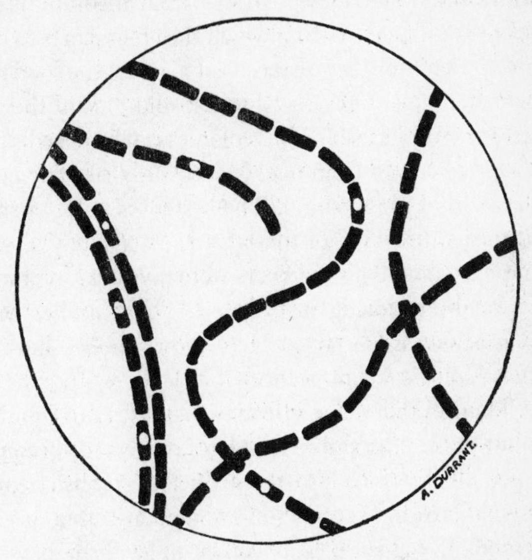

FIG. 177. Anthrax bacilli highly magnified, some showing spores.

Developing his technique, Koch made the two contributions that have enabled bacteriology to develop as an independent science; first he raised the microscopic visibility of bacteria; second he improved the method of cultivating them.

As regards visibility. The eye does not readily distinguish the shape and arrangement of very minute objects under the microscope. Stains had been occasionally used to raise their visibility. For the aniline dyes, which had become familiar in the textile trade (p. 457), bacteria have a special affinity. Koch introduced them into bacteriology (1877) and elaborated methods of staining suitable for a variety of organisms. Staining methods employed by bacteriologists have become very elaborate but are largely derived from those of Koch.

With a staining process of his own devising, Koch demonstrated the elusive tubercle bacillus in 1882.

As regards cultivation. Early workers on bacteria were constantly hampered and confused by preparations that contained more than one species—'impure cultures'. Elaborate methods were needed to obtain 'pure cultures', that is of one species only. Koch introduced the beautifully simple device of a transparent medium of jelly-like consistence (1881). The fluid from which the organism is to be grown can be diluted indefinitely before it is used to infect the solid medium. The very smallest quantities and highest dilutions of the infective substance can thereby be used. Thus colonies can be raised, each from very few bacteria or even from one. Moreover, anaerobic organisms can be grown within the medium. The system is still in general use and has led to the discovery of the organisms of many diseases.

Meantime Pasteur and his followers were investigating the powers of the body for protection against disease. These studies became of the highest practical importance. Here, however, we have only to consider their biological significance.

It was well known that many infectious diseases can commonly be contracted but once. The sufferer, having recovered, becomes 'immune'. Pasteur had before him the artificial establishment of immunity to smallpox by vaccination as demonstrated by Edward Jenner in 1798. Vaccinia is, in effect, a mild form of smallpox. Pasteur found a parallel in the disease 'fowl cholera'. He prepared cultures of its germs and found that, if kept too long, chickens inoculated with them developed only a mild attack. In such birds, on recovery, he inoculated an ordinary culture of the disease. The animals withstood it. They had become 'immune'. This discovery (1880) opened up yet another department of biology, Immunology.

Pasteur now turned to anthrax and found a somewhat similar series of events. An 'attenuated' culture of anthrax that he had prepared could be inoculated safely into susceptible animals which thereby become immune to the disease. He thus protected millions of cattle and carried out successful researches of a similar kind in other diseases.

The biological problem of the nature of immunity early arose. Two rival theories were advanced. One, associated specially with the name of ELIAS METSCHNIKOFF (1845–1916), may be called the 'solidist', the other the 'humoral' theory. According to Metschnikoff, infective organisms that find their way into the body are engulfed and digested by certain cells which he named 'phagocytes' (1884, Greek 'devouring cells'). He based his views primarily on invertebrate animals. This certainly takes place also in vertebrates but in them is concerned chiefly with local inflammation. The general nature of immunity is not explained by it.

The humoral view was forwarded by a discovery of EMIL VON BEHRING (1854–1917) in 1890. He produced immunity against a disease by injecting serum from an animal that had been infected and had recovered. The serum, in this case, is the carrier of the immunity. It bears with it the *antitoxin*—a word Behring introduced.

The problem of immunity has turned out to be enormously complex. The body, under various forms of excitation, is constantly throwing into its blood-stream substances which are somehow directly or indirectly inimical to the organisms of disease. In the last decade of the nineteenth century and later much work was done on immunity, largely by PAUL EHRLICH (1854–1915) who made many important practical advances. It must be admitted, however, that little light has been thrown on the composition or mode of action of the immunizing substances. Their discovery has so far added only to the wonder at the vast complexity of the physiological mechanism.

The practical results of the study of the microbic origin of disease are so impressive that the gaps in knowledge of the nature and life of micro-organisms are often forgotten. Biologically, the first question is 'How does an infectious disease originate?' There is no answer. All through the animal and vegetable kingdoms we meet the phenomenon of *parasitism*. The parasite lives in or on its host. In that limited habitat, its organs are lost or become 'degenerate'. The flea, on the surface of the body, loses its wings. The mistletoe, living on the apple tree, can no longer root in earth. Many parasitic plants have lost their chlorophyll. Intestinal worms have lost their sense-organs; some are without digestive canal and even without power of movement. It has been assumed that the bacteria are such degenerate organisms, but there is no knowledge of their ancestry.

How does such a form become parasitic? We are in the dark.

Diseases often attack only one particular species, but why? On this we have a little information but it can hardly be generalized.

Again, if a disease is traced through history, it is evident that its type changes. Scarlet fever, measles, influenza, are very different in one year from what they are in another. Why? Has the infective organism changed or has the host changed? We can but guess. In the face of such ignorance it is blind optimism to speak of our knowledge of disease as a conquest. It is part of the function of science to define the limits of our knowledge. Unjustified optimism may be an enemy of science as well as may unreasoning credulity.

At the very end of the nineteenth century, there came into view an aspect of infectious disease that bore on the most fundamental biological problems. It was demonstrated that certain diseases are conveyed by infectious material which does not lose its virulence even when passed through filters of minute mesh. Filters can be constructed which hold back not only bacteria but far smaller bodies. In 1892 it was demonstrated that the juices of tobacco plants suffering from 'mosaic disease', when thus filtered, can still convey the infection. In 1897, Friedrich Loeffler (1852–1913) showed that the same is true of the virus of the foot-and-mouth disease. This opened a new field. The investigation of these 'viruses' is a task of our time. The question of their relation to non-living substance and of their origin and fate are among the contemporary problems.

(ii) *Embryology, Embryogenesis, Sex*

(*a*) *Embryology and Embryogenesis*

The genesis of living things has always roused interest and speculation. The logical order would be to start with consideration of the nature of sex, especially in its cellular expression, then to follow with the earliest stages of development and to end with an account of the later process by which the embryo becomes adult. This, however, has not been the order of scientific advance. Technical difficulties delayed effective analysis of sex and of embryogenesis until those of embryonic development had become clear. We therefore follow something nearer to the historic than the logical order, treating first the embryo and embryogenesis and then sex. To do otherwise would obfuscate historic perspective.

Knowledge of embryogenesis effectively began with KARL ERNST VON BAER (1792–1876). He saw the mammalian ovum in 1827. Between 1828 and 1837 he illustrated the development of vertebrates, basing his account on that classical subject, the hen's egg. In doing this he compared the gill slits of fish to those which appear tempor-

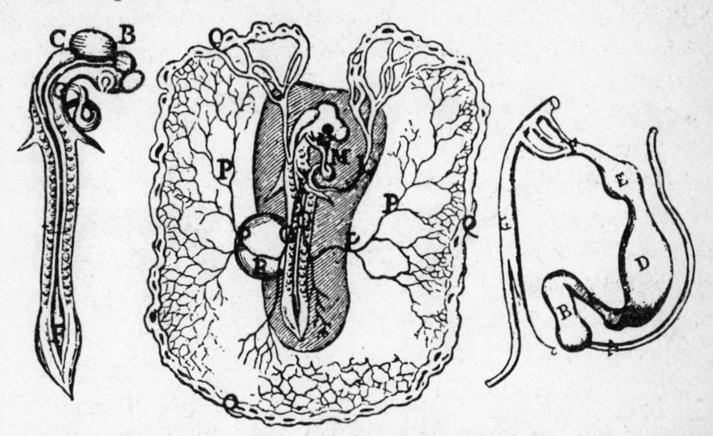

FIG. 178. Developing chick from Malpighi 1686. In the centre is the whole embryonic area, at about the end of the second day of incubation. The embryo itself is seen with its large head containing the three 'cerebral vesicles', which are the rudiments of the brain, the large eye, the protuberant coiled heart, M, from which arteries pass to the 'vascular area'. The segmented vertebral column is well seen, as well as the vessels forming a network as they meander over the vascular area PP. The vitelline veins which bring blood back to the heart are seen at QQQ. To the left is the embryo more enlarged. To the right is a further enlarged figure of the heart; the part D will ultimately form the ventricle, B the atrium, and E the aorta. At F the aorta sends forth three branches which unite again. The nature of these was not understood in Malpighi's time. They have been explained by embryologists, under the inspiration of evolutionary theory, as corresponding to the blood vessels that supply the gills of a fish-like ancestor.

arily in the chick. The vessels that supply their walls had been seen by Malpighi (pp. 283 f.) in the seventeenth century (Fig. 178) as well as by certain of von Baer's contemporaries who gave no meaning to them.

ROBERT REMAK (1815–65) and JOHANNES MÜLLER (1801–58) demonstrated (1843) that the animal ovum—which they recognized as a cell (p. 413)—when fertilized, divided into a mass of cells according to regular rules. The mass separated into three 'germ layers' which they named 'ectoderm', 'endoderm', and 'mesoderm' (Fig. 179). The first forms the skin and, by invagination, the nervous system;

the second the digestive organs; the third most of the remaining organs and tissues. In 1849 T. H. Huxley (1825–95) suggested that jellyfish and kindred forms, unlike higher organisms, have not three germ-layers but only two.

Many details of development suggested that, after the three germ-

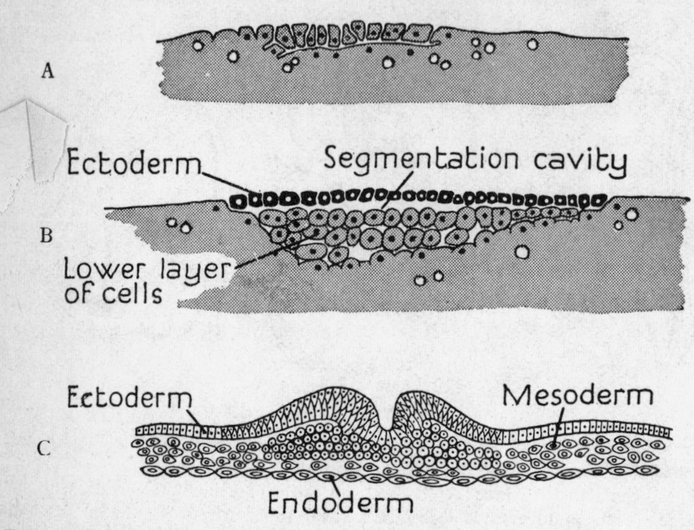

FIG. 179. Diagram of sections of three early stages in development of hen's egg. A. Section through germinal disk of egg about the middle of its stay in the uterus. The cells are shown segmenting off from the yolk. At the edges of the disk are seen nuclei still embedded in yolk. B. Section towards the end of segmentation. The ectoderm and endoderm are separated from each other and a segmentation cavity has made its appearance. This corresponds to the beginning of the gastrula stage (Fig. 181) greatly deformed by the mass of yolk. The ectoderm is represented as black. C. Transverse section of the embryo about the twentieth hour of incubation. The ectoderm and endoderm are clearly differentiated from each other. The neural groove has appeared. From the junction of ectoderm and endoderm the mesoderm is being given off.

layers have differentiated, the structure of the embryo comes to resemble successively organisms of an order lower than itself but becoming progressively higher as development advances. This 'recapitulation' view provided the main doctrinal basis of embryological research in the later nineteenth century.

In 1854 the gifted LOUIS AGASSIZ (1807–73) produced his *Lectures on Comparative Embryology*. They were intended for laymen and were reported in a Boston paper. The work contains the phrase 'In studying the metamorphoses of animals, from the first formation to the full-grown condition, we may find a natural scale by which we can measure the position we may ascribe to any animal belonging to this family.' In 1859 he went further and wrote: 'Embryology furnishes the best measure of the true affinities between animals.' Agassiz was an opponent of Darwinian evolution. Later, when organic evolution had become generally accepted, the recapitulation

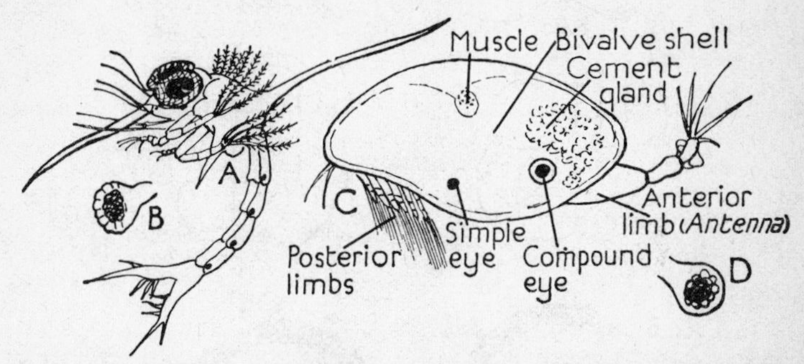

FIG. 180. Developing forms of Crustacea from Vaughan Thompson, 1828–30. A. 'Zoea' proved by Thompson to be the larval form of the shore crab. B. Stalked compound eye of slightly older 'zoea'. This type of eye is characteristic of certain groups of higher Crustacea. C. Free-swimming Crustacean, proved by Thompson to be the larval form of a fixed 'barnacle', an organism not at that time classed as a crustacean. The larva attaches itself by its anterior limb or 'antenna'. This becomes converted into the foot or peduncle of the adult barnacle. The foot becomes attached to a fixed object by cement from the cement gland. The larval form possesses eyes both simple and compound. These disappear in the adult barnacle. D. Compound eye of larval barnacle, which may be compared to B.

view was epitomized by Haeckel (pp. 475 f.) as 'Ontogeny [Greek "individual development"] is an epitome of Phylogeny' [Greek "race development"].

Nineteenth-century Darwinians greatly oversimplified the bearing of individual development on evolutionary history. An interesting case is Fritz Müller (1821–97), an eccentric pupil of Johannes Müller. His work (1864), though conversational, contains many new observations. Treating the larval forms of certain Crustacea as 'historical documents', he regarded their phases as closely resembling their ancestors. Examples of such phases had long before been seen by J. Vaughan Thompson (1779–1847) who had no evolutionary theories (Fig. 180). Fritz Müller's views roused in Haeckel (p. 475) an enthusiasm for embryology. This, in turn, inspired a series of magnificent studies by Haeckel's pupil, ALEXANDER KOWALEWSKY (1840–1901). Specially influential were those on Amphioxus

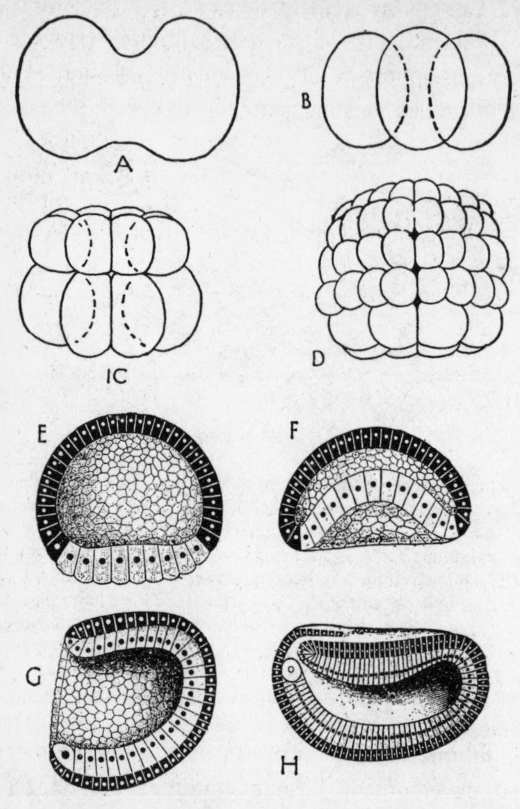

FIG. 181. Early development of Amphioxus. A–D. Segmentation of the egg. E–M. Later stages shown in section. The cells of the ectoderm are represented as black, with white nuclei, the cells of the endoderm and mesoderm as white, with black nuclei. E, F, and G. Formation of gastrula. H. Beginning of formation of neural canal, that is, of the central nervous system. In G and H a large posterior cell is seen which is the beginning of the mesoderm.

(1866–77). They have largely determined the modern view of the position of the vertebrates (Fig. 181–183).

Amphioxus is a small lancet-shaped marine creature first discovered in 1834 in British waters. It has many features in common with the lowest vertebrates and with all vertebrate embryos. Thus it has no backbone but in its place a stiff rod of peculiar structure, the *notochord* (Greek *nōton* 'back'). This organ is present in all vertebrate-embryos but in adults is more or less completely replaced by a chain of cartilages or bones, the vertebrae. Like a fish and like all vertebrate embryos, its gullet is pierced by a series of gill-slits (Fig. 183). The later embryonic stages of Amphioxus, like the adult

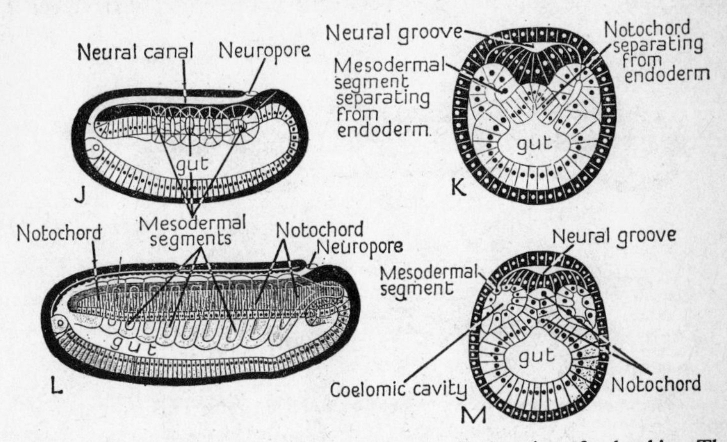

FIG. 182. J–M. Sections of Embryos of Amphioxus. J. Embryo just after hatching. The neural canal is now complete and opens anteriorly by a 'neuropore'. Segments of mesoderm have formed at the side of the gut. K. Transverse section through middle of J. The notochord is separating from the endoderm. The mesodermal segments, each with a hollow cavity ('coelom'), are also separating off from the endoderm, a somewhat unusual developmental relation. In most coelomates the mesoderm arises from the junction of endoderm and mesoderm and splits after separation. L. More advanced larva with nine mesodermal segments, the last of which is still connected with the point of junction of ectoderm and endoderm. Notochord well developed. M. Transverse section through the middle of L. The mesodermal segments are now quite separate and show each its own cavity.

organism, present many features characteristic of the most primitive fishes. Like them it is limbless and jawless, and has a regularly segmented muscular system arising from the mesoderm and a continuous notochord pinched off from the endoderm (Figs. 182, 183).

ERNST HAECKEL (1834–1919) was an extreme champion of evolutionary theory. He was an excellent artist, whose imagination was prone to rule his pen and brush. For more than a generation he purveyed a crude philosophy which grew into something resembling a religion. In science, his peculiar use of analogy resembled that of the scholastics whom he denounced. He confidently constructed genealogical trees of living things, which now raise only a smile. Most of his works rest undisturbed on the less accessible shelves of libraries. Nevertheless, he was a great biologist and his writings contain much that is still part of the fabric of biological thought.

Haeckel's pupil, Alexander Kowalewsky, when investigating the embryology of Amphioxus, observed that these creatures develop,

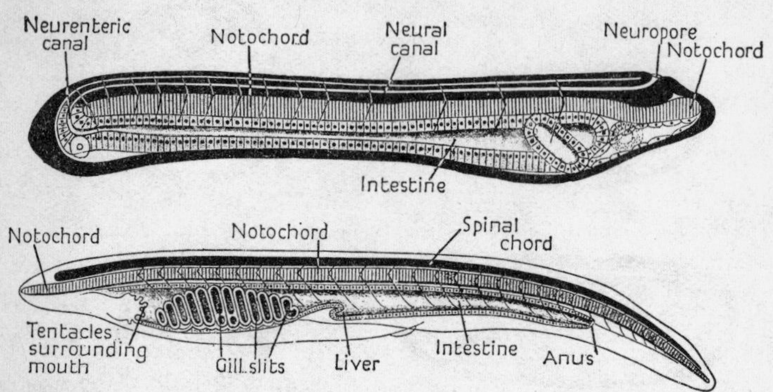

FIG. 183. *Above*, Late embryo Amphioxus with 14 body segments indicated by angular lines. *Below*, Young Amphioxus. Body segments indicated by angular lines. Gill slits in process of formation. Kowalewsky (1867).

by progressive division of the egg-cell, into a hollow sphere, the *morula* (Latin 'little mulberry'). This in due course invaginates into a cup, much as a hollow ball may be invaginated, the *gastraea* (Greek *gaster* stomach, womb). These were held by many naturalists to be ancestral forms of all multicellular animals (Fig. 181).

From about 1878 the gastraea theory was developed by pupils of Haeckel, notably by the brothers Oscar (1849–1922) and Richard (1850–1937) Hertwig, into the *coelom* theory (Greek *koilŏma* 'cavity'). This generalization distinguishes orders of animals, in whose development the mesoderm splits into two layers enclosing a cavity lined by a definite sheet of flat cells, from those in which this does not happen. This coelom and its varied fates and forms is still the basis of classification of animals. A ready example of a coelomic cavity is that containing the abdominal viscera of man. From about 1880 the classificatory scheme which follows, largely due to Haeckel and the Hertwigs, has been generally acceptable to biologists.

A. PROTOZOA (unicellular forms).

B. METAZOA (multicellular forms).

1. *Acoelomata* (without coelom).
   (*a*) Body of only 2 layers (e.g. Sponges, Jellyfish, Sea-anemones).
   (*b*) Body of 3 layers (e.g. Turbellaria, Flukes).

2. *Coelomata*.
   (*a*) Annelida (e.g. Segmented worms, e.g. Earthworms).
   (*b*) Arthropoda (e.g. Crustacea, Insects, Spiders).
   (*c*) Mollusca (e.g. Snails, Mussells).
   (*d*) Chordata (with notochord).
      (i) Protochordata (e.g. Amphioxus).
      (ii) Vertebrates.

In each division there are many other groups which are here omitted for the sake of simplicity.

There has always been a school that seeks to refer all the phenomena presented by living things to known physical laws. Such was the Swiss, WILHELM HIS (1831–1904), who described lucidly and exactly the development of the higher animals and especially of man. He was a main agent in introducing the systematic use of an instrument for cutting serial sections (1870), and such a 'microtome' became indispensable for biological research. He treated the various layers and organs of the embryo as tubes and plates, some more, some less elastic (1874). Local inequalities of growth and differences in the consistence of the tissues could, he held, account for the formation of the various organs and structures.

To accept such mechanical forces as themselves adequate to explain development involves the ancient fallacy that the originals (primordia or rudiments) of the various organs are represented in the germ-layers. The germ-layers themselves would then be similarly represented in the ovum. This is a thin disguise for the doctrine of preformation (pp. 384 f.). Apart from theoretical objections there is experimental evidence against this. But for a time, certain observations, notably those of WILHELM ROUX (1850–1924), seemed to be in its favour. Moreover it accorded with the general mechanistic teaching of the age.

Roux, another pupil of Haeckel, initiated experimental embryology. In 1881 he set out a mechanical basis for the functional adaptation of the parts to each other. He saw that development cannot be treated as a uniform process and distinguished two periods. In the first, the true *embryonic period*, the structures were formed that are predetermined[1] (Roux's own term). The second is the *period of functional development* when, according to Roux, the structure of the parts is influenced by their specific action. In this stage structure determines function, but function also determines structure. This is true but as thus analysed goes little further in illuminating either. A mystery does not cease to be mysterious either by limitation in space and time, or by re-expression in terms of the unknown. Nevertheless, the experimental approach of Roux was a real advance in embryological method.

Like his contemporary, Weismann (p. 483), Roux held that the fertilized egg-cell contains a very complex structure or machine which, after fertilization, is divided or disintegrated into its constituent but still complex parts by progressive division of the cells. A critical experiment raised in a new way the old antithesis between preformation and epigenesis (p. 384). In a frog's egg which had just divided into two cells, Roux destroyed one cell without injuring the other (1888). The living cell continued to develop but as a half-embryo. This would seem to imply preformation, to the extent, at least, that one cell contains the germ of one half of the frog, the other of the other half. Nevertheless, this view did not prove generally tenable (see below).

O. Hertwig had already shown the value of eggs of sea-urchins for experiment (1884). HANS DRIESCH (1867–1941), one of the few scientific anti-Darwinians of the late nineteenth century, began to experiment with these eggs in 1891. He succeeded in separating the two cells into which a sea-urchin's egg had segmented, and from each half-egg reared a complete larva, half the normal size (1900). He obtained complete embryos from yet later stages of division, even down to $\frac{1}{32}$nd of the normal size. It was later shown that in a frog's egg, such as that on which Roux had experimented, if the dead cell be skilfully removed, the remaining cell will develop as a whole embryo, though of course half the normal size. Such contradictory

[1] The term is significant as showing how even extreme mechanists fall into teleological language.

results focused attention on experiment and in this mood embryology began its course in the twentieth century.

We turn to consider the knowledge of the early divisions of the egg-cell at the end of the nineteenth century. The process had been found to have great differences in different organisms. A main factor was the presence of yolk. This provides nourishment to the embryo. Some eggs, like the hen's, contain masses of yolk, the protoplasm being collected at one spot, where division begins (Fig. 179). The yolk is gradually absorbed by the developing embryo and finally the yolk-sac itself is included in its alimentary canal. Other eggs, such as that of the Amphioxus, have no yolk and divide, at first, almost uniformly (Fig. 181). Yet other eggs, such as that of the frog, are intermediate between these two types. O. Hertwig showed (1897) that the distribution of the yolk itself is a mechanical cause of great differences in the mode of development. Thus he centrifugalized frogs' eggs so that the yolk collected at one pole and the protoplasm at the other. The egg then divides only at the protoplasmic pole as with the hen's egg. This has been confirmed in many other ways on many other animals.

Interesting results were elicited especially by HANS SPEMANN (1869–1947) in connexion with the power of healing or regeneration of lost parts, especially in embryonic and growing animals. It is remarkable that certain nerves are very faithful to certain structures, despite evolutionary changes in the form, position, and function of those structures. Thus the complexity of the nerve supply of an organ may tell of its evolutionary history. This affinity between an organ and its nerve is expressed by the refusal, as it were, of an ingrafted organ to co-operate with any but the nerves which normally supply it. Spemann, however, showed that it was also true that some organs, if suitably ingrafted, can be persuaded to form quite abnormal connexions. Thus the eye of an amphibian can be removed and engrafted in another individual in such a way as to acquire its innervation from the nerve that normally supplies the tongue. The subject was, however, hardly developed until after the period of which we treat.

(*b*) *Sex and the Cellular Phenomena of Reproduction*

By the sixties it had become evident that all development, growth, and tissue production could be expressed in terms of the rate and manner in which cells reproduce themselves. It was also recognized that the nucleus was in some sense the controller of cell-division. Simple or 'direct' division of the nucleus had been seen. With the improved techniques of stains, microtomes, and oil immersion lenses of the seventies and eighties, the much more complex 'indirect'

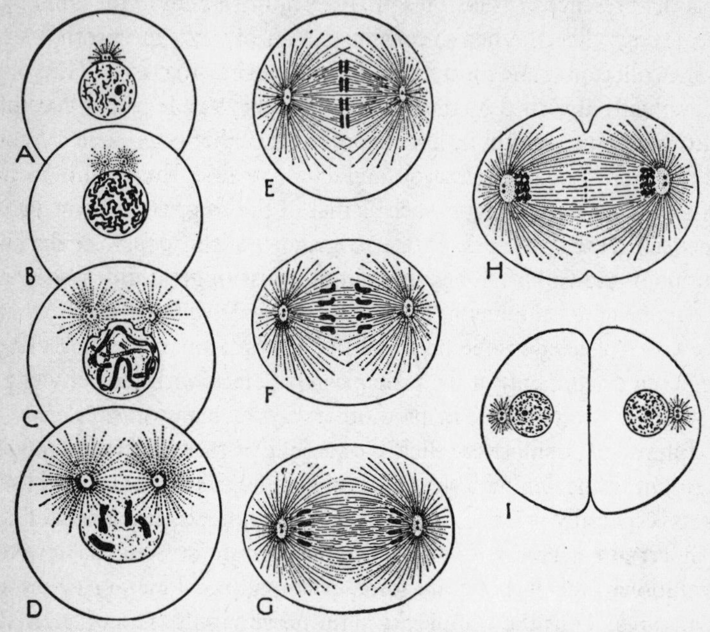

FIG. 184. Diagrams of process of mitosis in animals. A–D, Prophase. E. Metaphase. F, G. Anaphase. H, I. Telophase. Modified from Wilson.

nuclear division became recognized as normal. This remarkable phenomenon was elucidated in plants, notably by EDOUARD STRASBURGER (1844–1912) and in animals by WALTHER FLEMMING (1843–1915) in the early seventies. Indirect cell-division, as worked out by them, may be reduced to a schematic form (Fig. 184).

In an ordinary resting nucleus there can be distinguished a fine network of material that stains deeply with certain dyes. As division approaches, this becomes arranged into a long thread that is more or less regular, continuous, and spiral. The thread breaks up into a series of separate filaments, the chromosomes (Greek 'colour bodies') which stain deeply. By the time that the chromosomes are distinguishable, other changes have occurred. The delicate membrane which surrounds the nucleus has disappeared. On either side of the nucleus a minute body, the *centrosome*, has become apparent.[1] From each centrosome there radiates a series of lines forming a starlike *aster*.

The two asters extend until they meet around the equator of the nucleus. There the chromosomes collect. They then split along their length, and the halves pass, each in an opposite direction to its twin, along the lines of the asters toward opposite poles where they become clustered. Around each cluster a fine membrane forms. The daughter nuclei are now complete and the chromosomes lose their distinct outlines. In the protoplasm between the two daughter nuclei a partition appears. Thus arise two cells, each with its own nucleus. This process was called by Flemming (1882) *mitosis* (Greek *miton* 'thread'). It was suggested that this process is to effect an exact partition of mother nucleus between daughter nuclei. Its 'meaning'[2] has become more evident, as the chromosomes have been studied further, in connexion with the sexual process to which we now turn.

It had long been known that as the animal ovum becomes mature and ready for fertilization, it gave off two minute 'polar bodies', which play no further active part. In 1877 O. Hertwig showed that each of them is formed around a nucleus given off from the nucleus of the unfertilized ovum. Important additions to the knowledge of the reproductive process were made in 1887 by EDOUARD VAN BENEDEN (1845–1910). He found that the number of chromosomes was the same for each cell in a given body, and probably characteristic for each species. He further showed that the number of chromosomes was halved during production of polar bodies—a process known as maturation—and restored during the union of the sex-cells. For this work he used a species of intestinal round-worm because its chromosomes are large and few—four only in the body cells, two in the sex-cells (Fig. 186). The reduction in number of the chromosomes of the sex-cells became known as *meiosis* (Greek 'diminution').

Meanwhile, THEODOR BOVERI (1862–1918), a pupil of O. Hertwig, was investigating the centrosomes and their surrounding asters. He

[1] Most plant cells have no centrosomes.

[2] A teleological term often used by biological mechanists.

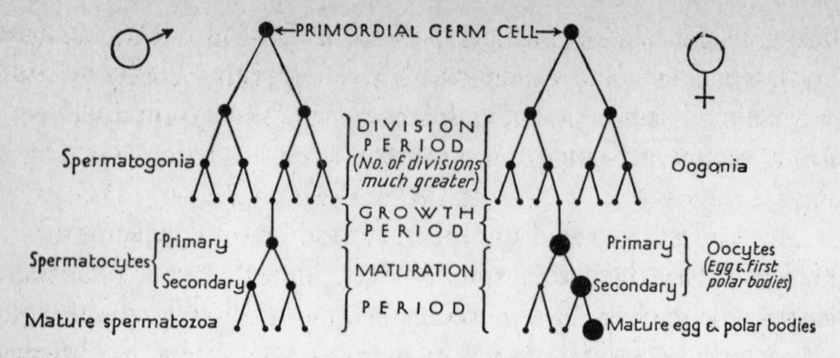

FIG. 185. Diagram of origin of spermatozoa and ova.

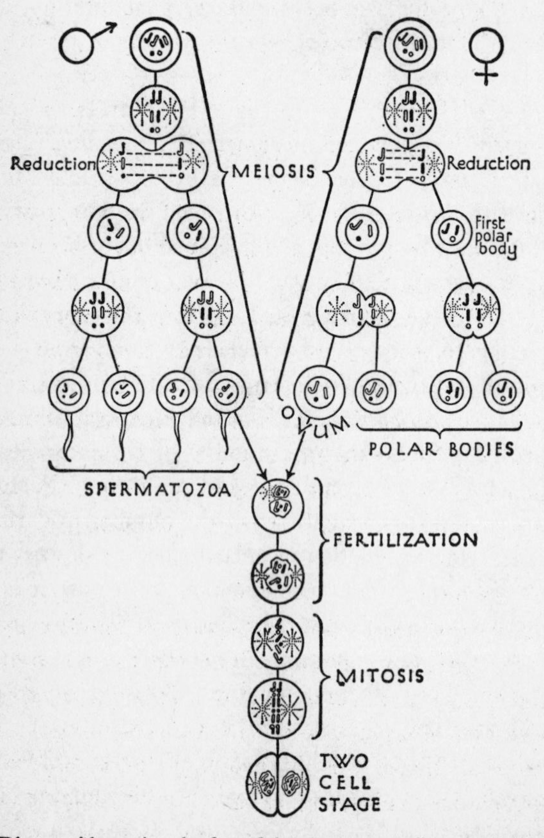

FIG. 186. Diagram illustrating the fate of the chromosomes in animals during meiosis and subsequent fertilization.

showed their relation to the processes of division (1887), treating them as the dynamic system of the cell. In 1890 Hertwig pointed out that the nuclear and cellular phenomena displayed in the formation of ovum and spermatozoon, though superficially widely different, are basically alike. In 1892 Boveri drew up the general diagram of spermatogenesis and oogenesis that is still current (Fig. 185).

In 1894 Strasburger in a searching examination of the development in higher non-flowering plants (ferns, &c.) observed a process of reduction in the number of chromosomes similar to that distin-

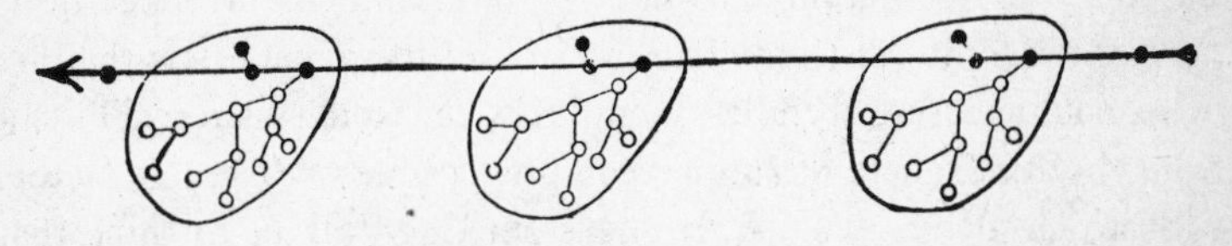

FIG. 187. Diagram to illustrate Weismann's conception of the continuity of the germ-plasm. Only the germ-cells (black) carrying the germ-plasm are continued from generation to generation. The body-cells (white) are destined for death.

guished by Van Beneden in the maturation of the egg of a round-worm. He inferred that, in these plants, the chromosome number of the sex-cells must be half that of the plant generation that bears them. This close parallelism between plants and animals has been worked out in much detail and is a most impressive fact.

In 1895 KARL RABL (1853–1917) made a valuable suggestion based on theory. He expressed the view that the chromosomes retain their individuality even when the nucleus is in the 'resting' stage, when the chromatin usually appears as a network. This persistence of the chromosomes is presupposed by modern genetics.

(iii) *Rise of Genetics*

The nuclear phenomena of sex were thus familiar by the last decade of the nineteenth century. Their relationships to the phenomena of heredity were not at all clear though they were in part forecast in 1892 by AUGUST WEISMANN (1834–1914).

Weismann developed the conception of the continuity of substance from parent to offspring. He gave to this substance the name *germ-plasm*. For him, the offspring resembles the parent simply because it is derived from the same substance, i.e. the germ-plasm, the essential germinal element of the germ-cells. The body he looked on as a vehicle for the conveyance of the germ-plasm. 'A hen', said Samuel Butler (1835–1902), 'is simply an egg's way of producing another egg.' The body-cells do not convey any of their own substance but merely nourish and carry the germ-cells, which become the succeeding generation.

The picture that Weismann raised is of a continuous stream of life (Fig. 184). These seemingly simple germ-cells transmit an inconceivably complex inheritance. The germ-plasm has passed through the countless generations that have been. It contains, wrapped in its mysterious folds, all the possibilities of all the generations that are to be. Such a picture of endless complexity and wonder is very different from the atmosphere of simple, comprehensible, measurable factors with which men of science are accustomed to deal, or to think that they deal. The vision is not very far from those mysteries associated with some religions.

The question arose, is the germ-plasm homogeneous or do only parts of it carry the hereditary qualities? The answer could be obtained only by examination of the structure and behaviour of the germ-cells and notably of their nuclei. This task occupied many observers during the late nineteenth century and has been continued with great activity to our time.

It soon became clear that it is the nuclei that carry most, if not all, the elements of heredity since the spermatozoon, genetically equal to the ovum, consists of little but nucleus. Moreover, it was found that part of the protoplasm of the egg could be removed without affecting the general course of development.

Weismann's stress on the importance of the germ-plasm led him to regard it as the sole transmitter of the heritage of the past. Could it be altered in the present? Could characters impressed on the body from without be passed on to offspring? His answer was 'No'. The matter seemed susceptible of experimental test. The experiments applied in Weismann's day were of the nature of mutilations. The evidence showed that characters thus modified are *not* passed on. But in the twentieth century 'mutations' (nearly always disadvantageous) have been artificially induced by direct action on the sex-cells themselves. Much information of this kind became more intelligible in the light of the discoveries of Mendel which were published in 1866 and 1869 and 'rediscovered' in 1900.

GREGOR MENDEL (1822–84) in examining pea-plants found two sharply marked races, the tall and the short. Mendel experimentally fertilized flowers of tall plants with pollen of short. The offspring, the *first filial generation*, were all tall plants. Tallness prevailed over shortness. Tallness is *dominant*, as we now say, over shortness which is *recessive*.

Mendel next let the flowers of this first filial generation be fertilized with their own pollen. In the following or *second filial generation*, shortness reappeared in some of the offspring. The opposed characters of tallness and shortness were, however, distributed not at random but in a definite, constant, and simple ratio of three dominant talls to one recessive short.

Moreover, Mendel showed that if the recessives of the second filial generation be self-fertilized, they invariably breed true. If, however, the dominants of the same generation be self-fertilized, they breed true only in a certain fixed proportion of cases. The proportion is thus expressed. One of every three talls breeds true, and the other two behave as the first filial generation in giving three talls to one short. In fact, so far as these characters of tallness and shortness are concerned, there are only three kinds of pea. These are:

(*a*) Shorts which breed true.
(*b*) Talls which breed true.
(*c*) Talls which give a fixed proportion of talls and shorts.

These results may be expressed in a genealogical tree:

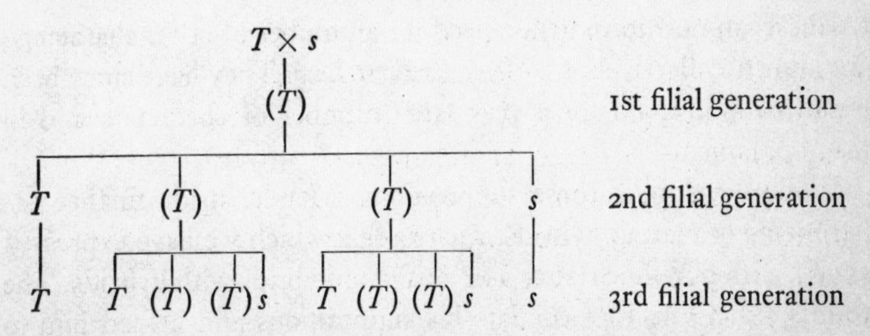

In this tree it is not necessary to exhibit both parents except those of the parental generation, since all other generations are self-fertilized. Thus, in each generation, except the first, the male and female elements carry the same inheritance. (*T*) is used here as a sign for a tall giving a fixed proportion of talls and shorts, *T* for a tall that breeds true, *s* for a short that breeds true, as all shorts do.

Mendel showed that the same constant numerical distribution of opposed characters persists throughout successive generations bred from a (*T*). He elicited an identical numerical distribution for a number of other opposed characters, among them being round form and wrinkled form and green and yellow colour of the peas themselves. Opposed characters that behave thus are now known as 'Mendelian'.

Now came the question of the interpretation of these phenomena. Mendel concluded that they could be explained, or rather re-expressed, by making the following four suppositions:

(*a*) Each of these pea plants carried within itself two height-influencing factors. These two are either one for tallness together with one for shortness, or again two for tallness, or again two for shortness.

(*b*) Of the two factors carried, one only passes to each offspring. Thus the offspring derives one of its two factors from each of its two parents.

(*c*) The chances of the combination of either factor from one parent with either factor from the other parent are equal.

(*d*) In the case of dominant factors, whether either one or two be carried, the individual will manifest the dominant character. In the case of recessive factors, the individual will bear the recessive character only if two be carried.

These suppositions, generalized for a number of other characters, are known collectively as *Mendel's First Law*. They have since been found to hold good for a very large number of characters and in many organisms.

While developing these suppositions, Mendel made further experiments of mating hybrids, such as that which we have expressed as (*T*), with pure lines (such as *T* or *s*) and hybrids with hybrids. The results aided him to formulate his suppositions and helped him to confirm them. These results may be expressed graphically in a simple manner. Let *T* represent a tallness-carrying dominant factor and the small letter *s* a shortness-carrying recessive factor. Then every possible type of parentage is covered by the formulae *TT*, *Ts*, *sT*, or *ss*. Of these four types, *Ts* and *sT* are indistinguishable and are, in fact the forms represented above by (*T*).

Let us now consider the offspring from all kinds of mating of these three types. Let each square represent one offspring which carried two height factors, one from each parent. Only four offspring are represented in each case, but for higher numbers the average results are statistically as indicated.

(*a*) Result expected and found when *TT* is mated with *ss*:

	T	*T*
s	*Ts*	*Ts*
s	*Ts*	*Ts*

i.e. 4 *Ts*

(*b*) Result expected and found when *Ts* is mated with *Ts*:

	T	*s*
T	*TT*	*sT*
s	*Ts*	*ss*

i.e. 1 *TT*, 2 *Ts*, 1 *ss*

(*c*) Result expected and found when *TT* is mated with *Ts*:

	T	*T*
T	*TT*	*TT*
s	*Ts*	*Ts*

i.e. 2 *TT*, 2 *Ts*

(*d*) Result expected and found when *Ts* is mated with *ss*:

	T	*s*
s	*Ts*	*ss*
s	*Ts*	*ss*

i.e. 2 *Ts*, 2 *ss*

The simple suppositions of Mendel are capable of exhibition for other pairs of opposed characters which do not interfere with each other. Thus tallness and shortness, wrinkledness and smoothness, greenness and yellowness, each behaves in a Mendelian fashion without regard to the other pairs.

Mendel also made experiments with two and more pairs of opposed characters. Thus he crossed peas whose seeds were yellow *Y* and round *R* with peas whose seeds were green *g* and wrinkled *w*, *Y* being dominant to *g* and *R* to *w*. The offspring of the first filial generation were all yellow and round.

The numerical results in the second filial generation may be thus expressed:

When the first filial generation, usually indicated as *F*1, was self-fertilized, the result in the progeny was as follows:

Yellow and Round . . .	9
Yellow and wrinkled . .	3
green and Round . . .	3
green and wrinkled . .	1

The meaning of this distribution can be grasped from the accompanying diagram (Fig. 188). It will be seen from the diagram that either one of the colour factors may meet equally often with either one of the shape factors, the manifestation in the offspring being, however, controlled by the operation of *dominance* and *recessiveness.* Comparable but more complex results can be obtained from three or more pairs of opposed characters.

The statement that a pair of opposed characters will behave independently of another pair and yield results on a numerical basis which illustrate this independence is known as *Mendel's Second Law.* It has needed much modification in the light of subsequent knowledge. Nevertheless, since Mendel's time an enormous number of simple Mendelian characters in many organisms, both animals and plants, has come to light.

In 1866 Mendel published his highly significant results in the clearest possible manner. Nevertheless they were almost ignored during his lifetime. The explanation lies perhaps in the orientation of the research of the age. Men seldom find that for which they do not look—that is the key to the value of hypothesis to science. The naturalists of the time—following Darwin—were looking always for series of variations which led gradually from one species to another. Their experimentation was always between varieties that differed

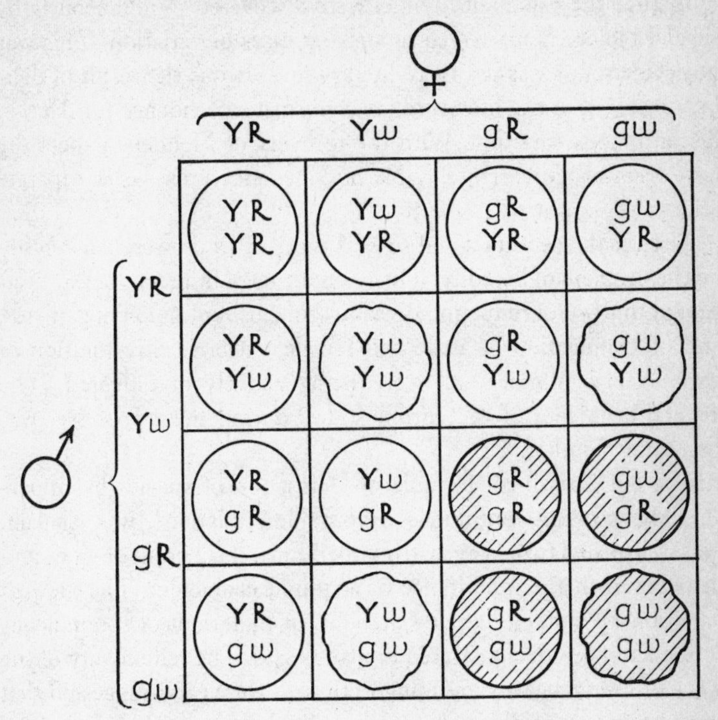

FIG. 188. Diagram illustrating the sixteen combinations that result when the four kinds of egg-cells are fertilized by the four kinds of pollen grains.

from each other in many characters. These differences concealed such simple numerical relationships as those elicited by Mendel.

Mendel succeeded because he simplified his problem so that at first only a single difference was under observation and he was fortunate in having selected a form that was readily self-fertilized. Darwin had dismissed large discontinuous changes as unlikely to be of service to the species. During the last years of the century several workers had nevertheless come upon striking cases of discontinuous variation, but the statistical methods of some influential investigators of the time, involving the use of large numbers of individuals, had the effect of concealing these rare natural 'mutations', as they were called from 1894. Thus the statistical method was eliminating discontinuous variation from the observer's range of vision.

In 1900 the papers of Mendel were discovered simultaneously by several biologists interested in striking cases of variation. This was no accident nor was it a mere literary find. It was the result of deliberate search of the literature for confirmatory evidence for data on discontinuous variation. With the recovery of Mendel's papers the new results became intelligible and genetics arose as a separate science.

Thus at the century's end several types of study were converging on the problem of heredity. The phenomena of indirect nuclear division or mitosis, of reduction division or meiosis, of union of gametes, and of conjugation of male and female nuclei as a restoration of chromosome number—all were being actively investigated. The general behaviour of the chromosomes during these processes was becoming familiar.

The old Darwinian formula of 'formation of species by imperceptibly minute continuous favourable variation' was shaken. Naturalists had turned away from the Lamarckian point of view, and disbelieved in the inheritance of acquired characters. The view of variation by sudden jumps, saltations, or mutations of spontaneous origin had been demonstrated in many cases. The rediscovery of the work of Mendel gave a meaning to these features and suggested that the chromosomes themselves were the carriers of heredity. The degree to which bodily characters could be analysed on a Mendelian basis was beginning to be realized as the twentieth century dawned.

(iv) *Physiological Synthesis*

(*a*) *Internal Environment*

With the parallel development of ideas on (*a*) the vital processes as a continuous elaboration and breaking down of living substance, (*b*) the living body as a structure composed of cells, and (*c*) the physical basis of life as protoplasm, there developed views of the organism as a physico-chemical mechanism. Mechanist views have never lacked critics, and it is significant that the most effective critic of mid-nineteenth-century mechanism, JOHANNES MÜLLER (1801–58), was himself an experimental physiologist of genius who is largely responsible for the picture of the body as a machine. His most important physiological investigations dealt with the action and mechanism of the senses, and were important starting-points for modern research.

The doctrine specially associated with Müller's name is the 'principle of specific nerve energies'. This teaches that the general character of the sensation, following the stimulation of a sensory nerve, depends not on the mode of stimulation, but on the nature of the sense organ with which the nerve is linked. Thus mechanical stimulation of the nerve of vision produces luminous impressions, and no other; stimulation of the nerve of hearing gives rise only to an auditory impulse, and so on. It is well to consider some of the implications of this very simple statement.

What do we know of the world in which we live? Only what our senses tell us. But how do our senses convey anything to us? *That* no man can answer. All we know is that certain external events initiate specific disturbances in certain sense organs, that the nerves from these organs convey disturbances to the brain or central nervous system, and that a sensation then arises. We can indeed picture a mechanism by which the external event may elicit a specific nerve-impulse, and we know something of the nature of the impulse and how it travels up the nerve and how it affects the brain. But how this chain of events produces something that *we* feel, and how that may induce *us* to take action—of these things we are not only ignorant but it is difficult to believe that we can ever be other than ignorant. Here perhaps is a veil which never can be rent by mortal man.

But consider further. External events are known to us only through our senses. Nevertheless from one and the same event we may receive completely different sensations. Thus, an electric stimulation of the optic nerve will give rise to a visual sensation; the same stimulation of the olfactory nerve yields a sensation of smell; of the auditory nerve a sensation of sound. Further, different events may give rise to the same order of sensation. Thus it matters not whether the optic nerve be stimulated by electricity, by heat, or mechanically, the sensation aroused will be visual. If our optic nerve were grafted to our auditory organ and our auditory nerve to our optic organ we should find ourselves transported to a world so strange that we cannot form the remotest conception of it. (Such an operation may actually be practicable in certain organisms.) To beings with senses different from ours the world would be utterly different.

The law of specific nerve energies is thus fundamental for our view as to the range of validity of scientific method, and indeed of experience as a whole. That law is a standing criticism of the 'common-sense' view that the world is as we see it, and that its contents, and particularly the living things in it, can be completely understood. Ultimate reality cannot be reduced to a conceptual form.

Müller laid emphasis on the existence of something in the vital process that, he thought, must remain, insusceptible of mechanical explanation or physical measurement. This doctrine occasionally misled him. Thus he held it impossible to measure the velocity of the nervous impulse. Yet it was measured by his own pupil, Helmholtz (1821–94), ten years later. Vitalistic views may be useful to philosophers, but the working man of science had best follow Claude Bernard's advice and forget them while he is at his appointed task.

The French physiologist CLAUDE BERNARD (1813–78), one of the greatest of all biological thinkers and experimenters, was the most effective contributor to the presentation of a concert of all the bodily processes as a chemico-physical mechanism. One of his greatest discoveries was that the liver builds up, from the nutriment brought to it by the blood, certain highly complex substances which it stores against future need, and that these substances, notably that known as *glycogen*, it subsequently modifies for distribution to the body according to its requirements.

It was already recognized that the source of bodily energy is the breaking down of nitrogenous substances, of which the final degradation product is urea (pp. 408 f.). Bernard, by his work on glycogen, demonstrated that the body not only can *break down* but also can *build up* complex chemical substances. This it does according to the requirements of its various parts.

Bernard thus destroyed the conception, then still dominant, that the body could be regarded as a bundle of organs, each with its appropriate and separate functions. He introduced a conception that the various forms of functional activity are interrelated and subordinate to the physiological needs of the body as a whole.

No less important, as bearing on this conception, was Bernard's work on digestion. Up to his time, an elementary knowledge of the facts of digestion in the stomach formed the greater part of digestive physiology. Bernard showed that this digestion is 'only a preparatory act' and that numerous other processes are involved. Thus the juice of the 'pancreas' or sweetbread, poured into the intestine near the lower opening of the stomach, emulsifies the fatty food-substances as they leave the stomach and splits them into fatty acids and glycerin. He showed further that the pancreatic juice has the power to convert insoluble starch into soluble sugar for distribution to the body in the blood, and that it has a solvent action on such proteins as have not been dissolved in the stomach.

A third great synthetic achievement of Bernard was his exposition of the manner of regulation of the blood-supply to the different parts of the body. This we now call the 'vaso-motor mechanism'. In 1840 the existence of muscle-fibres in the coats of the smaller arteries was discovered. Bernard showed that these small vessels contract and expand, thereby regulating the amount of blood supplied to the part to which they are distributed. This variation in calibre of the blood-vessels is, he showed, associated with a complex nervous apparatus. The reactions of the apparatus depend upon a variety of circumstances in a variety of other organs. Thus he provided another illustration of the close and complex interdependence of the various functions of the body upon each other.

Bernard's clear conception of the reciprocal relations of the organic functions led him to a very valuable generalization. He perceived that the characteristic of living things, indeed the test of life, is the preservation of internal conditions despite external change. 'All the vital mechanisms', he held, 'varied as they are, have only one object, that of preserving constant the conditions of life in the internal environment.' This phrase is the seal on Bernard's belief that the living organism is *sui generis*, something quite different from everything that is not living. The organism has an object, and it uses a mechanism for attaining that object. Is this conception infinitely removed from that of Aristotle? Is it not teleology undisguised ?

What is the internal environment of an organism? Bernard was thinking chiefly of the blood. But if we think of a part in terms of cells we see the environment of the cell made up of four main factors:

(*a*) The neighbouring cells and cell products.
(*b*) The substances that are brought to it by the blood.
(*c*) The substances that it throws off and that are removed from it by the blood.
(*d*) The nervous impulses that come to it.

The whole vast mass of physiological research since Bernard's time may be regarded as a commentary on these four factors of the internal environment.

(*b*) *Supremacy of the Nervous System*

It will be impossible to follow further all the factors of internal environment, but there is one upon which we must enlarge, since the whole standpoint with regard to it has altered fundamentally since the time of Bernard. It is the consideration of the nervous system and its relation to the body as a whole.

Between 1811 and 1822 FRANÇOIS MAGENDIE (1783–1855) and Sir CHARLES BELL (1774–1842) had been at work on the double spinal roots from which most of the nerves of the body arise. They showed that of these roots, one conveys only sensory elements, and the other only motor elements. Thus the investigation of the action of individual nerves became possible.

In the first half of the nineteenth century there appeared many comparative studies on the nervous system. Cuvier based his classificatory system in part upon the nervous system (p. 381) but not until the appearance of T. H. Huxley's *Manual of the Anatomy of the Invertebrated Animals* (1877) did full stress come to be laid on the ascendancy of the nervous system in all members of the animal series.

Despite the lead of Huxley, the nervous physiology of invertebrates remained neglected. But the internal structure of the nervous system of mammals had been investigated in very great detail. It has been found to be almost inconceivably complex. The investigations have been greatly helped by the introduction of new technique, at which we may now glance.

The early anatomists recognized that the central nervous system consists of two main parts—the grey and the white matter, and that in the brain the grey matter is mostly on the surface, while in the spinal cord it is mainly central. Soon after the foundation of histology as a special science it was observed that white matter consists of masses of enormous numbers of fibres while grey matter contains also numerous cells. These facts were known to Purkinje (1835, pp. 411–12) and were formally set forth by JAKOB HENLE (1809–85, p. 465). It was, however, more than forty years before KÖLLIKER (pp. 513–14) proved that all nerve-fibres are nothing more than enormously elongated processes given off from nerve-cells with which they retain continuity (1889). These nerve-cells are to be found either in the central nervous system itself or in the various ganglia.

In 1873 the Pavia professor, CAMILLO GOLGI (1844–1926), introduced a method of depositing metallic salts within various cell structures. These deposits are very evident under the microscope, and Golgi succeeded in applying this method to the central nervous system. He showed that the cells in that system tend to resemble irregular polygons from the angles of which project processes, *axons*, the essential parts of the nerve-fibres which ultimately end in a complicated system of branches, *dendrites*. The dendrites form twig-like 'arborizations' round other dendrites linked to other cells. Ultimately the system ends in terminal cells associated with sense organs, or glands, or muscles.

The method of Golgi was developed especially by RAMON Y CAJAL (1852–95) of Madrid. His researches stamped upon biology the conception of an immensely complex series of systems for the transport of nervous impulses. These systems, if intact and working well, determine the activities, the reactions, the whole life of the organism.

While the various nervous tracts were thus being traced, much work was in progress in the localization of the functions of the different parts of the nervous system. In 1861 PAUL BROCA (1824–80), demonstrated in a post-mortem room at Paris a relationship between loss of speech and injury to a definite area of the cortex. Broca made many contributions to the knowledge of the brains of men and of apes.

Others soon continued his work in the experimental field. In 1870 a versatile naturalist, GUSTAV FRITSCH (1838–91), and a student of insanity, EDUARD HITZIG (1838–1907), working together at Berlin, found that stimulation of certain parts of the cortex regularly produced contraction of certain muscles. DAVID FERRIER (1843–1928), followed this up by demonstrating that other areas of the cortex, which do not evoke muscular activity, are nevertheless functionally differentiated (1876).

Since then the surface of the brain has been mapped in great detail. Special areas have been associated with movements of different parts and different organs. Others are related to various forms of sensory discrimination such as sight, sense of position, weight, taste, and the like. Yet others are involved in the use of language, written and spoken. Their interrelationships are complex and often indirect, and are quite unlike those vainly imagined by the phrenologists.

Influential in determining modern views of the action of the nervous system have been researches on the nature of 'reflex action', that is, non-voluntary movement in response to a sensory stimulus. The conception may be traced from Descartes onwards (pp. 276 f.) but the term 'reflex action' was invented (1833) by MARSHALL HALL (1790–1857). The study of reflexes has resulted in the localization of functions in both the grey and white matter of the spinal cord much as with the cortex.

Since Hall's time there has been vast extension of the conception of reflexes. In addition to the simple nervous arc there are also more complex arcs which depend for their action on an elaborate mechanism. Beside 'spasmodic' events, as sneezing, coughing, scratching, &c., many of the ordinary acts of life, standing, walking, breathing, &c., are expressible as reflexes. The attempt has been made by IVAN PAVLOV (1849–1936) and others to press even 'instincts' into the same category, and the cortex has been shown to have the power of establishing new reflexes. This school has sought to explain all the reactions, and indeed most of the life of the higher organisms on a purely objective basis without reference to volitional elements.

If the simple reflexes of animal bodies are tested, it will be found that they clearly serve certain ends. Lightly touch the foot of a sleeping child and it will withdraw it. Tickle the ear of a cat and it will shake it. Exhibit savoury food to a hungry man and his mouth will 'water'. These examples might be multiplied a hundredfold. Such reflexes are admirably adapted to their ends. Many will continue in an animal in which the brain has been removed, provided that the spinal cord be still intact. Nevertheless, in the higher animals, and especially in man, the reflexes are controllable to a greater or less extent by the will.

But to leave the question at that would give a false idea of the extremely complex functions performed by the central nervous system. Thus the spinal cord which, to the naked eye, is a longitudinal and little differentiated nervous mass, is in fact a collection of nerve-centres which have historically, both in the individual and in the race, been formed by the union of a series of separate segments. Each segment in this system governs certain functions or movements of the body, and the activity of each segment is related in various ways to the activity of the other segments. There is thus a highly intricate process of 'integration' which runs right through the nervous system.

The growing knowledge of the bodily functions of chemical and physical nature gradually revealed that these activities are far more largely under nervous control and discipline than was formerly conceived. Thus the main factor in the activity of any part is its blood-supply, but the blood-supply is determined, as Bernard showed (pp. 492 f.), by the state of contraction of the vessels of supply which are in their turn under nervous control. Similar relations prevail for the state of nutrition of muscles, for the action of the sweat glands of the skin, for the mechanism of childbirth, and for a thousand bodily states. The regulation and control of all these events, processes, and states by the nervous system has since come to be called *nervous integration*, a subject specially associated with the work of Sir CHARLES SHERRINGTON (1861–1954).

During the nineteenth century there was an enormous extension of scientific interest in the analytical study of animal function through physical experiment. The exponents of this science of physiology applied themselves mainly to the higher animals. They devoted themselves to an examination of the parts or functions in the adult or developed state. The results were portentous in bulk, complexity, and interest, yet they went only a very little way to help us in considering the organism as a whole.

The animal body is, as it were, a vast and complex maze. The physiologist enters it, and he wanders there as long as he will. But his close and detailed report on its paths and walls helps but little toward the exposition of the design as a whole. The physiologist, in his special studies, is bound to consider isolated functions—wall by wall, path by path as it were. He selects respiration, nutrition, muscular movement, the action of the nervous system, or the like. But the performance of each of the functions of each of these systems is inextricably linked with the performance of the functions of all the other systems. A plan of the labyrinth still eludes us.

We are always looking for metaphors in which to express our idea of life, for our language is inadequate for all its complexities. Life is a labyrinth. But a labyrinth is a static thing, and life is not static. Life is a machine. But machines do not repair themselves, nor do they reproduce themselves. Life is a laboratory, a workshop. But it is a workshop in which a thousand processes go on within a single microscopic cell, all crossing and intercrossing and influencing each other, and it is a workshop which is constantly multiplying itself and producing its like.

It is but a metaphor. When we speak of the ultimate things we can, maybe, speak only in metaphors. Life is a dance, a very elaborate and complex dance. The physiologist cannot consider the dance as a whole. That is beyond his experimental power. Rather he isolates a particular corner or a particular figure. His conception of the dance, as thus derived, is imperfect in itself and, moreover, in obtaining it he has disturbed the very pattern of the dance. The shortcoming of his method becomes fairly evident when he seeks to relate his corner to another in a far distant part of the dance.

Moreover, even should he seek to treat the organism as a whole, he is still almost bound to consider it as an 'individual' complete and separate in itself, shut off from its environment and its history, born, as was Minerva, armed and fully equipped from the head of Jove. But living beings are not so. There is every degree of independence of their fellows among organisms. 'Individuality' too comes into prominence in the higher animal groups. The term is almost inapplicable to plants, in which physiology is, in effect, of a community, and that is a study not far, in its conceptions, from that of bionomics or 'ecology' as it is now called. The very idea of the 'individual' involves an historical record which physiology, alone among the biological sciences, almost completely ignored during the nineteenth century.

Physiology alone is of its nature incapable of presenting any picture of the mode of action of the organism as a whole, though modern doctrines of the workings of the nervous system have given some explanation of certain forms of animal behaviour. Yet the functions of the nervous system, like those of other systems, are relative to the other functions of the body. Not only is respiration, for example, regulated by the nervous system, but the nervous system itself is regulated by the character of the respiration. Raise the amount of carbon dioxide in the blood, and the respiratory movements are first stimulated and finally diminished via action on the respiratory centres. It would be possible to show that the same is true of any system or part of a system in relation to any other. What picture, then, can physiological processes give us of the interrelated complex of activities that we call an organism?

The physiologist has found that his science can be best prosecuted on the higher animals. Why? Because the functions of these creatures are best differentiated. If he wishes to study movement, respiration, nutrition, nervous action, he finds in the higher animals separate organs devoted to these processes. Such organs he cannot so easily, or cannot at all, find in the lower organisms. In certain of them, the Protozoa, every process is carried on in a minute single cell. Are we right to call such physiological marvels 'lower' organisms? Of such marvels there may be tens of thousands in a drop from a puddle.

But the most distinctly and clearly developed characteristic of the highest animals is their psychological status. To discuss these in the mechanistic nomenclature adopted by physiology is merely contradiction in terms. The one thing that we really know is our own thoughts, and external things—including the science of physiology—we know only in relation to these. How then can external things be said in any sense 'to explain' our thoughts? It is more intelligible to invert the process and to say that phenomena—including those of physiology—are parts of our thinking, than to say that our thinking can be built up of phenomena.

But if we emphasize the conception of science as dealing with phenomena—'things which appear'—we reach a *modus vivendi* both for a conception of mind, and for the findings of science. Having agreed that science shall deal only with phenomena, we expressly exclude our own mind, which is not an appearance at all, but that to which appearances happen. Science must keep to the phenomenal lev[illegible]ecute physiological study. But no am[illegible]present an entity in which any ele-me[illegible]nt of mind found in other organ-isms than myself? Unless the solipsist view be taken, this question must be answered in the affirmative.

(v) *Evolution*

(*a*) *The Word*

The leading contributions of the nineteenth century to the conception of a mechanical world are the twin doctrines of Energy and Evolution. As with most important scientific ideas, the enumeration of neither can be dated exactly or placed to one man's credit. To the doctrine of Energy it is convenient to attach the name of Joule, and the date 1842 (pp. 375 f.). The doctrine of Evolution has become so closely linked with the name of Darwin that 'Darwinism' is often taken as a synonym of this doctrine which is dated to 1859, the year of publication of the *Origin of Species*. The term 'Evolution' should, however, be retained for the philosophical view that the world attained its present form not by a single creative act, but by a slow process over long ages. This was held by a number of ancient thinkers such as Plato (p. 43), and by several unorthodox medieval thinkers, such as Averroes (pp. 155 f.). Of this view, the doctrine of Evolution of Organic Forms, or Darwinism proper, is a special case.

The Latin word *evolvere* means to unroll, to roll forth, to revolve. In classical usage its noun *evolutio* acquired the special meaning of the unrolling of a scroll in order to read it, 'the opening of the records' as we might say. In the Vulgate version of the Scriptures (*c.* 400), *evolvere* is used either in its literal sense or, most often, to designate passage of time as marked by the revolving heavens. Derivatives of *evolvere* had little application in the Middle Ages, since scrolls had been replaced by books with leaves, and no form of it occurs in the Authorized Version of the English Bible (1611). The word *Evolution* was given currency in modern literature by the group of seventeenth-century philosophers known as the 'Cambridge Neoplatonists'. They employed it to describe the unrolling, as of a scroll, of vast records of Time (cf. *Revelation* vi. 14; *Isaiah* xxxiv. 4). 'The whole Evolution of ages, from everlasting to everlasting, is represented to God at once', wrote (1667) their founder Henry More (1614–87), paraphrasing 'a thousand years in Thy sight are but as yesterday when it is past' (*Psalm* xc. 4).

Search of the writings of many philosophers of the eighteenth century, notably those of Leibniz (1646–1716), Diderot (1713–84), and Kant (1724–1804), reveals uses of the word evolution extended from that of the Cambridge Neoplatonists, and even adumbrations of the modern philosophical sense considered under heading (5) below. During the next century the word 'Evolution' was developed on lines comparable to those of the Cambridge Neoplatonists by the 'Naturphilosophen', and notably by Oken (pp. 389 f.), in connexion with their doctrine of 'ideas'. In this sense it was reimported into nineteenth-century English, probably by Samuel Taylor Coleridge (1772–1834). 'The sensible world', he wrote, 'is but the evolution of Truth, Love and Life or their opposites in Man' (1820).

In the course of its varied and adventurous career the word 'Evolution' thus acquired many different meanings and shades of meaning. It entered into the technical vocabulary of biological science—where we are here concerned with it—in at least five clearly distinguishable senses.

(1) Evolution naturally and conveniently designated the process, mainly an unfolding, of the parts of a bud opening into a flower; or again of the imago of an insect, such as a butterfly, in its final transformation from the pupa.

(2) There were two rival theories as to how living things develop. One held that the germ contained the living organism in a substantially complete state, folded on itself. This had to *unfold* in order to pass from the embryonic stage. The other held that the germ was at

first uniform, and that the form of the embryo was later generated in it. The philosophical biologist Bonnet (p. 384) gave wide currency to the former view under the name *Evolution* (1762), while the latter came to be known as *Epigenesis*. It is usually said that it is the epigenetic view that has prevailed. In the literal sense, but not in certain other senses, this is the case (pp. 477 f.).

(3) There has always been a philosophical problem of the relation of Being to Becoming. We need not follow this discussion in its vast divarications. St. Augustine posed the problem for the next millennium and a half: 'In the beginning God made Heaven and Earth, that is the *seeds* of Heaven and Earth, for the material of Heaven and Earth was yet in confusion; but since it was inevitable that from these seeds Heaven and Earth would be, therefore the material is thus called' (*De genesi contra Manichaeos*). These are the *seminales rationales* of the great medieval Christian thinkers who stressed *being* rather than *becoming*. These *seminales* in the mind of God were for them the ultimate reality. Bonnet is, in this sense at least, belatedly medieval, insisting that every being already is, and only seems to become. Seventeenth-century thinkers, startled by the changes newly revealed by the telescope in the heavens, and by the extraordinarily complex processes discerned by means of the microscope in the development of individuals on earth, directed attention to *becoming*. This was expressed by the scientific dilettante Matthew Hale (1609–76), for example, who writes of an 'ideal principle in the *evolution* whereof Humane Nature must consist'. Several eighteenth-century authors treat in a similar manner of the 'evolution of ideas', including 'ideas' in the technical sense of the *Naturphilosophie*.

(4) Great confusion has been caused by an early and still current misapplication of this last use of the word in biology. The process of *development* of the organism (*not* its unfolding) became called its 'evolution'. Thus Erasmus Darwin, grandfather of Charles, wrote of 'the gradual *evolution* of the young animal or plant from the egg or seed' (*Botanic Garden*, 1791), meaning its epigenetic development, and *not* its evolution in the sense of Bonnet (p. 384). This confusing usage has persisted to our time.

(5) Finally the word is used for a process (or the result of a process) by which, in long stretches of time, organic types develop (or have

developed) from other types. More or less definite expressions of this view can be traced far back, but no earlier exact use has been found of the word 'Evolution' to designate it than that of Lyell (1797–1875) in his *Principles*. There he discusses in detail the biological theories of Lamarck, and notably the view of that naturalist that 'certain organisms of the ocean existed first, until some of them by gradual *evolution*, were improved into those inhabiting the land' (1831).

The word 'Evolution' has been awarded numerous other technical meanings in departments other than biology, as for instance in mathematics, and in military tactics, where we do not follow it. It is necessary, however, to remind the reader that the biological meanings of the word all interdigitate, and that this fact is not without significance in the development of the philosophical conception of evolution. The word, in fact, carries with it all the trailing clouds of a confused and intricate past.

(*b*) *Early Evolutionists*

Among naturalists, the idea of the transformation of species was more or less overtly expressed by Hooke (1635–1703), Ray (1627–1705), Goethe (1749–1832), Oken (1779–1851), and many others. That it was much in the air is shown by the repeated insistence by Linnaeus, Haller, Bonnet, and many orthodox biological thinkers that species are not transformed from other species but exist in the form in which they were first created. (The difficulties that arose from the geological record in the minds of Cuvier and his contemporaries are reviewed on pp. 326–30 and 393.) The whole direction of biological activity in the period of Linnaean dominance was against discussion of variation or transformation, and favoured treatment of the world of life as something static. Nevertheless, a few eighteenth-century naturalists were able to break away from this view.

The first to give both form and substance to a conception of evolution of living things was BUFFON (1707–88, pp. 326, 399). His great *Natural History* (1749–1804), in forty-four volumes, sought to cover the whole area of natural knowledge. He himself regarded it as a sort of commentary on Newton's conception of a mechanical world. A new element in Buffon's work was its inclusion of living Nature which Newton had disregarded.

Buffon paid little attention to minor differences between organisms on which biological classificatory systems must necessarily be based. Thus the system of his contemporary Linnaeus did not appeal to him. He was interested rather in features that can be traced through very long series of organic forms. As regards the fixity of species he expressed himself variously, but he settled gradually into opposition to that view. Particularly he noted that animals possess parts which have no function as, for example, the lateral toes of the pig which, though perfectly formed, can never come into action. To explain these, he conceived that a species may alter in type from time to time, but retain marks of its previous form, as the pig retains its disused toes. Then, moving a little further, he concluded that some species are degenerate forms of others. Thus the ape is a degraded man, the ass a degraded horse, and so on. Buffon was so popular a writer that a search of the literature of his age would certainly reveal many repetitions and variations of these views (pp. 326 f.).

The ideas of Buffon were examined by ERASMUS DARWIN (1731–1802), grandfather of Charles Darwin. He, like Buffon, sought to indicate how living things had acquired their manifest adaptations to their environment. In his *Zoonomia; or the Laws of Organic Life* (1794–6), he sums up the difficulties among which Buffon had been groping and gathers together precisely those classes of facts that were most to impress his grandson.

When we revolve [writes Erasmus Darwin] first the *changes which we see naturally produced in animals after their birth*, as in the butterfly with painted wings from the crawling caterpillar, or the [air-breathing] frog from the [water-breathing] tadpole; secondly the *changes by artificial cultivation*, as in horses exercised for strength and swiftness, or dogs for strength, courage, or acuteness of smell, or swiftness; thirdly, the *changes produced by climate*, the sheep of warm climates being covered with hair instead of wool, and the hares and partridges which are long buried in the snow becoming white during the winter months; fourthly, the *changes produced before birth by crossing or mutilation*; fifthly, the *similarity of structure in all the warm-blooded animals, including mankind*, one is led to conclude that they have alike been produced from a similar living filament.[1] (Very greatly abbreviated.)

[1] This 'filament' is a spermatozoon which he regarded, following Buffon, as a sort of biological unit.

The mechanism by which such changes come about is, he believed, the transmission of characters acquired, sometimes at least, as an act of will. 'All animals undergo perpetual transformations; which are in part produced by their own exertions . . . and many of these acquired forms or propensities are transmitted to their posterity' (*Zoonomia*).

JEAN BAPTISTE DE MONET DE LAMARCK (1744–1829), was the greatest systematist of his age, but many of his views were so fanciful that he was lightly esteemed by most of his contemporaries. The interest of the theory by which Lamarck is remembered was not fully realized until after his death. It was discussed in detail by Lyell (1831, pp. 326 f.).

Lamarck held that no frontiers can ultimately be found between species. It seemed to him, therefore, intrinsically improbable that they are permanently fixed. In reaching this conclusion he too laid stress on the domesticated animals, which vary greatly from their wild originals. Who, seeing for the first time a greyhound, a spaniel, and a bulldog, would not think of them as different species? Yet all have a common ancestor. Their different characters have been produced by man's selective breeding. In Nature, too, variations comparable in kind to these are occasionally found within the same species. The agent that produces them is, according to Lamarck, the environment. Species, he thought, maintain their constancy only so long as their environment remains unchanged.

But how do changes of environment give rise to variation and produce new species? In answer Lamarck enunciated the 'law of use and disuse', inseparably connected with his name. He supposed that changes of environment lead to special demands on certain organs. These, being specially exercised, become specially developed. Such development, or some degree of it, is transmitted to the offspring. Thus a deer-like animal, finding herbage scanty, took to feeding on leaves of trees. It needed a longer neck to reach the leaves. In the course of generations, during which the poor creatures were always straining their necks to reach their food, long necks became an ever more accentuated feature of their anatomy. Thus emerged a beast recognizable as a giraffe. Conversely, useless organs, such as the eyes of animals that live in darkness, being unexercised, gradually became functionless and finally disappeared. The character of a longer neck or of defective eyes was acquired by the individual in its lifetime and transmitted, in some degree at least, to its descendants.

The great assumption is that acquired characters are inherited. Whether and in what sense acquired characters can be inherited is a matter of current discussion, but it is certain that in the sense suggested by Lamarck they are not. Nevertheless Lamarck directed attention to one of the most important biological problems. Unfortunately some of his early supporters set forth evolutionary schemes that were fantastic to the last degree. This resulted in biological speculation falling into disrepute for the first half of the nineteenth century.

Yet there was one writer of the time, whose work bore upon the subject, against whom the charge of reckless speculation could most certainly not be made. The Rev. T. R. MALTHUS (1766–1834) was a cautious and somewhat formal writer on mathematical and economic subjects. He produced anonymously in 1798 his *Essay on Population* At that time political theory was a matter of acute controversy in connexion with the French Revolution. Such topics as the 'rights of man', 'natural justice', and the like were in the public mind. The most flourishing school of thought in England was the 'utilitarian', the direct ancestor of that liberal philosophy on which Britain rose to industrial and imperial greatness during the nineteenth century. Adam Smith (1723–90), Joseph Priestley (1733–1804), and Jeremy Bentham (1748–1832) were the chief spokesmen of this movement. Many believed that a day was dawning when, amidst universal peace, all men would enjoy complete liberty and equality. Malthus brought out the difficulties that must arise in such a state from over-population, by his famous (but fallacious) principle that populations increase in geometric, but subsistence at best only in arithmetic ratio. He argued that a stage must be reached at which increase in population will be limited by sheer want. Thus he held that 'checks' on population are a necessity in order to avoid vice and misery.

Darwin read a later edition of the *Essay* of Malthus in 1838, and the *Principles* of Lyell in 1831. The one suggested to him the idea of the Struggle for Existence and the Survival of the Fittest, the other the general doctrine of Evolution. In the first half of the nineteenth century both these ideas were discussed by several biological writers, accessible to Darwin. None put the two ideas together, or at least none put them together adequately.

(*c*) '*The Origin of Species*'

It is the great achievement of CHARLES DARWIN (1809–82) that he persuaded the scientific world, once and for all, that many diverse organic forms are of common descent, that species are inconstant and in some cases impossible of definition, and that some mechanism must be sought to explain their evolution. In search of this mechanism, he directed attention to the occurrence of variation, to its persistence, and to the question of its origin and its fate.

In 1859 appeared Darwin's classic *Origin of Species*. He had opened a notebook on the subject in 1837, made a first draft of it in 1842, a second in 1844, and in 1858 published, simultaneously with ALFRED RUSSEL WALLACE (1823–1913), a preliminary sketch of his views. It is interesting that Wallace, like Darwin, seems to have derived ideas from Malthus. The *Origin* is one of the world's great books, and has proved significant for almost every human activity. But despite the conviction that it carried, and despite the fact that for the half-century after its publication its ideas provided the main stimulus for biological research, its arguments are frequently defective.

Darwin's basic claim is that organs and instincts have been 'perfected by the accumulation of innumerable slight variations, each good for the individual'. For this, he says, it is necessary to admit only three propositions. (*a*) 'That gradations in the perfection of any organ or instinct, either do now exist or could have existed, each good of its kind.' (*b*) 'That all organs and instincts are, in ever so slight a degree, variable.' (*c*) 'That there is a struggle for existence leading to the preservation of each profitable deviation of structure or instinct.' But this assumes that the 'profitable deviations' are inherited. Thus not three but at least four propositions are really needed.

Again, after discussing our knowledge of the distribution of species in time and space—which carries irresistible conviction of organic evolution as an historical process—he turns to discuss conditions under which a variation is perpetuated.

Man does not produce variability [in domestic animals]; he only exposes beings to new conditions, and then nature acts on the organization, and causes variability. But man can select variations, and accumulate them in any desired manner. He thus adapts animals and plants for his own benefit. He can influence the character of a breed by selecting, in each successive generation, individual differences so slight as to be quite inappreciable by an uneducated eye. That many of the breeds produced by man have to a large extent the character of natural species, is shown by the doubts whether many are variations or aboriginal species.

In the preservation of favoured individuals and races, during the Struggle for Existence, we see the most powerful means of selection. More individuals are born than can survive. A grain in the balance will determine which shall live and which die—which variety or species shall increase in number, and which shall decrease, or finally become extinct.

There will in most cases be a struggle between the males for possession of the females. The most vigorous individuals will generally leave most progeny. But success will often depend on special weapons or means of defence, or on the charms of the males; and the slightest advantage will lead to victory.

There are here, as we can now see, certain fallacies, e.g.:

(1) All domestic breeds have not been produced by selecting very slight individual differences. Some domestic breeds have certainly been produced by breeding from individuals which presented great deviations from the normal.

(2) That a natural variation should confer an advantage is not enough to secure its perpetuation. The advantage must be effective, and it must be transmissible. Now it is difficult to believe that the earlier stages of some developments are effective as, for example, a wing so little developed as to give no power of flight or of gliding.

(3) Darwin assumes that species differ from their nearer relatives in having some special advantages that enable them to adapt themselves to slightly different conditions. Closely allied species, however, often live in apparently identical areas and conditions. The characters by which species differ from their fellow species can seldom be shown to be advantageous.

4. Sexual selection acts only in a few higher animal groups.

Darwin's presentation of Natural Selection as an effective agent is probably at its weakest in dealing with the problem of disuse. Here he assumes the inheritance of acquired characters in a form hardly differing from that of Lamarck.

Disuse, aided sometimes by natural selection, will often tend to reduce an organ, when it has become useless under changed conditions of life; and we can clearly understand on this view the meaning of rudimentary organs. But disuse and selection will generally act on each creature, when it has come to maturity and has to play its full part in the struggle for existence, and will thus have little power of acting on an organ during early life; hence the organ will not be much reduced or rendered rudimentary at this early age. The calf, for instance, has inherited teeth, which never cut through the gums of the upper jaw, from an early progenitor having well-developed teeth; and we may believe that the teeth in the mature animal were reduced, during successive generations, by disuse or by the tongue and palate having been better fitted by natural selection to browse without their aid; whereas in the calf, the teeth have been left untouched by selection or disuse.

The full title of Darwin's book was *The Origin of Species by means of Natural Selection, or the Preservation of Favoured Races in the Struggle for Life*. Darwin himself compared the action of natural selection to that of a man building a house from stones of all shapes. The shapes of these stones, he says, would be due to definite causes, but the uses to which the stones were put in the building would not be explicable by those causes. The conception reveals the general weakness of Darwinistic thought which treats natural selection as though it were an active and directive agent. For when a man builds a house, there is the intervention of a definite purpose, directed towards a fixed end and governed by a clearly conceived idea. The builder *selects* in the proper sense of the word. But the acts of selection—mental events in the builder's mind—have no relation to the 'causes' which produced the stones. They cannot be compared with the action of Natural Selection. If a metaphor be sought for the action of Natural Selection, a better one might be the arrangement of stones on a sandy shore. Large stones are found high up on the beach. The stones become smaller as we descend towards the sea. On approaching the brink, we come upon a zone of sand. This arrangement is due to the forces of winds, waves, and tides acting, according to their nature, and according to the nature of the rocks of which the cliffs are built, over a long period of time. Provided that it be kept well in mind that it is a metaphor, and provided that no teleological view is implied, there can be no harm (and not very much good) in calling this a 'selective action' of the forces of wind, waves, and tides upon the disintegrated rocks.

Darwin repudiated teleology, but in his title, almost as though wishing to emphasize it, he repeats the teleological metaphor and speaks of the *Preservation of Favoured Races*. But how do we know that races are favoured? By their preservation! And what is preservation? A favour! And what is the favour? Preservation!

So, too, with the phrase *Survival of the Fittest*. In the sense in which some early Darwinians used the word, fittest was naïvely confused with physical or even athletic fitness, and to it an ethical corollary was sometimes forcibly adjusted. But the only kind of fitness implied in the Darwinian phrase was fitness for survival. It is doubtless a good thing, on an ethical level, to be brave as a lion, and a bad thing, on an ethical level, to be timid as a rabbit. But, on a biological level, either quality may indicate fitness. Lions survive because of their courage in seeking their prey, rabbits because of their timidity in fleeing from those that prey upon them. Both are tests of fitness. Those that survive are fit, and those that are fit survive; and survival is the test of fitness, and fitness the test of survival! He that fights and runs away, lives to fight another day.

Thus these phrases are, on analysis, devoid of ultimate meaning. This is very far from saying that no meaning can be extracted from the history of their use. Darwin was an investigator of the very first rank, but he was inexpert in the exact use of language and had little philosophical insight. Nor was his discovery quite of the nature that many of his followers thought it to be.

There is one species whose origin raised acute controversy. Anatomists had always drawn attention to the likeness of the structure of man to that of the apes. Darwin at first expressed no opinion on this point. Several of his supporters, notably T. H. HUXLEY (1825–95) devoted attention to it. The formal expression of Darwin's views was reserved till 1871, when at the opening of *The Descent of Man* he wrote: 'Huxley has conclusively shown that in every visible character man differs less from the higher apes than those do from the lower members of the same order.' This was very different from a demonstration of any intermediate form between man and the higher man-like apes. Nevertheless, evidence of this sort was gradually accumulating.

In 1856, three years before the publication of the *Origin*, the long bones and part of the skull of a man-like being had been unearthed in the small ravine of Neanderthal in Rhenish Prussia. They were all at first misinterpreted as pathological. Huxley ultimately recognized them as those of a human being, but the most ape-like yet found. The species to which these bones belong is now entitled *Homo Neanderthalensis*.[1]

Since the discovery of Neanderthal man, many other species of fossil man have been discovered. On the other hand, several fossil species of apes approaching nearer than living forms to the human stem have also been found. The ape-man series is now probably more complete than that of most comparable mammalian groups.

Even before Darwin, and still more after him, evolutionary doctrine was applied to human habits, language, social organization, and psychology. Thus arose a science of Anthropology, which owes a deep debt to JACQUES BOUCHER DE PERTHES (1788–1868). As early as 1830 he had discovered in the gravels of the Somme certain flints which he believed bore evidence of very ancient human workmanship. In 1846 he demonstrated the existence of such flints in company with the remains of elephant, rhinoceros, and other tropical or extinct forms. He established the existence of man from human products in Pleistocene and early Quaternary times. In 1863 de Perthes clinched this view by discovering near Abbeville, in a Pleistocene deposit, a human jaw associated with worked flints.

These conclusions were accepted, though with caution, by Lyell in his *Antiquity of Man* (1863). Since that time the study of the works and arts of Stone-Age man has developed parallel with the study of his physical structure. The succession of the cultures, crafts, and art of Palaeolithic Man and their emergence into those of Modern Man and notably into the culture known as Neolithic have now become

[1] A Neanderthal skull had been found at Gibraltar as early as 1848, but had not been brought to scientific notice.

familiar. It has been equated with geological and geographical change.

The subject of Organic Evolution has been pursued along many paths which pass the frontiers of biology and indeed of the sciences in the limited sense. In 1852—seven years before the publication of the *Origin*—the philosopher, HERBERT SPENCER (1820–1903), expounded doctrines of Evolution in a work where that word was used to describe a general process of production of higher from lower forms. He devoted the remainder of his long life to a highly elaborate exposition of what he regarded as the implications of evolution in every department of the inorganic and the organic world, in the structure of human society, and in the human mind. He eagerly adopted Darwinian principles as soon as the opportunity arose. Since his political philosophy of extreme 'individualism' fitted well the feeling of the age, his works were very widely read. The phrase 'Survival of the Fittest' was coined by him (1864).

That the philosophical system of Spencer is an object of derision is one of the few points on which all philosophers seem now to agree. There are few living who can claim to have studied all his works. But despite his extreme dryness as a writer, Spencer was a great phrase-maker. A surprising number of his dicta have obtained currency. A selection of passages from one section of one chapter of his first independent work *Social Statics* (1850) will suffice to indicate the mental atmosphere of the scientific public to which the *Origin* was delivered, nine years later.

'Progress is not an accident but a necessity. It is part of nature.' 'All perfection is a fitness to the condition of existence.' 'Evil tends perpetually to disappear.' 'Nature's rules have no exceptions.' 'In virtue of an essential principle of life, non-adaptation of an organism to its conditions is ever being rectified. Whatever possesses vitality obeys this law. We see it illustrated in the acclimatisation of plants, in the altered habits of domestic animals, in the varying characteristics of our own race. . . *Such changes are towards fitness for surrounding conditions.*' 'Civilisation instead of being artificial is a part of nature; all of a piece with the development of the embryo or the unfolding of a flower. . . . Man needed one moral constitution to fit him for his original state; he needs another to fit him for his present state; and he has been, is, and will long continue to be, in process of adaptation. By *civilization* we signify the adaptation that has already taken place. In virtue of this process man will eventually become completely suited to his mode of life.'

(*d*) *Reception of Evolutionary Doctrine*

The *Origin* came thus to a world well prepared. The central idea of the work was far from being new or even modern. Nevertheless, it created a revolution in biology, and indeed in almost every department of thought. It was the first work by a cautious, penetrating, highly competent, and experienced investigator that set forth a large and carefully sifted body of evidence on the subject of 'Evolution'. Darwin himself was not fond of using this word, but usually refers to it, very properly and modestly, as a man of science should, as 'the species question'. He thus clearly seeks to confine his discourse to the region in which he has evidence to offer. His great book, however, was the first that suggested a simple and apparently universally acting biological mechanism producing changes of form.

The story of the rise of Darwinism is one of the most familiar in the history of science. Among the opponents of Darwin were Owen, who occupied a very important scientific position, and Agassiz, a very accomplished naturalist. Both were still bemused by *Naturphilosophie*, as was also von Baer (p. 471), now in extreme old age. All opposed to evolution the '*idea*' or '*type*' of Goethe and Cuvier, a metaphysical conception and, of its nature, insusceptible of demonstration (pp. 385 ff.).

In Germany, then swept by 'liberal' ideas, Darwinism made rapid progress. Its ablest critic was the Swiss, ALBRECHT KÖLLIKER (1817–1905). Without denying the inconstancy of specific forms, and while fully accepting evolution within the limits of certain wider groups, he indicates several weaknesses in the Darwinian position:

(1) Absence of any experience of the formation of a species.

(2) Absence of any evidence that unions of different varieties (i.e. of *incipient species* on Darwin's view) are relatively more sterile than unions of the same variety.

(3) Extreme rarity of true intermediate forms between known species, whether living or fossil.

Kölliker and other critics claimed that the 'chance' element in Darwin's scheme was but a veiled teleology. Natural selection had been elevated to the rank of a 'cause' leading to an 'effect' and science has to deal not with causes but with conditions. In Kölliker's view, Darwin was dealing with the 'might' and 'may be' and not with any theory that could be tested by experience. Here Kölliker was right. Evolution was perhaps unique among major scientific theories in that the appeal for its acceptance was not the evidence for it, but that any other proposed interpretation of the data seemed wholly incredible.

In France the reception of Darwinism was on the whole hostile and its advance slow. The influence of Cuvier was still paramount. The ultimate victory was complete, though several very able biologists, such as Bernard (p. 492), remained unconvinced to the end. The movement led to a revival of interest in Lamarck, and *transformisme*, as evolution was called, received in France a Lamarckian tinge.

The battle of evolution is now a stricken field, and the whole of modern biology has been called 'a commentary on the *Origin of Species*'. Biologists are now at one in the view that living forms correspond to a limited number of common stocks, and tolerable agreement has been reached as to the evolutionary history of these stocks. It was not many decades, however, before doubt began to dawn as to the mechanism of evolution. During Darwin's active period, Gregor Mendel (1822–84) was at his unnoticed work (1857–69, pp. 485–90), the rediscovery of which (1900) introduced a particulate view of inheritance, of which Darwin and his century knew nothing.

In the twentieth century, with the rise of genetics and the mathematical treatment of natural selection, the 'species question' seems nearing an answer, though this cannot be simple or probably the same for all groups of living things. In the wider range of thought, whether in science, or philosophy, or religion, or economics, or literature, or even in the higher arts, the 'evolution idea' is triumphant though its identity with 'progress' has long ago passed into the limbo of wishful thinking. That Darwin's 'explanation' of organic evolution turns out to be a redescription is a charge against his philosophic but not against his scientific powers. Such redescription is the normal process of advance of scientific theory (p. 296).

(vi) *The Turn of the Century*

Thus, we part with our story of 'classical science'. Its task was to describe the world in mechanical terms in the hope of reaching a unitary view. Despite its triumphs there yet remained in the narrative inconsistencies so evident and breaks so definite that they could be ignored only by the most optimistic or the least philosophical. Thus the Doctrine of Ether remained highly metaphysical, and there were unbridged gulfs between Matter and Force on the one hand and between the Living and the Not-living on the other.

During the nineteenth century science was immensely successful in many and revolutionary directions. It had improved the human lot. It had provided an intellectual stimulus that was far more effective than those of some other and more fatigued disciplines. It had rendered many current philosophical and theological positions completely untenable. It had—despite modern misunderstanding—introduced a humaner spirit into human relations. It provided a new basis for education, and had made certain of the older bases more than a little ridiculous. Most of all, it had inseminated a hopeful and at least partially justified view of human potentiality. Nevertheless, the method has its limits which have been more readily recognized by working scientific men than by some who assumed the task of interpreting them.

Science, of its nature, is incapable of accomplishing or even of attempting the task of resolving all the various discrepancies of thought into one whole. For this reason, among others, a history of science is, in the strict sense of the word, hardly possible. Science cannot deal with the whole at all, but only with abstractions, with 'Departments of Scientific Inquiry' as we are accustomed to call them. But though it must perforce work in departments, it is by no means pledged to keep the boundaries of those departments fixed; it is committed to no doctrine of *status quo* for the frontiers on its maps. In changing those frontiers science must, at need, go back to its beginnings and question its own primary data: it must revise its metaphysic. In doing so it has come to presuppose a philosophy different from the classical materialistic plan. The world of science may come to be regarded as an evolutionary scheme in which will emerge patterns of *value*, precisely that type of pattern in fact that was so stoutly repudiated by the materialist philosophers of a previous generation.

The generation of philosophers that could ignore the great scientific conclusions is now at rest and is not likely to be disturbed. It seems probable that Science itself is now reaching a stage in which an adequate scientific equipment will involve some regard to the world as an interconnected whole, in other words, in which Science and Philosophy will dwell less apart. This does not mean that Science will abandon its method of abstraction—for then it would cease to be Science—nor does it mean that Science will seek a refuge in that tomb which has become the peaceful abode of an older philosophy based on ratiocination. But it does mean that the frontiers of scientific abstractions may be rendered more fluid and that philosophical methods may have some share in determining the nature of the change. Notably it seems probable that the conceptions of the separation of mind from mind and of mind from matter may need modification. There are many indications that the tendencies of science since the later nineteenth century have been working in these directions.

INDEX

Bold-face figures indicate important entries